E. Tolosa, R. Duvoisin, and F. F. Cruz-Sánchez (eds.)

# Progressive Supranuclear Palsy: Diagnosis, Pathology, and Therapy

Springer-Verlag Wien GmbH

Prof. Dr. E. Tolosa

Neurology Service, Hospital Clinic, Faculty of Medicine,
University of Barcelona, Spain

Prof. Dr. R. Duvoisin

Neurology Service, Robert Wood Johnson University Hospital,
New Brunswick, NJ, U.S.A.

Dr. F. F. Cruz-Sánchez

Neurological Tissue Bank, Faculty of Medicine,
University of Barcelona, Spain

© 1994 Springer-Verlag Wien
Originally published by Springer-Verlag Wien New York in 1994

Product Liability: The publisher can give no guarantee for information about drug dosage and application thereof contained in this book. In every individual case the respective user must check its accuracy by consulting other pharmaceutical literature. The use of registered names, trademarks, etc. in this publication does not imply, even in the absence of a specific statement, that such names are exempt from the relevant protective laws and regulations and therefore free for general use.

Typesetting: Best-set Typesetter Ltd, Hong Kong

Printed on acid-free and chlorine-free bleached paper

With 113 Figures

ISBN 978-3-211-82541-9          ISBN 978-3-7091-6641-3 (eBook)
DOI 10.1007/978-3-7091-6641-3

# Foreword

When Steele, Richardson and Olszewski described Progressive Supranuclear Palsy in Archives of Neurology in 1964, it was thought to be a rare disease. Recent pathological studies of large numbers of patients diagnosed as having Parkinson's disease in life have highlighted the fact that at least one in ten of such cases have some other condition. Progressive Supranuclear Palsy is one of the commonest alternative diagnoses. This book is therefore a timely review of present understanding of Progressive Supranuclear Palsy. Much has been learnt about this sporadic illness of middle and late life although its cause remains unknown, and its treatment continues to be difficult. The Editors have selected a team of authors who review the clinical aspects, neuro-imaging findings, neuropathology, neurochemistry, epidemiology, and therapy of Progressive Supranuclear Palsy. All are to congratulated on producing an excellent and detailed picture of contemporary knowledge of the condition. Anyone interested in Progressive Supranuclear Palsy cannot do better than to start by reading this book. It is to be hoped that it will prompt further investigation to establish its cause and cure.

C. D. Marsden, London

# Preface

Progressive Supranuclear Palsy (PSP) is a neurodegenerative disorder characterized clinically by the appearance of a supranuclear gaze palsy and extrapyramidal features which include bradykinesia and axial dystonia. Postural instability and falls and frontal-type behavioural and cognitive disturbances complete the clinical picture of this progressive disorder. Although it is not an uncommon disorder in movement disorder clinics, due to its rarity and its protean clinical manifestations it is frequently diagnosed correctly years after the onset of symptoms and not infrequently confused with other diseases in which dementia or parkinsonism are prominent.

PSP was defined as a separate nosological disorder by Steele et al. in 1964 and subsequent publications on the subject have confirmed the essential clinical and pathological features outlined by these authors and made important contributions towards the characterization of the disorder. These new publications have emphasized the clinical heterogeneity of PSP, not all patients, for example, develop supranuclear gaze palsy and others show profound dementia, and also the not infrequent difficulties pathologists also have in diagnosing the disorder. Also new information has emerged on the natural history and epidemiology of PSP and on its neuropathology, biochemistry, neurophysiology, and therapeutics. All authors in this volume are experts in different fields, basic or clinical, in the area of Movement Disorders and their contributions review and add new information on all of the areas outlined above. This book will therefore update the reader on modern concepts regarding PSP.

This book is intended primarily for a clinical audience. Neurologists in general and, in particular, Movement Disorder specialists, who are faced with the difficult task of diagnosing and treating patients with PSP will hopefully find in the text information that will be of help in managing their patients. Neuropsychologists, pathologists, radiologists, and other professionals interested in new developments in the field of neurodegenerative disorders will also find in this book, we belief, valuable information.

We wish to thank each of the authors of this volume for the valuable personal time they have spend writing their outstanding and timely contributions. We also wish to extend our thanks to the Fondo de Investigaciones Sanitarias de la Seguridad Social of the Spanish Ministry of Health for their help in organizing in Barcelona, in October of 1992, the symposium "Advances in Progressive Supranuclear Palsy". Many of the contributors to this book participated in this symposium, the first ever to be organized solely on PSP-related issues, reflecting the growing interest of the scientific community on this disorder.

Despite new developments PSP remains a progressive debilitating disorder for which we have no cure. We hope that this book will help early recognition and proper management of PSP and stimulate successful research that will soon enable us to dispose of an effective, curative treatment that we can offer our patients.

E. Tolosa
R. Duvoisin
F. F. Cruz-Sánchez

# Contents

X Contents

# Clinical aspects

# Historical notes

## John C. Steele

Umatac, Guam, U.S.A.

**Summary.** Progressive supranuclear palsy (PSP) is the name Dr. J. Clifford Richardson chose to designate an unusual clinical syndrome he first identified in the 1950s. Neurofibrillary degeneration is the hallmark of this fatal brain disease, and during our study of Richardson's patients, Professor Jerzy Olszewski and I also observed granulovacuolar degeneration, and widespread nerve cell loss and gliosis in subcortical and brain stem nuclei. The histopathological features bear a striking resemblance to those seen in postencephalitic parkinsonism after von Economo's epidemic encephalitis, and in the parkinsonism-dementia complex of Guam (PDC).

During the past 30 years, neurologists confirm that progressive supranuclear palsy is a universal, sporadic and not uncommon neurodegeneration of middle and late life. Many fine studies, as reported here, have advanced our understanding of PSP but its cause, and thereby its cure, is still to be revealed.

These historical notes tell of our observations from 1955 to 1975. We are pleased that colleagues remember these early descriptions and honor us by calling this disease, the Steele-Richardson-Olszewski (SRO) syndrome.

In Toronto, in 1955, neurologist J. Clifford Richardson was consulted by a 52 year old friend and business executive because of clumsiness, trouble in seeing, and mild forgetfulness. During the next 4 years, Richardson was puzzled as his friend progressively developed an unusual constellation of signs which included supranuclear ophthalmoplegia affecting chiefly vertical gaze, pseudobulbar palsy, dysarthria, dystonic rigidity of the neck, and mild dementia (Fig. 1).

As Dr. Richardson was observing the evolution of this remarkable illness, he was astonished to identify similar symptoms in three middle-aged Canadian veterans who were admitted to the Neurology Service at Sunnybrook Military Hospital for long term care. The first was a West Indian laborer, who immigrated to Canada in 1913 and developed unsteady walking in 1954. Next was a truck driver who came from England when he was a child and who was entirely well until 1953 when his personality altered and he was confused. The third veteran had come from the Ukraine in 1913 and worked as a laborer until 1956 when he developed difficulty with vision, his speech became slurred, and he had trouble in swallowing.

**Fig. 1.** First case of PSP identified by Richardson in 1955. Loss of downward gaze in attempting to look at his watch (Steele et al., 1964)

Dr. Richardson recognized that these four patients had the same disease. It was unfamiliar to him, and to other senior neurologists who visited Toronto and examined them.

In 1961, when I (Fig. 2) began my neurology residency after graduation from the University of Toronto, Dr. Richardson encouraged me to investigate and describe this disease with him and Professor Olszewski, Chief of Neuropathology at the Banting Institute. I was honored by their invitation.

Clifford (Ric) Richardson (Fig. 3) graduated from the University of Toronto Medical School and when he decided to specialize in neurology in 1935, he travelled to the National Hospital for Nervous Diseases, Queen Square, London. After six months, he became one of the few Canadians to be appointed House Physician spending two years apprenticed to Drs. Holmes, Wilson, Martin, Symonds, Critchley and Denny-Brown. He returned to Canada and had a distinguished clinical and teaching career at the Toronto General Hospital. In 1960 he became Head of the Division of Neurology, and in 1972, when the special role of neurology within the Department of Medicine was recognized through the establishment of a Chair, Dr. Richardson became the first Professor of Neurology. After his retirement from the University in 1975 he was Professor Emeritus and he continued in active practice until just months before his death in June, 1986 at the age of 77 (Wherrett, 1986).

Professor Jerzy Olszewski (Fig. 4), a graduate in Medicine from Wilno University in Poland, came to the Montreal Neurological Institute (MNI) in

**Fig. 2.** John C. Steele (1934–present)

**Fig. 3.** J. Clifford Richardson (1909–1986)

1948 by the invitation of Dr. Wilder Penfield. He brought with him a remarkable knowledge of the cytoarchitecture of the nervous system, and his insights as an anatomical microscopist (despite being color blind) are documented by three atlases he published while working at the MNI. In 1956 he came to realize how productively his cytoarchitectural skills could be applied to neuropathology and in that year he moved to the University of Saskatchewan to become an experimental neuropathologist. His accomplishments there were substantial. In 1959 he left Saskatoon for a professorship in neuropathology at the University of Toronto. He quickly established collaborative ties with other neuroscientists and his Division of Neuropathology flourished at the Banting Institute. He was an exceptional teacher who inspired his own love of knowledge in young doctors and, as I found, no resident was too junior for his personal attention. His charity was renowned and he was the most humble of great men I have known. He died

**Fig. 4.** Jerzy Olszewski (1913–1964)

in 1964 from myocardial infarction. It was untimely but his influence continues still through his residents and his work (Baxter et al., 1987).

From 1961 through 1963, Dr. Richardson and I studied the evolution of the illness and its clinical aspects in 8 patients. Then in 1963, when I was an assistant resident in neuropathology, Professor Olszewski and I described the histopathology of the disease in seven fatal cases and the detailed localization of lesions in four of these in which sufficient material was available for comprehensive study. It was his expertise in brain stem cyto-architecture that crystalized our thoughts on the disorder.

In June 1963, at the American Neurological Association meeting, Dr. Richardson presented the first clinical report of this disease (Richardson et al., 1963). He observed: "In recent years in Toronto we have seen these few cases of progressive degenerative cerebral disease with a common syndrome of ocular, motor and mental symptoms which we had not previously recognized and which seems to have escaped clear identification and description in the medical literature. Though the component features have varied in preponderance, the cases have all shown defects of ocular gaze, spasticity of facial musculature with dysarthria and sometimes dysphagia, extensor rigidity of neck with head retraction, and dementia. There have been less constant symptoms of unsteady gait, truncal apraxia with difficulties in movements such as sitting, lying down and turning in bed, and some pyramidal spasticity and related reflex changes in the limbs.

Seven of the cases have been followed clinically in Toronto and five of these have died and have been studied pathologically. We are indebted to

Dr. Francis McNaughton of Montreal for providing clinical notes and pathological material from the eighth case. Neuropathological observation in these six fatal cases has supported our clinical impression that they do not belong to any of the well recognized categories of classical degenerative disease such as paralysis agitans, motor neuron disease, olivo-ponto-cerebellar degeneration or Creutzfeldt-Jakob disease. None were explicable on a basis of cerebral arteriosclerosis.

The cases have all been male with symptoms starting between ages 52 and 62. The six fatal cases have all shown a steady progression of cerebral symptoms leading to death within seven years. Usually there was a terminal bedridden state of helpless rigidity, pronounced dysarthria and dysphagia, and considerable dementia. The earliest case, a business executive, had the typical upturned head and gaze with staring stiff face and severe dysarthria. He was grossly disabled by his inability to look down and by considerable truncal apraxia. His dementia was only mild and he had only slight pyramidal spasticity of his legs with fairly normal motor functions in the upper limbs. He died at home by suffocation-having turned in bed to a helpless prone position with his face in the pillow.

The ophthalmoplegia which was present in all these cases was always a defect of willed vertical gaze, though several cases later showed some faults of lateral conjugate ocular movement. The vertical gaze palsy was predominantly in depressor movement in four of the cases, and the usual posture was of elevation of the head and eyes in varying degree. In three of the cases there was about equal impairment of upward and downward gaze and in one surviving case the defect is chiefly in elevation of the eyes. In all cases there was loss of following vertical movements as well as willed gaze. Parinaud's syndrome would be an applicable term for most of the cases, but because of uncertainly about exact definition (of that syndrome), the eponym has been avoided. There was defective ocular convergence in three of the cases and diplopia was noted intermittently in two. The pupils were unaffected. Two of the cases showed some retraction of upper eyelids; there was no ptosis. In six of the eight cases in whom full testing was carried out, there was retention of full passive ocular movements when the gaze was fixed and the head moved by the examiner. Two cases showed a tendency to spasmodic ocular fixation with a doll's head effect. Bell's phenomena was absent in four of the five cases in which it was fully tested. Vestibular function was checked by caloric testing in three cases and showed normal responses.

Stiff spastic facial muscles with a rather staring expression and often an open mouth, were characteristic. Hyperactive jaw and facial jerks could be easily demonstrated in most. Dysarthria was constant and was severe in five cases, a gross slurring and slowing of speech in two cases progressing to anarthria. Forced laughter and crying was not encountered in any of these cases.

The motor changes were featured by a this stiff, rubbery contraction of posterior cervical and upper back muscles, causing an upturned posture of head and eyes. In later stages three cases showed more severe plastic rigidity of limbs, but none developed characteristic parkinsonian features and none showed tremor. One case at a late stage developed a peculiar athetoid posturing and distortion of purposeful movements of his upper limbs. Pyramidal tract involvement was exceptional and minor but two cases were noted to have spasticity and Babinski reflexes in later stages.

The gait was mildly unsteady and broad based in most cases. This was partially explained by the ocular defect with inability to look downwards, but there was some added disturbance of equilibrium.

Dementia occurred in some degree in all of the cases. It was a severe and prominent feature in two cases. Most of the patients had presenting symptoms of personality change and impaired intellectual ability but usually the dementia remained mild."

Drs. Houston Merritt, Robert Schwab, and Derek Denny-Brown were the discussants at this presentation. They congratulated Dr. Richardson for his detailed observations about this interesting condition which Dr. Denny-Brown thought must be quite rare since none of those attending, except Dr. McNaughton, knew of similar cases. Because there were no hereditary factors, Dr. Merritt thought it was likely that some toxic influence was responsible. Dr. Leonard Kurland commented on the unusual concentration of cases from Toronto and he also wondered if they could have been exposed, in common, to some regional environmental factor. When responding, Dr. Richardson observed that the group of cases reported from Toronto was a very small one. The patients came from various parts of Ontario and there was nothing peculiar about their diet. One patient lived in Montreal and Van Bogaert's case, which seemed to be the same disease, came from Belgium (Chavany et al., 1951) (Fig. 5).

Dr. Richardson concluded: "I doubt very much that there is any local geographic incidence. I expect that a good many cases of the same disease will be identified in other areas."

In a companion report at the meeting of the American Association of Neuropathology (Olszewski, 1963), Professor Olszewski observed: "Neuropathological studies have revealed nerve cell loss and gliosis in the pallidum, subthalamic and red nucleus, substantia nigra and locus coeruleus, the superior colliculi, periaqueductal gray matter, and pretectal regions, the vestibular nuclei, various nuclei of the reticular formation and the dentate nuclei. In these regions neurofibrillary tangles without senile plaques were present. Granulovacuolar changes were seen in the red nucleus and nuclei pontis. Glial nodules and perivascular cuffing by lymphocytes were observed infrequently. There was patchy loss of Purkinje cells and torpedo-like expansions were present on axis cylinders.

These cases are considered to belong to the same group as those reported by Van Bogaert (Chavany et al., 1951), Verhaart (1958), Neumann (1961) and Brusa (1962) and to represent a distinct clinical and pathological entity. For convenience we have retained Verhaart's designation of 'heterogeneous system degeneration'. Future investigations may, however, reveal a different term to be more appropriate."

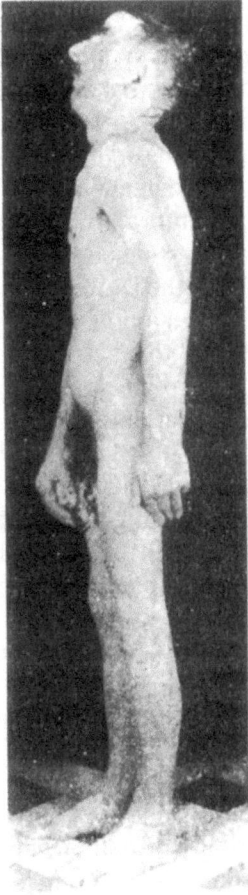

**Fig. 5.** Nuchal dystonia in PSP (Chavany et al., 1951)

In the discussion, Dr. Asao Hirano, a neuropathologist at Montefiore Hospital observed: "The paper presented by Dr. Olszewski is especially interesting in view of the current investigations of amyotrophic lateral sclerosis (ALS) and the parkinsonism-dementia complex (PDC) occurring among Chamorros on the island of Guam (Hirano et al., 1961). About 1 month ago, at the invitation of Dr. Olszewski, I had the opportunity to examine a patient and study slides of this very fascinating material in Toronto. One can recognize obvious differences between cases in Canada and those from Guam with respect to the main clinical manifestations, as well as to the location of the major histological involvement. The most severely involved areas of the nervous system in the Canadian cases are the subthalamic nuclei, superior colliculi and the vestibular and dentate-red nuclei systems, with demyelination of the brachium conjunctivum. In the

Guam series, on the other hand, although these structures may be slightly affected in certain cases, the most severe lesions are in other systems, namely the motor neuron-pyramidal system in amyotrophic lateral sclerosis and the cerebral cortex and substantia nigra, as well as other structures of the extrapyramidal system in the parkinsonism-dementia complex. The striking supranuclear ocular palsy observed here has not been seen so far in our Guam series, and the marked tendency to fall backward with forced extension of the neck is not a feature in our cases."

"On the other hand, it is interesting and important to note that there are certain features common to both the Toronto and Guam series. The blank and expressionless face in each, superficially resembles parkinsonism. In both there is a delayed response to questions, severe akinesia, advanced dysarthria, increased plastic tone of all extremities without obvious tremor, hyperreflexia, and dementia of varying proportion. The globus pallidus, substantia nigra, locus coeruleus, periaqueductal gray matter and various regions of the reticular formation are affected to a moderate or to a severe degree in both disorders. Histologically, there is neuronal loss and gliosis in all involved areas, as well as abundant neurofibrillary changes and granulo-vacuolar bodies. These histological and cytological features are essentially similar to those found in all of the Guam cases. The striking similarity of tissue response in these 2 disorders, occurring in 2 different geographical locations, certainly deserves attention not only in the clinical and patholo-gical sense but also from the standpoint of their familial and epidemiological features."

In April of 1964, our report of nine patients, seven of whom had died, was published in the Archives of Neurology (Steele et al., 1964).

We were not able to establish a definite etiology but we suggested a degenerative process or a chronic and late viral infection. The histopatholo-gical features bore a striking resemblance to those seen in postencephalitic parkinsonism and in the parkinsonism-dementia complex of Guam, but the distribution of lesions was distinctive, and differed in all three conditions.

In our 1963 reports we had called the condition heterogeneous system degeneration, a term first applied by Verhaart to a similar case he described in 1958. But since we were not certain the disease was a primary degenera-tion, or that it was truly related to other system degenerations, we agreed we should choose another term. In the summer of 1963, Dr. Richardson suggested we call it progressive supranuclear palsy (PSP), a clinical de-signation and the name by which the disease is now known. Some authors also refer to it as the Steele-Richardson-Olszewski syndrome, an eponym which was first used by Dr. Andre Barbeau in 1965.

By 1975, Dr. Richardson and I were able to conclude that the nosolo-gical entity which he had first identified as an unusual syndrome in the 1950s was clearly a unique, universal and not uncommon neurological disease of middle and later life (Steele, 1970, 1972, 1975). But the condition was not new, and the earliest clinical account seemed to be in the writings of Posey (1904) and Spiller (1905). Figure 6, reproduced from Spiller's paper, shows the gaze paralysis when the patient tried to look up, widened palpebral fissures and spastic facies. The patient was demented, mildly hypertonic and ataxic. No late follow-up or pathological study was reported.

Of 73 patients with PSP reported between 1951 and 1972, 51 had been male and of diverse ethnic and racial origin (Steele, 1972). With one exception (David et al., 1968), no similar disorder existed in their families and their past health had usually been good. Careful enquiry seemed to exclude anything in their life habits or circumstances which might have predisposed them to the disease. In none had there been known exposure to noxious or toxic agents or any antecedent illness to suggest encephalitis. The many reports then available indicated that the disease was not restricted to geographic or climatic regions.

In our orginal description, we observed widespread neuronal degeneration in diverse nuclei of the cerebellum, brainstem and diencephalon, and we were therefore surprised by the remarkable constancy of clinical signs and the common pattern of their emergence and progress. We recognized that the clinical illness depended upon the similar localization of lesions which occured sequentially. In view of the heterogeneity of nuclei affected, we anticipated that further observations would broaden the clinical spectrum of the disease, as the same or other nuclei became affected to different degrees or at different times. Contrary to our expectations, the 59 patients reported by others between 1964 and 1972 had, with minor exceptions, shown the same symptoms and pursued a similar course.

In 1972, I visited Micronesia and was captivated by the beauty of the islands and harmony of the Pacific lifestyle. I left Toronto and my fledgling career in academic neurology to work among native people as a general physician, helping to develop a health care system for the Trust Territory of the Pacific Islands. It was a second career of adventure and delight, and it began a romance with this region that continues. I maintained my interest in neurology and often crossed Micronesia's watery vastness by plane and steamer to search for neurological illnesses in its small and far flung populations. I observed that neurodegenerative diseases were quite uncommon among the 125,000 Micronesians and even those of advanced age seldom suffered dementia, parkinsonism or ALS (Steele, 1984).

When the Trust Territory dissolved in 1983, I stayed in the Pacific Islands and came to Guam, southernmost of the Mariana Islands, to establish a clinic for indigenous Chamorro veterans, to be the neurologist at the U.S. Navy Hospital, and to study the disease of Guam I first learned of from Dr. Hirano.

Although my friend Dr. Carleton Gajdusek had told me that amyotrophic lateral sclerosis and the parkinsonism-dementia complex on Guam were disappearing (Garruto et al., 1985), I immediately began to see many Guamanian Chamorros who suffered progressive neural disease and had varying degrees of dementia, parkinsonism and amyotrophy (Lavine et al., 1990). Of particular excitement for me was a 52 year old Chamorro military retiree I began to see in 1983. He suffered a rapidly advancing illness entirely like that of progressive supranuclear palsy, as I had observed in Toronto twenty years before (Fig. 7). He died in 1984 and neuropathological

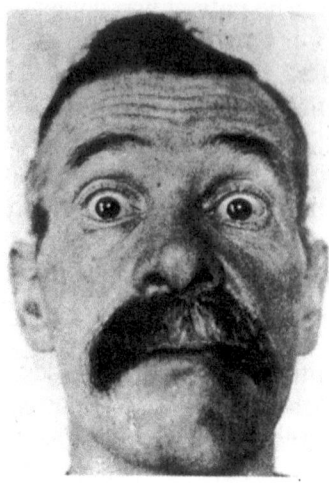

**Fig. 6.** Early case report of probable PSP. Paralysis of upward movement of the eyes when the patient tries to look up (Spiller, 1905)

**Fig. 7.** PSP features in a Chamorro with the parkinsonism-dementia complex of Guam (PDC). He has blepharospasm and is unable to look down at a pen (Bergeron and Steele, 1990)

studies by Dr. Catherine Bergeron showed changes of the parkinsonism-dementia complex (PDC), entirely like those described by Dr. Hirano (Bergeron and Steele, 1990).

In subsequent studies with Drs. Tanner, Lepore, Calne, and Duvoisin, we have shown that supranuclear disturbances of ocular and eyelid motility are very common in ALS/PDC of Guam and we know now that many patients with PDC show vertical gaze palsy, pseudobulbar palsy and dystonia as occurs in PSP (Tanner et al., 1987; Lepore et al., 1988). These findings are of particular interest because they are new observations which further relate the two diseases and were not known when the clinical and pathological interrelationships between PDC and PSP were being considered in 1963 (see Dr. Hirano's discussion above).

In 1963, PSP and PDC were diseases half a world apart. I have been fortunate to participate in studies of both. In the 1960s, in Toronto, we recognized and reported progressive supranuclear palsy as a separate disease. Now, in the 1990s, on Guam, we work to discover the cause of the amyotrophic lateral sclerosis/parkinsonism-dementia complex. When we succeed, I expect this knowledge will provide important insights about the pathogenesis of PSP and other related diseases of the nervous system which affect millions of the world's elderly and aging population (Monmaney, 1990).

## Acknowledgements

This article was prepared for Drs. Irene Litvan and Yves Agid and is the Introductory Chapter for their monograph, titled "Progressive Supranuclear Palsy: Clinical and Research Approaches" published and copyrighted by Oxford University Press in 1992. (ISBN 0-19-507229-4). Dr. Steele thanks the editors and Oxford University Press for their permission to reprint that article.

## References

Barbeau A (1965) Degenerescence plurisystematisee du nevraxe. Syndrome de Steele-Richardson-Olszewski. Un Med Can 94: 715–718
Baxter DW, Buettner-Ennever JA, Sharpe JA, Leigh RJ (1987) Jerzy Olszewski: cartographer of the brain stem reticular formation. Neurology 37: 1881–1882
Bergeron C, Steele JC (1990) Guam parkinsonism-dementia complex and amyotrophic lateral sclerosis. A clinico-pathological study of seven cases. In: Rose FC, Norris F (eds) ALS. New advances in toxicology and epidemiology. Smith-Gordon, Chapter 13, pp 89–98
Brusa A (1961) Degenerescence plurisystematisee du nevraxe, de charactere sporadique, a debut tardif et evolution subaigue. Rev Neurol 104: 412–429
Chavany JA, Van Bogaert L, Godlewski S (1951) Sur un syndrome de rigidite a predominance axiale avec perturbation des automatismes oculo-palpebraux d'origine encephalitique. Presse Med 50: 958–962
David NJ, Mackey EA, Lawton Smith J (1968) Further observations in progressive supranuclear palsy. Neurology 18: 349–356

Garruto RM, Yanagihara R, Gajdusek DC (1985) Disappearance of high-incidence amyotrophic lateral sclerosis and parkinsonism-dementia on Guam. Neurology 35: 193–198

Hirano A, Malamud N, Kurland LT (1961) Parkinsonism-dementia complex, an endemic disease on the island of Guam. II. Pathological features. Brain 84: 662–679

Lavine L, Steele JC, Wolfe N, Calne DB, O'Brien PC, Williams DB, Kurland LT, Schoenberg BS (1991) Amyotrophic lateral sclerosis/parkinsonism-dementia complex in southern Guam: is it disappearing? In: Rowland LP (ed) Advances in neurology. Amyotrophic lateral sclerosis and other motor neuron diseases. Raven Press, Chapter 56, pp 271–285

Lepore FE, Steele JC, Cox TA, Tillson G, Calne DB, Duvoisin RC, Lavine L, McDarby JV (1988) Supranuclear disturbances of ocular motility in Lytico-Bodig. Neurology 38: 1849–1852

Monmaney T (1990) Annals of science (An epidemic of brain disease). The New Yorker, October 29: 85–113

Neumann MA (1961) Heterogeneous system degeneration with particular involvement of the reticular substance: correlation with neurologic disorders and the concept of the Creutzfeld-Jakob disease. J Neuropathol Exp Neurol 20: 316

Olszewski J, Steele J, Richardson JC (1963) Pathological report on six cases of heterogeneous system degeneration. J Neuropathol Exp Neurol 23: 187–188

Posey WC (1904) Paralysis of upward movement of the eyes. Ann Ophthalmol 13: 523–529

Richardson JC, Steele J, Olszewski (1963) Supranuclear ophthalmoplegia, pseudobulbar palsy, nuchal dystonia and dementia: a clinical report on eight cases of "heterogenous system degeneration". Trans Am Neurol Assoc 88: 25–29

Spiller WG (1905) The importance in clinical diagnosis of paralysis of associative movement of the eyeballs (Blick-Lahmung) especially of upward and downward associated movements. J Nerv Ment Dis 32: 417–488, 497–530

Steele JC (1970) Progressive supranuclear palsy: report of a Thai patient. J Med Assoc Thailand 53: 364–368

Steele JC (1972) Progressive supranuclear palsy. Brain 95: 693–704

Steele JC (1975) Progressive supranuclear palsy. In: Vinken PJ, Bruyn GW, DeJong JMBV (eds) Handbook of clinical neurology. North-Holland, pp 217–229

Steele JC (1984) Micronesia: health status and neurological diseases. In: Chen KM, Yase Y (eds) Amyotrophic lateral sclerosis in Asia and Oceania. National Taiwan University, pp 173–183

Steele JC, Richardson JC, Olszewski J (1964) Progressive supranuclear palsy: a heterogeneous degeneration involving the brain stem, basal ganglia and cerebellum with vertical gaze and pseudobulbar palsy, nuchal dystonia and dementia. Arch Neurol 10: 333–359

Tanner CM, Steele JC, Perl DP, Schoenberg BS (1987) Parkinsonism, dementia, and gaze paresis in Chamorros on Guam: a progressive supranuclear palsy-like syndrome. Ann Neurol 22: 174

Verhaart WYC (1958) Degeneration of brainstem reticular formation, other parts of brain stem and cerebellum: example of heterogeneous system degeneration of central nervous system. J Neuropathol Exp Neurol 17: 382–391

Wherrett JR (1986) In memoriam: John Clifford Richardson. Can J Neurol Sci 13: 292–293

Author's address: Dr. J. C. Steele, Micronesian Health Study, University of Guam Health Sciences Bldg., Mangilao, Gu 96923, U.S.A.

# Clinical diagnosis and diagnostic criteria of progressive supranuclear palsy (Steele-Richardson-Olszewski syndrome)

## E. Tolosa, F. Valldeoriola, and M. J. Marti

Neurology Department, Hospital Clinic, Faculty of Medicine, University of Barcelona, Spain

**Summary.** Progressive supranuclear palsy (PSP) is characterized clinically by supranuclear gaze palsy, neck dystonia, parkinsonism, pseudobulbar palsy, gait imbalance with frequent falls and frontal lobe-type dementia. In the advanced typical case, when supranuclear gaze palsy and other main features are present diagnosis is relatively easy. Diagnostic problems, though, are frequent in the early stages due to the variable clinical presentation and in those atypical cases in which gaze palsy does not develop or that present as a severe dementig disorder or as an isolated akinetic-rigid syndrome. In this review we summarize the clinical features of PSP and emphasize those aspects helpful in the differential diagnosis with Parkinsnon's disease and other motor and cognitive disorders that can pose difficult diagnostic problems. Clinical diagnostic criteria are also discussed and modifications of those currently in used are proposed.

## Introduction

Although some reasonably convincing cases of what is now called progressive supranuclear palsy (PSP) had been described before 1964 it was in this year when Steele et al. (1964) clearly established PSP as a separate nosological entity different from other disorders presenting with supranuclear ophtalmoplegia, pseudobulbar palsy and parkinsonism and from those associated neuropathologically with neurofibrillary tangles, such as post-encephalitic Parkinson disease and the Parkinson-Dementia complex of Guam. Steele et al. (1964) called the disorder progressive supranuclear palsy, a term that vaguely describes the patients problems but which has survived, particularly in its abreviated form, PSP, the test of time, unlike other proposed terms. The cardinal manifestations of PSP were clearly outlined by Steele et al. in their original paper (Steele et al., 1964) and reviewed by Steele (1975) and others (Golbe et al., 1988; Kristensen, 1985; Lees, 1987) and include supranuclear gaze palsy, pseudobulbar palsy, prominent neck dystonia, parkinsonism, poor equilibrium and falls and subcortical dementia. The clinical picture is fairly uniform in the advanced, typical case and while each individual feature can be encountered in other neurological conditions the presentation on a given patient of several of these cardinal features is almost unique for PSP.

Difficult diagnostic problems can occur, though, when trying to diagnose PSP in the early stages. Indeed, the mean duration of PSP before diagnosis in Kristensen's review (1985) of 325 cases collected from the literature was 3.9 years and other authors, such as Pfaffenbach et al. (1972) report an even longer average delay in the diagnosis of PSP after symptom onset (4.5 years). This dealy in diagnosis is undoubtedly due to lack of familiarity of physicians with the clinical syndrome, but also to the heterogeneous presentation of the disorder in the early stages and in particular to the fact that the supranuclear ophtalmoplegia which remains the most characteristic clinical sign of the disease, tends to appear late after the onset of symptoms. Brusa et al. (1979), for example, in their review of 75 reported cases with neuropathological confirmation found that in half of these cases, slowness and limitation of both conjugate voluntary and vergence movements ocurred 2 to 4 years after the onset of any other symptom. In the early stage then a firm clinical diagnosis is difficult, frequently impossible and one is often satisfied in rulling out other neurological disorders if possible and establishing a diagnosis of possible or probable PSP which only the passage of time will clarify. Unfortunately there are no biological markers or neuroimaging techniques that offer great help with this difficult diagnostic task.

In the brief review that follows we summarize the main clinical features of PSP emphasizing their relative diagnostic value and insisting upon those clinical signs that are helpfull in the differential diagnosis with Parkinson disease (PD), and other motor and cognitive disorders that can pose on occasion diagnostic problems. A list of other less common or clinically significant features is also provided in Table 1. We also briefly comment on existing diagnostic criteria and propose a new set of them.

## Clinical manifestations

### Oculomotor disturbances

The neurological abnormality considered most important for the diagnosis of PSP is the presence of a supranuclear ophtalmophegia (SNO). Down gaze palsy has to be present for the gaze palsy to be considered of significance since upward gaze impairment can be encountered in PD and other neurological disorders and is present in some normal elderly individuals. Gradually the gaze disturbance involves both vertical and horizontal movements and a complete supranuclear ophtalmoplegia will eventually occur in more than half of the patients. Typically movements to command are more affected than those to pursuit and by definition the oculocephalic movements are preserved. Other oculomotor and eyelid abnormalities occur in PSP (Kristensen, 1985; Lepore et al., 1988; Troost and Daroff, 1977) and are summarized in Table 2. Some of them, such as lid lag, with markedly reduced blink rate or apraxia of lid opening, are characeristic of PSP.

**Table 1.** Clinical manifestations of PSP

*A. Main clinical features*
Supranuclear gaze palsy
Axial dystonia
Bradykinesia
Pseudobulbar palsy
Postural instability and falls
Behavioural and cognitive disturbances

*B. Uncommon and less prominent features*
Pyramidal tract signs
Focal or segmental dystonia
Abnormal sleep patterns
Inspiratory gasps; periods of involuntary tachypnea
Seizures
Action myoclonus
Paroxismal kinesigenic dystonia
Muscle atrophy, fasciculations and weakness
Rest tremor
Chorea
Major depression
Schizophreniform psychoses
Autonomic dysfunction

**Table 2.** Neuro-ophtalmologic features of PSP

*A. Ocular abnormalities usually preceding gaze palsy*
Defective visual supression of the vestibuloocular reflex
Loss of the fast component of the optokinetic nystagmus
Ocular fixation instability with square-wave jerks
Impersistence of gaze
Hesitancy when initiating an eye movement to command, usually a command
   to look down
Slow and hypometric saccades (initially on the vertical plane)

*B. Supranuclear gaze palsy*
Vertical supranuclear gaze palsy (saccadic movements affected before smooth-
pursuit movements, and downgaze usually more disturbed than upgaze;
   oculocephalic movements are preserved)
Reduced or absent convergence
Global (vertical and horizontal) supranuclear ophtalmoplegia

*C. Other uncommon oculomotor disturbances*
Internuclear ophtalmoplegia
Loss of Bell's phenomenon
Loss of oculocephalic movements and vestibuloocular reflexes, in advanced phases
   of the illness

*D. Eyelid disturbances*
Eyelid retraction (Cowper's sign) with wide-eyed staring look
Blepharospasm (eyebrow elevation sometimes associated)
Palpebral ptosis (due to levator inhibition)
Apraxia of lid opening
Apraxia of lid closure
Decreased blink rate (may result in exposure keratitis)

In contrast to the appreciation of Steele et al. (1964) that SNO was usually early and constant, later studies showed that it tends to occurs relatively late and is usually detected up to 4 years after disease onset (Pfaffenbach et al., 1972). Almost always, though, saccade movements are slow even though their magnitude may be normal and hypometric refixation occurs eventually with square wave jerks (Troost and Daroff, 1977). In these early stages the fast component of the opticokinetic nystagmus (OKN) is frequently abolished making this test a valuable diagnostic maneuvre. Still, loss of OKN is a frequent finding in olivopontocerebellar atrophy (OPCA) (Duvoisin, 1987) and in Parkinson-Dementia syndrome of Guam (Lepore et al., 1988). Failure in suppressing the vestibuloocular reflex is another initial oculomotor disturbance in PSP, but this sign can also be found in PD (Rascol et al., 1989) and OPCA (Duvoisin, 1987) and can not be considered as specific of PSP.

While SNO as described above constitutes the hallmark of PSP a number of features that characterize the gaze palsy of PSP make it particularly difficult at times to evaluate and probably contribute to diagnostic errors. On the one hand SNO may not only appear late but not appear at all. Numerous reports with pathological confirmations have insisted on this point (Dubas et al., 1983; Jellinger et al., 1980; Probst, 1977). Another problem is the changing characteristics of the gaze disorder that can vary from one day to another or even in consecutive exams in the same day. Severe cervical dystonia with neck flexion has made it impossible for us in some patients to determine how much down gaze palsy is present and whether voluntary movements were more affected than reflex ones. Finally, other non-familial multisystem degenerative disorders of the nervous system can present with a SNO indistinguishable from the one encountered in PSP (Lees, 1987).

Because of these difficulties strict criteria for defining the supranuclear gaze palsy of PSP have been proposed (Golbe and Davis, 1993). They include the presence of voluntary downward gaze of less than 15° degrees and preserved oculocephalic reflexes (except in the terminal phases of the illness) or alternatively require the presence of the following abnormalities: 1) hesitancy of voluntary gaze, 2) impaired OKN with stimulus moving downwards and 3) poor suppression of vertical vestibuloocular reflex. Except for specific research pourposes we belief that the presence of a supranuclear gaze palsy with down-gaze abnormality sufficiently defines the SNO that occurs in PSP.

*Parkinsonism*

Steele et al. (1964) in their original description emphazised that none of the cases that they reported had been considered as parkinsonism by any of the numerous neurologists that had examined them. Subsequent authors though, have considered differently the importance of parkinsonism in the clinical syndrome of PSP.

Duvoisin considers the presence of bradykinesia essential in the diagnosis of PSP (Duvoisin et al., 1987) while others like Kristensen (1985), have considered parkinsonism as a less constant and minor finding. Brusa et al. (1979), reviewed the clinical features of 75 published pathologically proven cases of PSP and mentions that bradykinesia was present in only 22% of cases during the first year from onset of symptoms. Still, mostly due to the frequent presence of bradykinesia, parkinsonism has been considered the major differential diagnosis of PSP (Duvoisin et al., 1987) and PSP has been encountered frequently among populations of parkinsonian patients. Jackson et al. (1983) for example, found that 16 (3.9%) among 415 patients with parkinsonism had findings of PSP and Agid et al. (1986) reports that 7% of patients admitted to their hospital with the diagnosis of parkinsonism turned out to have PSP. Duvoisin et al. (1987) have estimated that the true prevalence of PSP among patients diagnosed on clinical grounds as having parkinsonism must be in the neighbourhood of 12%. In part these discrepancies result from a different use of the term parkinsonism. Some authors include dysequilibrium and falls as part of the parkinsonism of the patients, together with bradykinesia and rigidity (Brusa et al., 1979). Others, when describing parkinsonism in PSP are describing bradykinesia and rigidity (Blin et al., 1990) and still others mostly or exclusively bradykinesia (Duvoisin et al., 1987). Contributing to this problem is the inherent difficulty that exists in diagnosing parkinsonian signs in patients with other neurological signs. Bradyphrenia, for example, interferes with the appreciation of bradykinesia, and dystonia of the neck can make it impossible to diagnose neck rigidity.

While both, in the early and late stages, PSP differs considerably from PD it is true that in a minority of cases both disorders can be very difficult to differentiate. In such cases an akinetic-rigid syndrome without ophtalmoplegia marks the onset and subsequent progression of the illness for a number of years. In these patients the diagnosis of PSP can be suspected on the basis of the clinical differences between both parkinsonian syndromes (see Problems in Differential Diagnosis) and the poor response to levodopa encountered in PSP patients but a firm diagnosis will not be possible until supranuclear ophtalmoplegia, or additional signs typical of PSP such as neck hyperextention develop. A sustantial beneficial reponse to levodopa can occur ocasionally further complicating the differential diagnosis. A recent clinico-pathological study (Hughes et al., 1992) points to the difficulties in separating clinically PD from PSP. Among hundred brains received in the PD London Brain Bank from patients diagnosed as PD, the authors established the neuropathological diagnosis of PSP in 7 cases, typical in 2 and atypical in 5. Unfortunately further details on the clinical characteristics of these seven patients are not given in the paper.

Bradykinesia and increased muscle tone and gait difficulties and falls are frequent in PSP and it is because of these signs that these patients are cared for by parkinsonologist and later perhaps published as cases of atypical parkinsonism. This does not meen that the differentiation between these two disorders is particularly difficult. The onset of both disorders is quite different and the fully developed picture of PSP differs considerably from PD and other forms of parkinsonism. The face, while inmobile is deeply furrowed with an astonishing look, the body is not flexed and posturing of the hands as it occurs in PD does not occur. Further the gait is not a "petit pas" and associated movements of the arms are present. Rest tremor is a very infrequent feature.

Even though the clinico-pathological correlation of PSP is not well defined, the presence of parkinsonism is to be expected in PSP considering that the zona compacta of the substantia nigra is invariably involved by the pathological process and the marked dopamine loss (80 to 90%) generally present in the caudate and putamen of these patients (Jellinger et al., 1980; Kish et al., 1985). PET studies with 18F-dopa have found reduced striatal (caudate greater than putamen) 18F-dopa uptake in PSP with 90% of individual cases showing reduced caudate 18F-dopa influx. In PSP, though, unlike in PD, striatal 18F-dopa uptake does not correlate to locomotor status (see Brooks, in this volume, p. 119).

### Dystonia posture and gait

Dystonia of the neck muscles, "nuchal dystonia", is an important feature of PSP. In contrast to the flexed posture of the limbs, neck and trunk in Parkinson disease, the neck is held in extension and on passive motions it offers a rubbery resistance to anterior-posterior movements, whereas tone on rotation is normal. Frequently neck extention is more pronounced in the erect posture. Further due to dystonic involvement of the upper trunk muscles PSP patients tend not infrequently to stand with the trunk hyperextended. These postural abnormalities aggravate the gait disorder of the patient and contribute to the falls backward that these patients frequently experience. The prominent dystonic rigidity of the neck contrasts with the absence or only mild increase in tone encountered in the limbs.

Neck dystonia in extension is rarely the initial manifestation of PSP but if it appears early it is as distinctive as the ophtalmoplegia. While not as characteristic nuchal dystonia is not infrequently in flexion or neutral. In some instances, like in severe PD patients, the trunk is markedly bend forward due to axial dystonia in flexion. Rarely nuchal rigidity remains mild until an advanced stage.

It is unclear which lesion of the central nervous system are responsible for axial dystonia. Steele et al. (1964) thought it was caused by the peculiar combination of lesions in the substantia nigra, pallidum, subthalamic nuclei and red nuclei. Behrman et al. (1969) suggested that the neck posture might be a manifestation of segmental descerebrate rigidity caused by partial denervation of motor neurons in the upper cervical cord. Messert and Van Nuis (1966) and others have attributed it to lesions of the vestibular nuclei.

Besides the neck, dystonic spasms can be uncountered in PSP patients involving a number of other different regions. The staring facial expression referred by Steele as "spastic" is essentially dystonic in the sense of a sustained abnormal posture of the face. Prominent trismus starting four years after disease onset has also been reported in a patient by Anzil (1969). Although limb dystonia is not common and mentioned only seldom, recent reports indicate that it can be found more frequently than previously suspected. Rafal and Friedman (1987) found limb dystonia in 8 out of 30 patients with PSP and observed that it did not correlate with severity of neck dystonia. Other reports (Léger, 1987), describe limb dystonia as an early sign of PSP.

Gait is always abnormal in PSP and is usually affected in the early stages with a symmetrical bilateral involvement. Although slowness, shortness of stride, shuffling, festinantion and freezing may be seen later, as in Parkinson's disease, gait tends to be broad-based, and associated arm movements are preserved. Dysequilibrium due to posture instability is also a prominent and initial feature, and usually the finding that best explains the frequent falls that these patients suffer.

*Pseudobulbar function*

Signs of bulbar dysfunction of varying degrees of severity are constant findings in PSP and can be the earliest manifestation of the disease and slurred speech and difficulty swallowing frequently preceed the onset of gaze disturbances. Facial and jaw jerks are exagerated and tongue movements are slow. The speech is slurred from spastic dysarthria and generally slow. It can have a strained strangled quality and can be hypophonic as in Parkinson disease. It gradually worsens and may become unintelligible. Dysphagia also can occur early and tends to progress. Initially patients experience difficulty in handling solids or fluids and drooling can be prominent. In advanced stages patients frequently choke on their own saliva and explosive coughing is not uncommon. Most patients eventually require semi-liquid food that has to be administered via nasogastric tube. Recently Sonies (1992) reported on the investigation of swallowing mechanisms in 22 patients with PSP. All patients reported having difficulty swallowing and most were aware that they had problems during both oral and pharyngeal phases of swallowing. Many reported coughing to clear their throats and thus avoid aspiration. Some patients had deficient control of the timing of initiation of swallowing since they could not swallow on command. Triggering of a single swallow stimulated additional swallows even when no bolus was in the mouth, further indicating a deficient control in terminating the swallowing response. When compared to Parkinson disease, PSP patients were highly aware of their difficulties swallowing and rarely had episodes of silent aspiration common to Parkinson disease. According to the authors of this study head hyperextension may be partially responsible for the dysphagia in PSP since it delays the triggering of the swallowing reflex, inhibits elevation of the hyoid bone and can be another cause of the lengthy delay in initiation of swallowing characteristic in this disorder.

Uncontrollable crying and laughter are common, although it is said less so then in other forms of pseudobulbar palsy. Such responses can be overlooked since they become manifested only when the patient tries to engage in some form of conversation. These exacerbated emotional responses are frequently a source of major disability since they interfere with communication, particularly when already altered by dysarthria, palilalia, drooling and explosive coughing.

## Mental status

Patients with PSP frequently have behavioural and cognitive problems. They complain of forgetfulness and their families indicate that they have become innatentive, disinterested and indifferent to their surroundings. They may have lost the capacity to take initiatives and have difficulty in making decisions. Another frequent complain is that the patient speeks little and that his or her mental responses are slow. Generally though both the patient and their relatives express their belief that the mental status of the patient is otherwise acceptably normal an opinion which frequently agrees with the physician impression that the patients responses are very slow but accurate.

The degree and nature of cognitive impairment in PSP has been a matter of debate. Dementia was present in 7 of the 9 patients originally described by Steele et al. (1964) but in later series it has been a variable finding. This is so in part because in some studies clinical criteria have been employed to define dementia while others have based the diagnosis on the results of neuropsychological test batteries. Further difficulties encountered in assesing the cognitive functions of PSP are related to the methodological difficulties present in evaluating cognitive functions both at bedside and in the neuro-psychological laboratory in patients with prominent gaze abnormalities, dysarthria and slowness of thought. It is frequently striking when caring for these patients that they "look demented", more than what they eventually turned out to be when one takes the time to talk to them and to their relatives inquiring about their level of mental funtioning. The masked face with staring look, lack of spontaneous verbal output, severe dysarthria and explosive speech with bouts of incontrollable crying or laughter, presence of palilalia and long delay in answering when questioned, all of them common manifestations of the illness, frequently convey the wrong impression about the presence of dementia or its degree.

In 1974 Albert et al. (1974) introduced the term *subcortical dementia* to describe the characteristic profile of cognitive dysfunction occuring in PSP patients and which they attributed to the well known subcortical lesions occuring in the disorder. They characterized the dementia of PSP by slowness of thought (bradyphrenia), forgetfulness, changes in personality with apathy or depression and impaired ability to manipulate acquired knowledge. They drew attention to the differences between this clinical syndrome and the so-called cortical dementia seen in disorders with prominent cortical pathology such as Alzheimer disease, where afasia, apraxia and agnosia are prominent. A number of studies using comprehensive neuropsychological testing (Cambier et al., 1985; Grafman et al., 1990; Litvan et al., 1989; Maher et al., 1985; Pillon et al., 1986) have since shown, that a substantial intellectual impairment occurs in most patients with PSP and also that these patients have 1) cognitive slowing, 2) particular difficulties with tests of frontal lobe functions and 3) relative preservation of memory and instrumental funtions such as language and praxis. The *abnormalities in frontal lobe-like funtions* or so-called executive functions (Lezak, 1983), explain the behavioural and cognitive alterations such as impaired attention, poor abstract thinking and reasoning, mild to moderate memory loss and impaired elaborate linguistic abilities. They occur frecuently in association with the classical frontal lobe signs of forced hand grasping, groping and increased oral reflexes and the presence of utilization behaviour and motor perseveration. Pillon and Dubois (1992) have recently presented evidence that the frontal lobe like sndrome of PSP evolves as the disease progresses but that the neuropsychological deficits remain sufficiently specific even in late stages of the disease to conform of the characteristic pattern of dementia described by Albert et al. (1974).

It has been recently pointed out that some of the neuropsychological disturbances in PSP could be in part related to the visual disorder, particularly those detected in tests with a strong visual scanning requirement (Fisk et al., 1982). Pirozollo et al. (1986) for example, found that their patients performed less well then controls only in those cognitive functions requiring visual orientation or visual memory. Rafal et al. (1988) found that PSP patients have difficulty with attention with vertically displaced stimuli but this was independent of motor gaze palsy. Clinical impression suggests that the eye abnormalities do not play a significant role in the cognitive functioning of these patients.

While a prominent dementia is diagnosed clinically in a minority of patients with PSP at least in the early stages, PSP has been diagnosed at autopsy in patients in which a progressive dementia had been the dominant clinical manifestation throughout the course of the illness. These patients with a *dementing form of PSP* are generally diagnosed in life as Alzheimer disease or other dementing disorder. In these cases the correct diagnosis of PSP can only be made in life if the cardinal features of the illness appear in conjunction with the dementing state (Davis et al., 1985; Masliah et al., 1991; Olson et al., 1992).

The anatomical substrate and pathophysiology of the cognitive and behavioural abnormalities of PSP have not been clearly defined. Albert et al. speculated that the intellectual dysfunction in PSP might be due to impaired timing, alerting and activating mechanisms resulting in damage to the frontolimbic connections (Albert et al., 1974). Several authors have recently attributed most of these intellectual changes to frontal dysfunction due to "deafferentation" or "weak activation" of the prefrontal cortex secondary to the lesions in the subcortical structures. PET images of energy metbolism showing hypometabolism of 18 F-fluoro-deoxyglucose in the frontal lobe in

PSP are consistent with the presence of frontal dysfunction in PSP (Brooks, this volume, p. 119), and (Goffinet et al., 1989). Frontal hypometabolism, relative to the remainder of the cerebral cortex, has been shown to correlate with scores on neuropsychological tests specific for frontal lobe functions. While the origin of the frontal syndrome and the underlying hypometabolism are presumably caused by deafferentation of the frontal cortex due to subcortical lesions, it is also possible that cortical involvement with neurofibrillary tangles or with neuropil threads, as recently shown by Verny et al. (this volume, p. 179) may contribute to the intellectual disorder of these patients.

## Problems in differential diagnosis

The clinical spectrum of PSP may take years to develop fully. Once the main features are present it is a relatively easy disorder to diagnose since few other neurological disorders progress gradually to a state of bradyphrenia, supranuclear ophtalmoplegia, neck hyperextension, pseudo-bulbar palsy and marked postural abnormalities with frequent falls. Rarely, though, other neurological disorders can mimic PSP clinically since they manifest several or all of the cardinal features of PSP. Such disorders include diffuse Lewy body disease (De Bruin, 1992; Fearnley et al., 1991; Lewis and Gawel, 1990), Alzheimer disease, progressive subcortical gliosis (Foster et al., 1992) corticobasal degeneration (Gibb et al., 1989) and multisystem atrophy (both striatonigral and olivopontocerebellar types) (Al-Din et al., 1990; Koepper and Hans, 1976). Difficulties may also arise occasionally with multiinfarct states (see Winikates and Jankovic, this volume, p. 189), normo-pressure hydrocephalus (Morariu, 1979), rigid forms of Huntington disease, Jakob-Creutzfeld disease (Ross-Russell, 1980), the Parkinson-Dementia complex of Guam (Steele and Guzman, 1987) and other disorders such as Wilson's disease, Machado-Joseph's disease (Rosenberg et al., 1976), Gaucher's disease (Tripp et al., 1977), juvenile dystonic lipidosis (Neville et al., 1973), Pick's disease and the punch drunk syndrome (Roberts, 1969). Also, the presence of pseudobulbar palsy, exaggerated DTR's and gaze abnormalities with the presence of respiratory rythm disorders can resemble postencephalitic Parkinson's disease (Perkin et al., 1978).

The *early diagnosis* of PSP when most of the characteristic signs are yet to become manifested, can be specially difficult and description of pathologically proven cases indicates frequently early misdiagnosis. In particular, the diagnosis of PSP can not be established, in the absence of supranuclear gaze palsy. Considering the frequency of gait instability, parkinsonian-like, visual and pseudobulbar symptoms and memory and behavioural problems as presenting symptoms (Golbe et al., 1988; Jackson et al., 1983; Maher and Lees, 1986) and the different rate of appearance of the different neurological manifestations of the illness (Brusa et al., 1979) it is not surprising that initial diagnosis of PSP patients includes PD as well as other parkinsonian syndromes, the sporadic ataxias, Alzheimer disease or other dementing disorders as well as psychiatric diagnosis such as depression. Corticobasal ganglionic degeneration can occasionally present with supranuclear gaze palsy, parkinsonism and gait disturbances and also mimic PSP. Early prominent bulbar findings can be particularly difficult to separate from the bulbar forms of Amyotrophic Lateral Sclerosis and can also resemble a multiinfarct state. The ocular manifestations can also lead to a mistaken diagnosis such as progressive external ophtalmoplegia on a myopathic basis or myasthenia gravis, particularly if the presentation is atypical, e.g. with internuclear ophtalmoplegia (Mastaglia and Grainger, 1975; Perkin et al., 1978).

Particularly difficult from a diagnostic point of view are the problems that PSP patients pose when they present with atypical clinical manifestations. Jellinger et al. (1980), for example, described four pathologically proven cases of PSP in which the clinical manifestations developed before age 34 in mentally subnormal individuals who presented a prominent neuropsychiatric disorder which included violent behaviour with outbursts of irritability and aggressiveness, restlessness and disorientation causing early admission to psychiatric hospitals. Other *atypical forms* are those without ophtalmoplegia (Dubas et al., 1983; Jellinger et al., 1980; Probst, 1977), those presenting mostly as a severe dementing disorder (Davis et al., 1985) or as an isolated akinetic-rigid syndrome with limited or no response to levodopa (Lees, 1987) and those presenting as a pure akinetic syndrome (Matsuo et al., 1991; Yamamoto et al., 1990). Clinical features helpfull in differentiating PD from the akinetic-rigid variant of PSP are shown in Table 3.

The difficulties that clinicians can encounter in reliably diagnosing PSP are exemplified by a recent clinicopathological study of 13 autopsied cases

Table 3. Parkinsonism in PSP: differences with PD

| Progressive supranuclear palsy | Parkinson's disease |
| --- | --- |
| Symmetric parkinsonian signs | Asymmetric at onset |
| Gait impaired early | Little impairment of gait at onset |
| Falls early in the course | Falls occur late in the course |
| Early impairment of postural reflexes | Postural reflexes normal in early stages |
| Wide based gait | Marche a petit pas |
| Trunk posture in extension | Body flexion when walking |
| Arm swing when walking present | Early loss of arm swing |
| Astonished facial expression | Facial amimia |
| Blink rate 3–5 per minute | Blink rate 10–14 per minute |
| Rest tremor is uncommon | Rest tremor is frequent |
| More prominent axial dystonia/rigidity than limb dystonia | Limb dystonia/rigidity more common than axial |
| Absence of hand deformity | Characteristic hand deformity |
| Absent or poor response to levodopa | Good response to levodopa |
| Levodopa induced dyskinesias are rare | Levodopa induced dyskinesias are frequent |
| Wearing-off and "on-off" phenomena are unusual | Wearing-off and "on-off" phenomena are frequent |

of PSP (Olson et al., 1992). Most patients had clasical features of the disorder, including ophtalmoplegia, postural instability and extrapyramidal signs. Still, five of the thirteen patients were diagnosed clinically as having Alzheimer disease. Neuropathological diagnosis was also heterogeneous since besides typical PSP changes a diagnosis of Alzheimer disease was also made in 6 and changes typical of PD were present in 1.

## Clinical diagnostic criteria

In an attempt to improve diagnostic accuracy to give appropiate prognosis to patients and their families and to select correct cases for drug studies or other research pourposes, diagnostic criteria for PSP have been developed (Blin et al., 1990; Golbe and Davis, 1993; Lees, 1987). The requirements for a diagnosis of PSP according to these different criteria are quite variable. Some, such as those proposed by Lees, include the presence of SNO as essential for the diagnosis of PSP (Lees, 1987). Others require as essential SNO and bradykinesia (Golbe and Davis, 1993) while still others include SNO, bradykinesia and axial rigidity (see Duvoisin, this volume, p. 51). SNO itself is quite differently defined (see Section on oculomotor disturbances). According to these proposed criteria a number of additional items have to be present to complement these essential criteria in order to establish a diagnosis of PSP. Which ones and how many, again, depends on the opinion of the different authors consulted. As a result according to one set of criteria patients with a progressive neurological disorder, starting in middle or old age, can be diagnosed as having PSP if they have SNO and only bradykinesia, limb rigidity and poor response to levodopa as additional signs (De Bruin, 1992), while according to another set of criteria SNO plus bradykinesia have to be present and in addition any three of the following manifestations: a) dysarthria or dysphagia; b) axial rigidity greater than limb rigidity; c) neck in a posture of extension; d) minimal or absent tremor; e) frequent falls of gait disturbance occurring early in disease course and f) poor or absent response to levodopa.

We propose here based on previous data and our personal experience, a set of diagnostic criteria (Table 4) that we belief are easy to use by most practicing neurologists. They do not include certain clinical maneuvres, such as the vestibuloocular reflex or the assesment of utilization behaviour which are likely to be evaluated quite variably by different neurologists, nor results of laboratory tests, such as CT scans or neuropsychological testing, since the practical diagnostic value of such tests has not been clearly established. Since only few neurological disorders other than PSP present with the characteristic combination of supranuclear gaze palsy, pseudobulbar palsy, parkinsonism, disequilibrium and falls, bradiphrenia and frontal lobe signs and neck dystonia the use of the criteria we propose or of other similar ones will exclude the conditions most commonly misdiagnosed as PSP and most likely will yield only a small rate of false positive diagnosis when patients with the full blown syndrome are evaluated. Since we have con-

**Table 4.** Criteria for clinical diagnosis of PSP

---

Non familial disorder
Onset after age 40
Progressive course
Supranuclear gaze palsy with involvement of downward gaze

*Plus the following manifestations:*
(**probable PSP** if *three or more* are present; **possible PSP** if *two* are present)
Postural instability and falls
Bradykinesia with poor or transient response to levodopa
Pseudobulbar palsy
Prominent bradyphrenia with frontal lobe signs (e.g. grasping, perseveration)
Greater axial than limb dystonia/rigidity

*Plus absence of all of the following:*
Prominent cerbellar signs
Unexplained polineuropathy
Prominent, unexplained dysautonomia
Focal sensory deficits-primary or cortical
Unilateral limb apraxia

---

sidered, unlike others (Blin et al., 1990), the presence of SNO as a necessary criteria to diagnose "probable" or "possible" PSP our criteria will unavoidably leave undiagnosed the occasional case without ophtalmoplegia. Other atypical cases such as those presenting with prominent dementia or as a purely akinetic syndrome will also be excluded. Only well designed prospective clinicopathological studies will determine the value of these and other clinical criteria.

## Conclusions

PSP presents clinically with a rather distinct picture that generally allows differentiation with other neurological disorders. Still diagnostic difficulties may arise in the early stages when only few symptoms and signs such as poor equilibrium and falls or behavioural abnormalities are present or still in those atypical cases when the disorder manifests itself predominantly or exclusively as a parkinsonism or dementia and in those cases when a clear supranuclear ophtalmoplegia does not develop. In these cases the degree of diagnostic certainty can probably be improved taking advantage of diagnostic tests which complement the clinical diagnosis. These procedures include CT-scan, NMR, 123-iodobenzamide-SPECT (Schwarz et al., 1992) and PET studies, sleep polisomnography, neuropsychological evaluation and pharmacological tests such as the apomorphine test (Hughes et al., 1991). While the practical diagnostic values of such tests remains to be determined their judicious use undoubtedly helps in supporting the clinically-based diagnosis of PSP. Unfortunaly the value of these tests is probably less in the early stages of the illness.

While PSP is clinically heterogeneous and may present a serious diagnostic challenge to the clinician, the neuropathological diagnosis can also be difficult. It is straightforward to diagnose the typical case with neuronal loss and neurofibrillary tangles in the appropiate distribution. Judging from reports in the literature and from conversations with experienced pathologists though, cases are not infrequently encountered in which the histological changes are not typical, mostly because of the severity and distribution of such changes. The neuropathologist may have difficulties in classifying such cases which are frequently labeled as atypical (Hughes et al., 1992; Olson et al., 1992; see "The neuropathology of progressive supranuclear palsy" by P. Lantos, this volume, p. 137). Adding to these diagnostic difficulties is the presence in some PSP brains of Lewy bodies and Alzheimer-like pathology which may have to be considered as age-related or diagnostic of an associated neurological disorder such as Parkinson disease or Alzheimer disease (Cruz-Sánchez et al., 1992; Milder et al., 1984; Mori et al., 1986; Sasaki et al., 1991). Such neuropathological co-morbidity was present in 7 of the 13 cases reviewed by Olsen et al. (1992). Obviously both clinical and neuropathological markers are urgently needed to reliably diagnose PSP.

## References

Agid Y, Javoy-Agid F, Rubert M, et al (1986) Progressive supranuclear palsy: anatomo-clinical and biochemical considerations. Adv Neurol 45: 191–206

Albert ML, Feldman RG, Willis AL (1974) The "subcortical dementia" of progressive supranuclear palsy. J Neurol Neurosurg Psychiatry 37: 121–130

Al-Din ASN, Al-Kurdi A, Al-Salem MK (1990) Autosomal recessive ataxias, slow eye movements, dementia and extrapyramidal disturbances. J Neurol Sci 96: 191–205

Anzil AP (1969) Progressive supranuclear palsy: case report with pathological findings. Acta Neuropathol (Berl) 14: 72–76

Behrman S, Carroll JD, Janota I, Matthews WB (1969) Progressive supranuclear palsy. Clinico-pathological study of four cases. Brain 92: 663–678

Blin J, Baron JC, Dubois B, Pillon B, Cambon H, Cambier J, Agid Y (1990) Positron emission tomography study in progressive supranuclear palsy: brain hypometabolic pattern and clinicometabolic correlations. Arch Neurol 47: 747–752

Brusa A, Mancardi GL, Bugiani O (1979) Progressive supranuclear palsy 1979: an overview. Italian J Neurol Sci i: 205–222

Cambier J, Masson M, Viader F, Limodin J, Strube A (1985) Le syndrome frontal de la paralysie supranucléaire progressive. Rev Neurol (Paris) 141: 528–536.

Cruz-Sánchez FF, Rossi ML, Cardozo A, Deacon P, Tolosa E (1992) Clinical and pathological study of two patients with progressive supranuclear palsy and Alzheimer's changes. Antigenic determinants that distinguish cortical and sub-cortical neurofibrillary tangles. Neurosci Lett 136: 43–46

Davis PH, Bergeron C, McLachlan DR (1985) Atypical presentation of progressive supranuclear palsy. Ann Neurol 17: 337–343

De Bruin VMS, Lees AJ, Daniel SE (1992) Diffuse Lewy body disease presenting with supranuclear gaze palsy, parkinsonism and dementia: a case report. Mov Disord 7: 335–358

Dubas F, Gray F, Escourelle R (1983) Maladie de Steele-Richardson-Olszewski sans ophtalmoplégie: six cas anatomo-cliniques. Rev Neurol (Paris) 139: 407–416

Duvoisin RC (1987) The olivopontocerebellar atrophies. In: Marsden DC, Fahn S (eds) Movement disorders, vol 2. Butterworth, London, pp 249–271

Duvoisin RC, Golbe LI, Lepore FE (1987) Progressive supranuclear palsy. Can J Neurol Sci 14: 547–544

Fearnley JM, Revesz T, Brooks DJ, Frackowiak RSJ, Lees AJ (1991) Diffuse Lewy body disease presenting with a supranuclear gaze palsy. J Neurol Neurosurg Psychiatry 54: 159–161

Fisk JD, Goodale MA, Burkhart G, Barnett HJM (1982) Progressive supranuclear palsy: the relationship between ocular motor dysfunction and psychological test performance. Neurology 32: 698–705

Foster NL, Gilman S, Berent S, Sima AAF, D'Amato C, Koeppe RA, Hicks SP (1992) Progressive subcortical gliosis and progressive supranuclear palsy can have similar clinical and PET abnormalities. J Neurol Neurosurg Psychiatry 55: 707–713

Gibb WRG, Luthert PJ, Marsden CD (1989) Corticobasal degeneration. Brain 112: 1171–1192

Goffinet AM, De Volder AG, Gillain C, Rectem D, Bol A, Michel C, Cogneau M, Labar D, Laterre C (1989) Positron tomography demonstrates frontal lobe hypometabolism in progressive supranuclear palsy. Ann Neurol 25: 131–139

Golbe LI, Davis PH (1993) Progressive supranuclear palsy. In: Jankovic J, Tolosa E (eds) Parkinson's disease and movement disorders. William & Wilkins, Baltimore, pp 145–161

Golbe LI, Davis PH, Schoenberg BS, Duvoisin RC (1988) Prevalence and natural history of progressive supranuclear palsy. Neurology 38: 1031–1034

Grafman J, Litvan I, Gómez C, Chase TN (1990) Frontal lobe function in progressive supranuclear palsy. Arch Neurol 47: 553–558

Hughes AJ, Lees AJ, Sterm GM (1991) Challenge tests to predict the dopaminergic response in untreated Parkinson's disease. Neurology 41: 1723–1725

Hughes AJ, Daniel SE, Kilford L, Lees AJ (1992) Accuracy of diagnosis of idiopathic Parkinson's disease: a clinico-pathological study of 100 cases. J Neurol Neurosurg Psychiatry 55: 181–184

Jackson JA, Jankovic J, Ford J (1983) Progressive supranuclear palsy: clinical features and response to treatment in 16 patients. Ann Neurol 13: 273–278

Jellinger K, Riederer P, Tomonaga M (1980) Progressive supranuclear palsy: clinico-pathological and biochemical studies. J Neural Transm [Suppl] 16: 111–128

Kish SJ, Chang LJ, Mirchandani L, Shannak K, Hornykiewicz O (1985) Progressive supranuclear palsy: relationship between extrapyramidal disturbances, dementia and brain neurotransmitter markers. Ann Neurol 18: 530–536

Koepper A, Hans M (1976) Supranuclear ophtalmoplegia in olivopontocerebellar degeneration. Neurology 26: 764–768

Kristensen MO (1985) Progressive supranuclear palsy — 20 years later. Acta Neurol Scand 71: 177–189

Lees AJ (1987) The Steele-Richardson-Olszewski syndrome (progressive supranuclear palsy). In: Marsden DC, Fahn S (eds) Movement disorders, vol 2. Butterworth, London, pp 272–287

Léger JM, Girault JA, Bolgert F (1987) Deux cas de dystonie isolée d'un membre supérieur inaugurant une maladie de Steele-Richardson-Olszewski. Rev Neurol (Paris) 143: 140–142

Lepore FE, Steele JC, Tilson G, Calne DB, Duvoisin RC, Lavine L, McDarby JV (1988) Supranuclear disturbances of ocular motility in Lytico-Bodig. Neurology 38: 1849–1853

Lewis AJ, Gawel MJ (1990) Diffuse Lewy body disease with dementia and oculomotor dysfunction. Mov Disord 5: 143–147

Lezak MD (1983) Neuropsychological assessment. Oxford University Press, New York

Litvan I, Grafman J, Gómez G, Chase TN (1989) Memory impairment in patients with progressive supranuclear palsy. Arch Neurol 46: 765–767

Maher ER, Smith EM, Lees AJ (1985) Cognitive deficits in the Steele-Richardson-Olszewski syndrome (progressive supranuclear palsy). J Neurol Neurosurg Psychiatry 48: 1234–1239

Maher ER, Lees AJ (1986) The clinical features and natural history of the Steele-Richardson-Olszewski syndrome (progressive supranuclear palsy). Neurology 36: 1005–1008

Masliah E, Hansen LA, Quijada S, DeTeresa R, Alford M, Kauss J, Terry R (1991) Late onset dementia with argyrophilic grains and subcortical tangles or atypical progressive supranuclear palsy. Ann Neurol 29: 389–396

Mastaglia FL, Grainger KMR (1975) Internuclear ophtalmoplegia in progressive supranuclear palsy. J Neurol Sci 25: 303–308

Matsuo H, Takashima H, Kishinawa M, Kinoshita I, Mori M, Tsujihata M, Nagataki S (1991) Pure akinesia: an atypical manifestation of progressive supranuclear palsy. J Neurol Neurosurg Psychiatry 54: 397–400

Messert B, Van Nuis C (1966) A syndrome of paralysis of downward gaze, dysarthria, pseudobulbar palsy, rigidity of neck and trunk and dementia. J Nerv Ment Dis 143: 47–54

Milder DG, Elliott CF, Evans WA (1989) Neuropathological findings in a case of coexistent progressive supranuclear palsy and Alzheimer's disease. Clin Exp Neurol 20: 181–187

Morariu MA (1979) Progressive supranuclear palsy and normal-pressure hydrocephalus. Neurology 29: 1544–1546

Mori H, Yoshimura M, Tomonaga M, Yamanouchi H (1986) Progressive supranuclear palsy with Lewy bodies. Acta Neuropathol 71: 344–346

Neville BRG, Lake BD, Stephens R, Sanders MD (1973) A neurovisceral storage disease with vertical supranuclear ophtalmoplegia, and its relationship to Niemann-Pick disease. Brain 96: 97–120

Olson DA, Gearing M, Watts RL, Mirra SS (1992) Clinical-pathological heterogeneity in progressive supranuclear palsy. Ann Neurol 32: 244 (abstract)

Perkin GD, Lees AJ, Stern GM, Kocek RS (1978) Problems in the diagnosis of progressive supranuclear palsy. Can J Neurol Sci 5: 167–173

Pfaffenbach DD, Layton OD, Kearns TP (1972) Ocular manifestations in progressive supranuclear palsy. Am J Ophtal 74: 1174–1184

Pillon B, Dubois B, Lhermitte F, Agid Y (1986) Heterogeneity of cognitive impairment in progressive supranuclear palsy, Parkinson's disease and Alzheimer's disease. Neurology 36: 1179–1185

Pillon B, Dubois B (1992) Cognitive and behavioural impairments. In: Litvan I, Agid Y (eds) Progressive supranuclear palsy. Clinical and research approaches. Oxford University Press, New York, pp 223–239

Pirozzolo FJ, Jankovic J, Levy JK (1986) Progressive supranuclear palsy: are cognitive and motor deficits related. Neurology 36 [Suppl] 1: 308

Probst A (1977) Dégénérescence neurofibrillaire sous-corticale sénile avec présence de tubules contournés et de filaments droits: form atypique de la paralysie supranucléaire progressive. Rev Neurol (Paris) 133: 417–428

Rafal RD, Friedman JH (1987) Limb dystonia in progressive supranuclear palsy. Neurology 37: 1546–1549

Rafal RD, Posner MI, Friedman JH, Inhoff AW, Bernstein E (1988) Orienting of visual attention in progressive supranuclear palsy. Brain 111: 267–280

Rascol O, Clanet M, Monstastruc JL, Simonetta M, Soulier-Esteve MJ, Doyon B, Rascol A (1989) Abnormal ocular movements in Parkinson's disease. Brain 112: 1193–1214

Roberts AH (1969) Brain damage in boxers: a study of prevalence of traumatic encephalopathy among ex-professional boxers. London, Pitman

Rosenberg B, Nyhan WL, Day C, Shore P (1976) Autosomal dominance striionigral degeneration. Neurology 26: 703–714

Ross-Russell R (1980) Supranuclear palsy of eyelid closure. Brain 103: 71–82

Sasaki S, Maruyama S, Toyoda Ch (1991) A case of progressive supranuclear palsy with widespread senile plaques. J Neurol 238: 345–348

Schwarz J, Tatsch K, Arnold G, Gasser T, Trenkwalder C, Kirsch CM, Oertel WH (1992) 123-Iodobenzamide-SPECT predicts dopaminergic responsiveness in patients with de novo parkinsonism. Neurology 42: 556–561

Sonies BC (1992) Swallowing and speech disturbances. In: Litvan I, Agid Y (eds) Progressive supranuclear palsy. Clinical and research approaches. Oxford University Press, New York, pp 240–253

Steele JC, Richardson JC, Olszewski J (1964) Progressive supranuclear palsy. A heterogeneous degeneration involving the brainstem, basal ganglia and cerebellum with vertical gaze and pseudobulbar palsy, nuchal dystonia and dementia. Arch Neurol 10: 333–358

Steele JC (1975) Progressive supranuclear palsy. In: Vinken PJ, Bruyn GW, deJong JMB (eds) Handbook of clinical neurology, vol 22. System disorders and atrophies, part II. North-Holland, Amsterdam, pp 217–229

Steele JC, Guzman T (1987) Observations about amyotrophic lateral sclerosis and the Parkinsonism-Dementia complex of Guam with regard to epidemiology and etiology. Can J Neurol Sci 14: 358–362

Tripp JH, Lake BD, Young E, et al (1977) Juvenile Gaucher's disease with horizontal gaze palsy in 3 siblings. J Neurol Neurosurg Psychiatry 40: 470–478

Troost BT, Daroff RB (1977) The ocular motor defects in progressive supranuclear palsy. Ann Neurol 2: 297–403

Yamamoto T, Kawamura J, Hashimoto S, Nakamura M, Iwamoto H, Kobashi Y, Ichijima K (1990) Pallido-nigro-luysian atrophy, progressive supranuclear palsy and adult onset Hallervorden-Spatz disease: a case of akinesia as a predominant feature of parkinsonism. J Neurol Sci 101: 98–106

Authors' address: Prof. E. Tolosa, Neurology Department, Hospital Clinic, Villarroel 170, E-08036 Barcelona, Spain.

# Eyelid motor abnormalities in progressive supranuclear palsy

## F. Grandas[1] and A. Esteban[2]

[1] Servicio de Neurología, and [2] Sección de Neurofisiología, Hospital General "Gregorio Marañón", Madrid, Spain

**Summary.** Eyelid motor abnormalities found in progressive supranuclear palsy are reviewed. Electrophysiological correlates of blepharospasm, levator inhibition (blepharokolysis) and supranuclear paralysis of lid closure are presented.

Disorders of eyelid motility are not uncommon in progressive supranuclear palsy (PSP). They may be found in about one third of patients with this syndrome (Jackson et al., 1983; Golbe et al., 1989). This is not surprising since ocular and eyelid movements are highly coordinated, mainly in the vertical plane (Gordon, 1951; Kennard and Smith, 1963; Kennard and Glaser, 1964), and a supranuclear ophthalmoplegia with down gaze impairment is a cardinal feature of PSP (Steele et al., 1964). The spectrum of eyelid motor disorders described in PSP includes blinking abnormalities, lid retraction, blepharospasm, levator inhibition and supranuclear palsy of eye closure.

### Blinking abnormalities

A profound reduction of blink rates has been observed in patients with PSP (Karson et al., 1984; Golbe et al., 1989).

It has been suggested that eye-blink rates reflect the activity of the central dopaminergic systems (Karson, 1983). Indeed patients with either schizophrenia or Huntington's chorea show higher blink rates than normal subjects, while rates are decreased in Parkinson's disease and are even lower in PSP (Karson, 1983; Karson et al., 1984; Golbe et al., 1989). Furthermore, in Parkinson's disease blink rate abnormalities appear to be related to the degree of dopamine depletion, showing the lowest rates the most affected patients (Karson, 1988).

However, dopaminergic dysfunction may not be the only cause responsible for the low blinking rate in PSP. Rigidity and akinesia occur in both Parkinson's disease and PSP, but the neuropathological basis of the two disorders is different. In PSP, in addition to neuronal loss, gliosis and neurofibrillary tangles in the pars compacta of substantia nigra, the pallido-subthalamic complex, superior colliculus, periaqueductal grey matter and pretectal areas are also specially damaged (Lees, 1987). Lesions of the superior colliculus may be relevant in the pathogenesis of low blink rates in PSP since it receives projections from basal ganglia and visual cortical centers and might be involved in eye and eyelid movements related to visual stimuli (Asher and Gachelin, 1967; Karson, 1988).

Golbe et al. (1989) described "slow blinks" in a patient with PSP, whose eyeblinks consisted of a 0.5 sec. closing phase, a 4 sec. closed phase and a 0.5 sec opening phase, and ocurred one or two times per minute.

Patients with PSP increase their low blink rates during versional gaze tasks (Golbe et al., 1989), perhaps in an attempt to facilitate changes of ocular fixation as blinks can influence saccadic initiation, velocity and accuracy as a synkinetic phenomenon (Gordon, 1951; Zee et al., 1983).

## Lid retraction

Lid retraction is an eyelid disturbance observed in patients with PSP which, in combination with low blink rates and ophthalmoplegia, contributes to give the patient a fixed and staring appeerence (Pfaffenbach et al., 1972; Kristensen, 1985; Maher and Lees, 1986; Lees, 1987).

Lid retraction might be a consequence of levator palpebrae superioris and Müller's muscles dysfunction. It has been suggested that lesions of pretectal area and posterior commisure nuclei play an important role in the patho-genesis of this lid disorder, as damage in these areas induced eyelid retrac-tion in monkeys (Pasik et al., 1969; Schmidtke and Büttner-Ennever, 1992).

## Blepharospasm

Blepharospasm is a focal cranial dystonia characterized by spasms of con-traction of the orbicularis oculi muscles (Grandas, 1989).

Blepharospasm has been noted in patients with PSP (David et al., 1968; Pfaffenbach et al., 1972; Mastaglia et al., 1973; Sing et al., 1974; Jackson et al., 1983), usually in combination with other supranuclear lid disorders (Golbe et al., 1988), with a prevalence ranging in different series from 10 to 28% (Brusa et al., 1980; Jankovic et al., 1988).

Electromyographic recordings of blepharospasm (Fig. 1) show bursts of EMG activity in the orbicularis oculi muscle with persistent activity in the levator palpebrae superioris, coinciding with the clinical spasms of eye closure. Co-contraction of these antagonist muscles reflects a dysruption of their physiological reciprocal innervation (Van Allen and Blodi, 1962; Esteban and Salinero, 1979) and may be regarded as an equivalent of the reciprocal inhibition abnormalities found in other forms of dystonia (Rothwell et al., 1983).

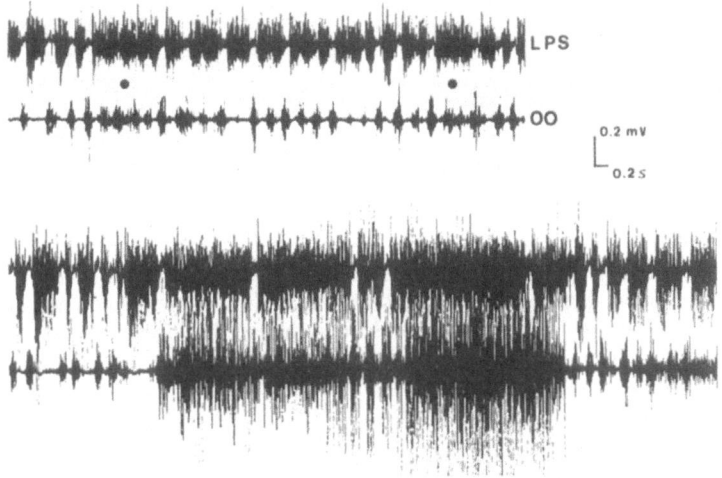

**Fig. 1.** Blepharospasm. Top: Spontaneous irregular bursts in the orbicularis oculi (OO) muscle, which ocasionally tend to fuse, coinciding with activity in levator palpebrae superioris (LPS) (dots). Bottom: Co-contraction is specially manifested during closure of the eyes on command

In idiopathic blepharospasm an enhanced excitability of brainstem interneurons which mediate the late response of blink reflex has been found, reflecting an abnormal excitatory drive, perhaps from the basal ganglia, to the facial motoneurons and brainstem interneurons involved in facial reflexes (Berardelli et al., 1985). The role of brainstem and basal ganglia in the pathophysiology øf blepharospasm is stressed by the reports of several cases of blepharospasm secondary to rostral brainstem or basal ganglia structural lesions (Jankovic and Patel, 1983; Powers, 1985; Keane and Young, 1985; Leenders et al., 1986; Gibb et al., 1988). A similar pathophysiological mechanism may account for blepharospasm in PSP.

### Involuntary inhibition of levator palpebrae superioris muscles (Blepharokolysis)

Goldstein and Cogan (1965) described four patients with "a nonparalytic motor abnormality characterized by the patient's difficulty in initiating the act of lid elevation". They coined the term of "apraxia of lid opening" for this disorder and differentiated it from paralytic ptosis for its transient appearence at the start of lid opening and from blepharospasm because no forceful contractions of orbicularis oculi were present. One of the patients had a supranuclear ophthalmoplegia suggestive of PSP.

Lepore and Duvoisin (1985) found a similar lid dysfunction in six patients with parkinsonism. One of them had PSP and two others showed supranuclear vertical gaze palsies. These authors described this lid opening disorder as consisting of: a) transitory inability to initiate lid opening; b) no

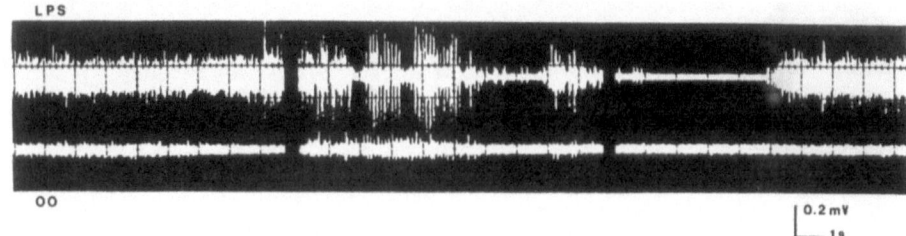

**Fig. 2.** Blepharokolysis. Contraction of levator palpebrae superioris (*LPS*) is recorded while eyes are open. This is followed by intermittent EMG activity and finally complete inhibition of LPS with absence of muscle contraction during several seconds. The orbicularis oculi (OO) remains silent except for a slight activation at the onset of LPS inhibition. EMG records are discontinued to show the different parts of this phenomenon

evidence of ongoing orbicularis oculi contraction, such as lowering of the brows beneath the superior orbital margins; c) vigorous frontalis contraction during periods of inability to raise eyelids and d) no oculomotor or ocular sympathetic nerve dysfunction and no ocular myopathy.

Esteban and Gimenez-Roldán (1988) reported electrophysiological evidence of prolonged inhibition of the levator palpebrae muscles in nine patients with parkinsonism and involuntary closure of the eyelids, three of whom had PSP. Electromyographic recordings during episodes of involuntary eyelid closure showed abrupt interruption of levator palpebrae superioris muscles activity, which normally contract tonically while the eyes are open. These periods of levator inhibition lasted from less than one second to 30 seconds and no appreciable concomitant orbicularis oculi activity was observed (Fig. 2).

The type of lid opening abnormality just described has been reported in PSP by other authors (Dehane, 1984; Jankovic, 1984; Golbe et al., 1989) and this disorder is presently considered as its commonest etiology (Lepore, 1988).

Goldstein and Cogan (1965) classify this impairment of lid opening as an apraxia and this terminology has been also used by several authors (Dehane, 1984; Lepore and Duvoisin, 1985; Golbe et al., 1989). However, all reported patients with 'apraxia of lid opening" had either extrapyramidal or pyramidal motor dysfunction (Lepore, 1988) and therefore this disorder cannot be regarded as a true apraxia since their motor system was not intact. Thus, the term "apraxia of lid opening" seems inappropiate. Esteban and Gimenez-Roldan (1988) proposed the term blepharokolysis (from Greek blepharon, eyelid; and kolysis, inhibition) to designate those episodes of involuntary closure of the eyes caused by abnormal levator palpebrae inhibition.

Although this disorder of lid opening has been attributed to akinesia of lid function (Case records of Massachusetts General Hospital, 1975), it is at present regarded as an involuntary levator palpebrae inhibition of supranuclear origin (Lepore and Duvoisin, 1985), which might represent an exagerated period of electrical inactivity in both levator and orbicularis muscles at the end of a blink, like that seen in patients with Parkinson's disease (Loeffler et al., 1966).

The anatomic basis of levator inhibition remains unclear. Sporadic loss of volitional eye opening with a stereotactically placed thermal lesion of the prerubral fields of Forel H and red nucleus has been reported in a patient so treated for a hyperkinetic movement disorder (Nasholl and Gills, 1967). Lesions of dorsal midbrain areas involving superior colliculus, periqueductal gray matter and pretectal areas may be relevant to the pathogenesis of blepharokolysis in PSP.

Blepharokolysis may be the cause of the ptosis reported in patients with PSP (Dix et al., 1975; Kristensen, 1985).

## Supranuclear palsy of voluntary eyelid closure

Other dysfunction of voluntary lid control is the inability to close the eyes either willfully or on command with preservation of reflex blinking. This eyelid disorder has been rarely reported in patients with PSP (Jankovic, 1984; Golbe et al., 1989), usually in combination with eye opening abnormalities.

Figure 3 shows electromyographic records of levator palpebrae superioris and orbicularis oculi muscles of a patient with PSP who had, in addition to a complex ophthalmoplegia with prominent downgaze palsy, inability to close the eyes voluntarily or on command, but preserved reflex blinking to visual threat, glabellar and corneal stimulation. There was no blepharospasm and eye opening was unimpaired. On attempting to close the eyes only EMG activity similar to spontaneous blinking was observed. When a sustained lid closure was obtained after repetitive eyelashes stimulation a moderate activity of both levator palpebrae superioris and orbicularis oculi was observed. The blink reflex elicited by electrical stimulation of the supraorbital nerve showed normal latencies of R1 and R2 components.

Thus, a combination of inadequate inhibition of levator palpebrae superioris and insufficient activation of orbicularis oculi, with an intact blink reflex arc, is the electrophysiological correlate of supranuclear palsy of voluntary eye closure in PSP.

The anatomical substrate of this eyelid dysfunction is not well known. A similar disorder has been described in bilateral cortical frontal lesions of different etiologies such as infarctions (Lessel, 1972; Ghika et al., 1988), Creutzfeldt-Jakob disease (Russell, 1980) and motoneuron disease (Esteban et al., 1978; Younger et al., 1988; Sunohara et al., 1989; Nishimura et al., 1990). In PSP, although neuropathological abnormalities are chiefly subcortical and cerebral cortex including frontal areas are largely spared (Lees, 1987), positron emission tomography studies with $F^{18}$-fluorodeoxyglucose or oxigen 15 have demonstrated a relative hypometabolism in the frontal lobes (Blin et al., 1990), specially in motor and premotor areas (Goffinet et al., 1989).

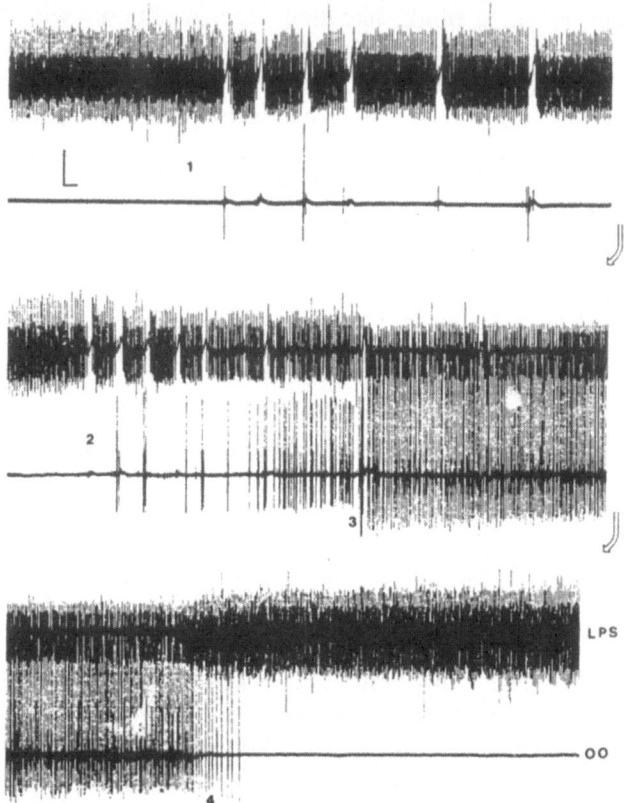

**Fig. 3.** Impairment of voluntary eyelid closure. The strong voluntary efforts to close the eyes on command were only followed by several irregular blinks (*1*). After a rapid, repetitive cilia stimuli (*2*), a weak but sustained eye closure was obtained (*3*). In this situation, a moderate activity was recorded in orbicularis oculi (OO) coinciding with a partial inhibition of levator palpebrae superioris (LPS) activity. Finally impersistence of eyelid closure occurs (*4*) and OO activity subsides with recovery of LPS basal tonic contraction. Calibration signals, 0.2 mV (vertical) and 0.2 s (horizontal)

    This lid abnormality may be due to a deficit of frontal cortical inhibition of levator palpebrae superioris rather than of orbicularis oculi activation (Schmitdtke and Büttner-Ennever, 1992) since in normal voluntary blinks inhibition of levator palpebrae superioris precedes orbicularis oculi activation (Björk and Kugelberg, 1953; Van Allen and Blodi, 1962; Loeffler et al., 1966; Esteban and Salinero, 1979).

    The inability to voluntarily close the eyes with normal blink reflexes was first described by Lewandowsky (1907), who regarded this defect as an eyelid apraxia. However this term, as the above mentioned "apraxia of eyelid opening", seems innapropiate because the motor system of patients reported with this eyelid dysfunction was not intact. Supranuclear paralysis of lid closure, as named by Lessel (1972) and Rusell (1980), is likely a more suitable term.

## References

Ascher P, Gachelin G (1966/1967) Rôle du colliculus supérieur dans l'elaboration de réponses motrices á des stimulations visuelles. Brain Res 3: 327–342

Berardelli A, Rothwell JL, Day BL, Marsden CD (1985) Pathophysiology of blepharospasm and oromandibular dystonia. Brain 108: 593–608

Blin J, Baron JC, Dubois B, Pillon B, Cambon H, Cambier J, Agid Y (1990) Positron emission tomography study in progressive supranuclear palsy. Brain hypometabolic pattern and clinicometabolic correlations. Arch Neurol 47: 747–757

Björk A, Kugelberg E (1953) The electrical activity of the muscles of the eye and eyelids in various positions and during movement. Electroencephalogr Clin Neurophysiol 5: 595–602

Brusa A, Mancardi GL, Bugiani D (1980) Progressive supranuclear palsy 1979: an overview. Ital J Neurol Sci 4: 205–222

Case records of the Massachusetts General Hospital (1975) N Engl J Med 293: 346–352

David NJ, Mackey EA, Smith JL (1968) Further observations in progressive supranuclear palsy. Neurology 18: 349–356

Dehane I (1984) Apraxia of eyelid opening in progressive supranuclear palsy. Ann Neurol 15:115

Dix MR, Harrison MJG, Lewis PD (1971) Progressive supranuclear palsy (The Steele-Richardson-Olszewski syndrome). A report of 9 cases with particular reference to the mechanism of the oculomotor disorders. J Neurol Sci 13: 237–256

Esteban A, De Andrés C, Giménez-Roldán (1978) Abnormalities of Bell's phenomenon in amyotrophic lateral sclerosis. J Neurol Neurosurg Psychiatry 41: 690–698

Esteban A, Giménez-Roldán S (1988) Involuntary closure of the eyelids in parkinsonism. Electrophysiological evidence for prolonged inhibition of the levator palpebrae superioris muscles. J Neurol Sci 85: 333–345

Esteban A, Salinero E (1979) Reciprocal reflex activity in ocular muscles: implications in spontaneous blinking and Bell's phenomenon. Eur Neurol 18: 157–165

Ghika J, Regli F, Assal G, Bogousslavsky J (1988) Inability to voluntarily close the eyes. Discussion of supranuclear disorders in palpebral closure based on 2 cases, with a review of the literature. Schweiz Arch Neurol Psychiatr 139: 5–21

Gibb WR, Lees AJ, Marsden CD (1988) Pathological report of four patients presenting with cranial dystonias. Mov Disord 3: 211–221

Goffinet AM, De Volder AG, Guillain C, Rectem D, Bol A, Michel C, Cogneau M, Labar D, Laterre C (1989) Positron tomography demonstrates frontal lobe hypometabolism in progressive supranuclear palsy. Ann Neurol 25: 131–139

Golbe LI, Davis PH, Lepore FE (1989) Eyelid movement abnormalities in progressive supranuclear palsy. Mov Disord 4: 297–302

Goldstein JE, Cogan DG (1965) Apraxia of lid opening. Arch Ophthalmol 73: 155–159

Gordon G (1951) Observations upon the movements of the eyelids. Br J Ophthalmol 35: 339–351

Grandas F (1989) Blepharospasm: clinical aspects and therapeutic considerations. Disorders of movement: clinical, pharmacological and physiological aspects. Academic Press, London, pp 287–294

Jackson JA, Jankovic J, Ford J (1983) Progressive supranuclear palsy: clinical features and response to treatment in 16 patients. Ann Neurol 13: 273–278

Jankovic J (1984) Apraxia of eyelid opening in progressive supranuclear palsy. Reply. Ann Neurol 15: 115–116

Jankovic J, Friedman D, Rey GJ, Levy J, Pirozzol FJ (1988) Progressive supranuclear palsy: clinical and neuroophthalmological findings in 87 patients. Book of abstracts. Ninth international symposium on Parkinson's disease, Jerusalem, p 82

Jankovic J, Patel SC (1983) Blepharospasm associated with brainstem lesions. Neurology 33: 1237–1240

Karson CN (1983) Spontaneous eye-blink rates and dopaminergic systems. Brain 106: 643–653

Karson CN (1988) Physiology of normal and abormal blinking. Adv Neurol 49: 25–37

Karson CN, Burns RS, Lewitt PA, Foster NL, Newman RP (1984) Blink rates and disorders of movement. Neurology 34: 677–678

Keane JR, Young JA (1985) Blepharospasm with bilateral basal ganglia infarction. Arch Neurol 42: 1206–1208

Kennard DW, Glaser GH (1964) An analysis of eyelid movements. J Nerv Ment Dis 139: 31–48

Kennard DW, Smyth GL (1963) Interation of mechanisms causing eye and eyelid movement. Nature 197: 50–52

Kristensen MD (1985) Progressive supranuclear palsy—20 years later. Acta Neurol Scand 71: 177–189

Leenders KL, Frackowiak RS, Quinn N, Brooks D, Summer D, Marsden CD (1986) Ipsilateral blepharospasm and contralateral hemidystonia and parkinsonism in a patient with unilateral rostral brainstem-thalamic lesion: structural and functional abnormalities studied with CT, MRI and PET scanning. Mov Disord 1: 51–58

Lees A (1987) The Steele-Richardson-Olszewski syndrome (progressive supranuclear palsy). In: Movement disorders, vol 2. Butterworths, London, pp 272–287

Lepore FE (1988) So-called apraxias of lid movement. Adv Neurol 49: 85–90

Lepore FE, Duvoisin RC (1985) "Apraxia" of eyelid opening: an involuntary levator inhibition. Neurology 35: 423–427

Lewandowsky M (1907) Über Apraxie des Lidschlusses. Berl Klin Wschr 44: 921–923

Lessel S (1972) Supranuclear paralysis of voluntary lid closure. Arch Ophthalmol 88: 241–244

Loeffler JD, Slatt B, Hoyt WF (1966) Motor abnormalities of the eyelids in Parkinson's disease. Arch Ophthalmol 76: 178–185

Maher ER, Lees AJ (1986) The clinical features and natural history of the Steele-Richardson-Olszewski syndrome (progressive supranuclear palsy). Neurology 36: 1005–1008

Mastaglia FL, Grainger K, Kee F, Sadka M, Lefroy R (1973) Progressive supranuclear palsy (The Steele-Richardson-Olszewski syndrome): clinical and electrophysiological observations in eleven cases. Proc Aust Assoc Neurol 10: 35–44

Nishimura M, Tojima M, Suga M, Hirose K, Tanabe H (1990) Chronic progressive spinobulbar spasticity with disturbance of voluntary eyelid closure. J Neurol Sci 96: 183–190

Nashold BS, Gills P (1967) Ocular signs from brain stimulation and lesions. Arch Ophthalmol 77: 609–618

Pasik P, Pasik T, Bender MB (1969) The pretectal syndrome in monkeys I. Disturbances of gaze and body postures. Brain 92: 521–534

Pfaffenbach D, Layton DD, Kearns TD (1972) Ocular manifestations in progressive supranuclear palsy. Am J Ophthalmol 74: 1179–1184

Powers JM (1985) Blepharospasm due to unilateral diencephalon infarction. Neurology 35: 283–284

Rotwell JC, Obeso JA, Day BL, Marsden CD (1983) Pathophysiology of dystonias. Adv Neurol 39: 851–864

Russell RRW (1980) Supranuclear palsy of eyelid closure. Brain 103: 71–82

Schmidtke K, Büttner-Ennever JA (1992) Nervous control of eyelid function. A review of clinical, experimental and pathological data. Brain 115: 227–247

Sing S, Smith BL, Lal A (1974) Progressive supranuclear palsy: report of 4 cases with particular reference to blepharospasm and levodopa therapy. Neurol India 22: 65–71

Steele JC, Richardson JC, Olszewski J (1964) Progressive supranuclear palsy. Arch Neurol 10: 333–359

Sunohara N, Mukoyama M, Funamoto H, Kamei N, Tomi H, Satoyoshi E (1989) Supranuclear paralysis preventing lid closure in amyotrophic lateral sclerosis. Jpn J Med 28: 515–519

Van Allen MB, Blodi FC (1962) Electromyographic study of reciprocal innervation in blinking. Neurology 12: 371–337

Younger DS, Chou S, Hays AP, Lange DJ, Emerson R, Brin M, Thompson H, Rowland P (1988) Primary lateral sclerosis: a clinical diagnosis reemerges. Arch Neurol 45: 1034–1037

Zee DS, Chu FC, Leigh RJ, Savind PJ, Schatz NJ, Reingold DB, Cogan DG (1983) Blink-sacade synkinesis. Neurology 33: 1233–1236

Authors' address: Dr. F. Grandas, Servicio de Neurologia, Hospital General "Gregorio Marañón", Doctor Esquerdo, 46, E-28007 Madrid, Spain.

# The auditory startle response in progressive supranuclear palsy

## J. C. Rothwell, M. Vidailhet, P. D. Thompson, A. J. Lees, and C. D. Marsden

Institute of Neurology, MRC Human Movement and Balance Unit, London, United Kingdom

**Summary.** The EMG characteristics of the normal auditory startle response in man are compatible with an origin in the pontine reticular formation and with conduction down the spinal cord in a slowly conducting, possibly reticulo-spinal pathway. The startle was reduced or absent in patients with progressive psupranuclear palsy, consistent with loss of neurones in the lower pontine reticular formation. In contrast, the startle was present and of normal form in patients with Parkinson's disease. However, it was delayed in onset. This result was not influenced by treatment with L-dopa. The late auditory startle in Parkinson's disease might be related to withdrawal of facilitatory input to brainstem centres from the basal ganglia.

## Introduction

In normal subjects, unexpected auditory stimuli produce a startle response. This usually involves flexion of the neck and trunk, and brief closure of the eyes. EMG recordings from many muscles show that although the onset latency can be quite variable, the muscles are activated in a very constant order once the response is underway. An example of a single startle response in a normal subject after an unexpected noise (120 dB, 1 kHz, 50 ms) is shown in Fig. 1.

The first muscle to contract is the orbicularis oculi, with a latency of about 40 ms. This is followed by activity in other face and neck muscles, and then by responses in the arms and legs. Colebatch et al. (1990) and Brown et al. (1991) have argued that the initial part of the orbicularis oculi response is not part of the startle reaction proper, but simply represents a blink reflex to auditory stimulation. The reason for saying this is that the startle response itself adapts very rapidly to repeated presentation of the stimulus. In most experiments, the stimuli are given at intervals of 20 *minutes* or more in order for a startle to be observed, and even then, the response may only occur on the first 3–4 trials. The blink reflex in the orbicularis oculi, in contrast, habituates much more rapidly: it is still present (without any other part of the startle response) if stimuli are given every

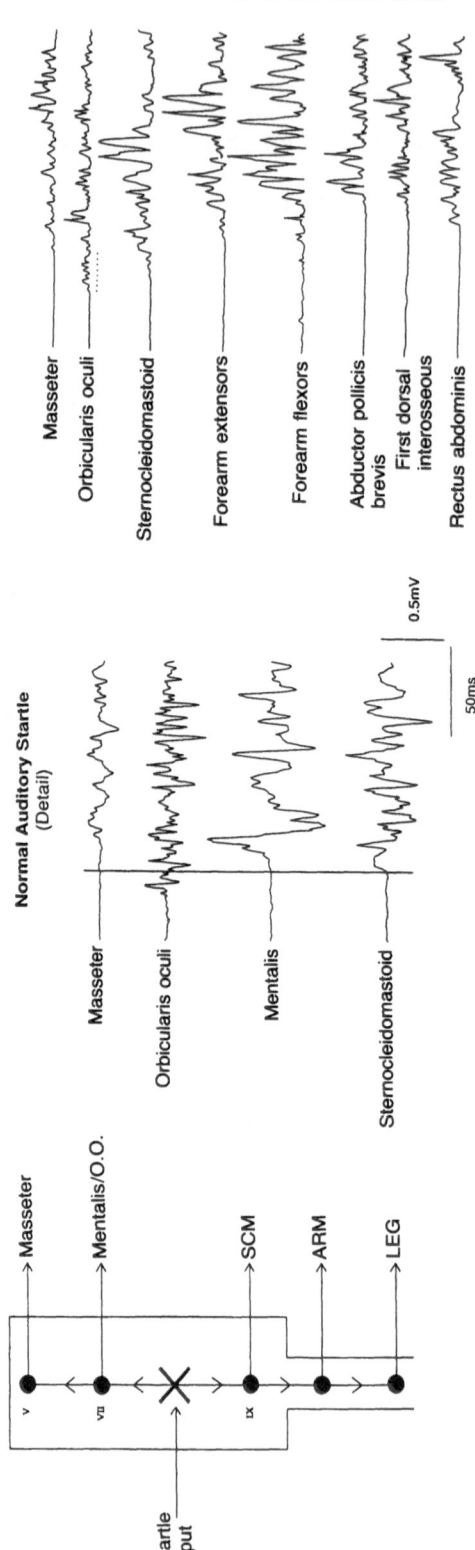

**Fig. 1.** Normal auditory startle response. The left panel shows the proposed origin of the startle in the lower brainstem. Input activates this centre, which then innervates cranial nerve nuclei in a caudo-rostral order, with sternocleidomastoid (XII) being innervated first. The middle panel shows a detail of the cranial nerve innervation in EMG recordings of a single startle response of a normal subject. The stimulus was given at the start of the sweep. There is an initial (blink reflex) response in the orbicularis oculi at about 35 ms. The main startle begins in the sternocleidomastoid (vertical line) at about 50 ms, and is closely followed by activity in mentalis and masseter. The right panel shows a typical startle in a different subject, with EMG recordings from many muscles. The stimulus was given at the start of the sweep (calibration bar, 50 ms). The blink reflex in orbicularis oculi is indicated by dotted underlining. The startle response begins in sternocleidomastoid, and spreads upwards to masseter, and down to the arms and trunk. Note the excessively long delay to responses in the intrinsic hand muscles compared to the latency in the forearm extensors or the rectus abdominis. (From Brown et al., 1991, with permission)

**Table 1.** A comparison of the efferent pathways subserving the normal auditory startle reflex with corticobulbar and pyramidal pathways in man

|  | Auditory startle reflex | Magnetic stimulation of motor cortex |
|---|---|---|
| Excess latency of masseter over SCM | 0.7 ms | −2.9 ms (masseter precedes SCM) |
| Excess latency of RA over SCM | 24.0 ms | 8.0 ms |
| Excess latency of APB over biceps | 22.4 ms | 10.5 ms |

The differences in the latencies to onset of EMG activity between individual muscles in the normal auditory startle reflex are calculated from the median latencies from 12 healthy seated subjects. The stimulus was 1 kHz, 124 dB tone of 50 ms duration delivered to both ears at intervals of about 20 min. The differences in the latency to onset of EMG activity between masseter and sternocleidomastoid (SCM), sterno-cleidomastoid and rectus abdominis (RA), and between biceps and abductor pollicis brevis (APB), following magnetic stimulation of the motor cortex, are calculated from the mean latencies reported by Cruccu et al. (1989) and Thompson (1991) in normal subjects. The pattern of recruitment of cranial nerves seen following magnetic stimulation of the motor cortex is reversed in the normal auditory startle reflex. The differences in latency to onset of EMG activity between sternocleidomastoid and rectus abdominis, and between biceps and abductor brevis, are longer in the auditory startle reflex than following magnetic stimulation of the motor cortex

20 *seconds*. Thus, it seems to involve a different mechanism than the startle proper, and is usually considered separately (see also Chokroverty et al., 1992).

If we assume that the early response in the orbicularis oculi is an auditory blink, rather than part of the startle, then the first muscle to contract is usually the sternocleidomastoid. This is followed by activity in muscles of the face, trunk, and limbs. The pattern of activity gives some information on the origin and conduction velocity of the pathway involved in the response.

The interval between the onset of activity in sternocleidomastoid and other trunk or limb muscles is quite long (see Table 1). For comparison, when transcranial magnetic stimulation is used to activate the large diameter component of the corticospinal tract, the interval between the onset of activity in sternocleidomastoid and rectus abdominis is of the order of 8 ms. In the startle, the interval is about 24 ms. If we assume that the conduction velocity of the peripheral efferent pathway from spinal cord to muscle is the same in each case, then the conduction velocity of the central descending tract responsible for the startle must be considerably slower than that of the corticospinal tract.

Precisely which pathways are activated by the startle is not clear. However, some clues come from the pattern of activation of muscles in the face and neck. Brown et al. (1991) initially showed that close inspection of the cranial nerve innervated muscles suggested an origin of the startle in the lower brainstem. Thus, sternocleidomastoid (XIIth nerve) was innervated first, and was followed by activity in the mentalis (VIIth nerve), and then the masseter (Vth nerve): the cranial nerve nuclei were innervated in a caudo-rostral order (see Fig. 1 and Table 1). The interval between the onset of activity in sternocleidomastoid and other muscles of the face is only very short (about 1 ms on average), and is often difficult to measure. In a recent study Chokroverty et al. (1992) could find no reliable difference in onset latency between the muscles of the cranial nerves. However, even a 0 ms difference between sternocleidomastoid and masseter is different from the 2–3 ms lead of masseter over sternocleidomastoid which occurs after magnetic stimulation of the motor cortex (see Table 1).

A final feature of the startle response indicates that the descending pathway involved is quite different from the corticospinal tract activated by motor cortex stimulation. Magnetic stimulation of motor cortex preferentially evokes activity in the hand and forearm muscles. In the startle, responses are largest in the axial and proximal muscles. Responses in hand muscles are small and are less frequently observed. Indeed in the startle, hand muscle activity begins some 20–25 ms after that in biceps, compared with 10 ms or so after magnetic stimulation of the brain. This extra latency cannot be due to slow conducting central pathways since the innervation level of hand and upper arm muscles is quite close. Instead it seems to reflect a specific pattern of organisation in the startle with excessively delayed and small responses in distal muscles.

The conclusion from these observations on normals is that the startle response probably originates in the lower brainstem and is conducted down the spinal cord by a slowly conducting pathway with preferential projections to proximal muscles. Such a pathway might be one of the reticulospinal tracts. These characteristics of the human startle pattern are very similar to those reported previously in extensive studies of the auditory startle response in rats and rabbits (Davis et al., 1982; Hori et al., 1986; Wu et al., 1988). The ventral pontine reticular formation, in particular the nucleus reticularis pontis caudalis was identified as one of the crucial zones in the generation of the auditory startle response in animals (Grillner et al., 1968; Hammond, 1973; Groves et al., 1974; Davis et al., 1982). Reticulospinal pathways were thought to be the main pathway conveying excitation to the spinal cord.

In the present chapter we describe the startle pattern in two groups of patients: eight patients with progressive supranuclear palsy, and 11 patients with idiopathic Parkinson's disease. In progressive supranuclear palsy there are widespread pathological changes in the brainstem, including degeneration of the pontine reticular formation (Steele et al., 1964) with severe neuronal loss (Malessa et al., 1991), which might be expected to be associated with substantial changes in the startle pattern.

### Material and methods

Details of the patients and methods are given in detail in the paper by Vidailhet et al. (1992). Briefly, the responses of 12 elderly normal subjects (mean age 51 years) were compared with those of 11 treated patients with idiopathic Parkinson's disease (mean age 55 years), and 8 patients with a clinical diagnosis of Steele-Richardson-Olszewski syndrome (mean age 67 years). Five of the parkinsonian patients were studied both *on* and 16 hrs *off* their normal L-dopa medication.

Startle reponses were recorded using surface electrodes in relaxed, seated individuals with eyes lightly closed. The stimulus was a 50 ms burst of 1 kHz at 90 dB given binaurally through headphones every 20 min or so.

## Results

Patients with Parkinson's disease had relatively normal startle responses, whether *on* or *off* drug therapy. The percentage of trials on which the response occurred and the size of the response was similar to normal, and the pattern of muscle activation the same (see Fig. 2). However, the onset latency of the startle was slightly longer in the patients than in the normals (onset of activity in sternocleidomastoid: normals, median 61 ms, range 42–125: patients, median 77 ms, range 58–102). L-dopa administration had no effect on the latency of the startle in the 5 patients tested both *on* and *off* therapy.

In contrast to the patients with Parkinson's disease, the startle was grossly abnormal in patients with progressive supranuclear palsy. In 3 of the

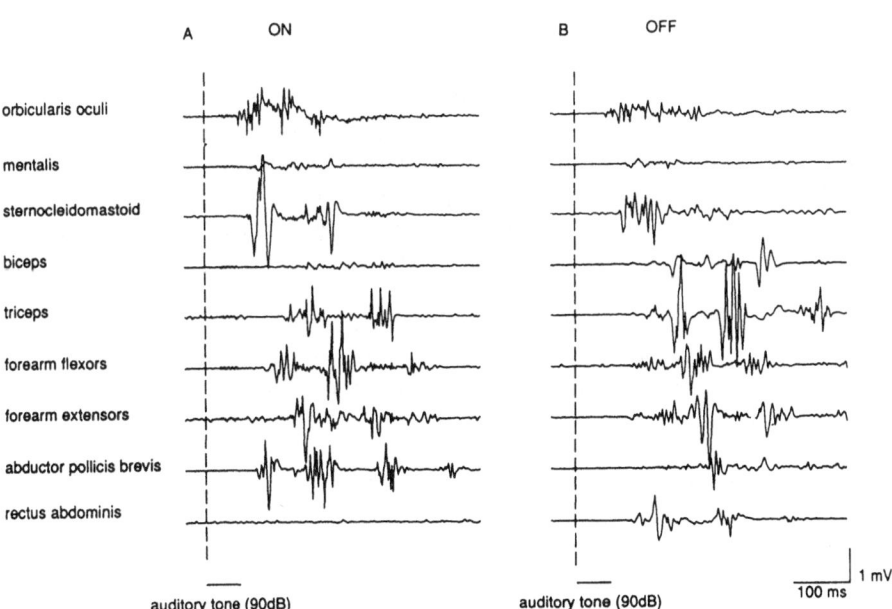

**Fig. 2.** Auditory startle responses in a patient with idiopathic Parkinson's disease when *on* medication and normally mobile (**A**), and when *off* medication and akinetic and rigid (**B**). Single trials are shown. There is no difference in the latency, amplitude or distribution of the responses in either state. However, the onset of activity in the orbicularis oculi and sternocleidomastoid is delayed in comparison with normal age matched subjects (see text for details). (From Vidailhet et al., 1992, with permission)

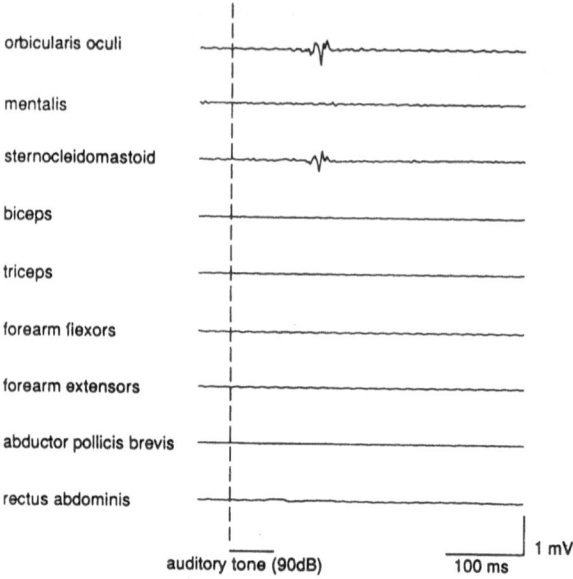

orbicularis oculi

mentalis

sternocleidomastoid

biceps

triceps

forearm flexors

forearm extensors

abductor pollicis brevis

rectus abdominis

auditory tone (90dB)    100 ms    1 mV

**Fig. 3.** An auditory startle response in a patient with the clinical diagnosis of progressive supranuclear palsy to an auditory tone delivered at the time of the vertical dotted line. A single trial is shown, The response is limited in its distribution with a small delayed response in only the orbicularis oculi and sternocleidomastoid muscles. (From Vidailhet et al., 1992, with permission)

patients the response was completely absent. In the remaining five, it was present, but severely attenuated. An example is shown in Fig. 3. The extent of muscle recruitment was limited compared to normal, and varied both from trial to trial and from patient to patient. In only 3 patients was a startle obtained in both the sternocleidomastoid and orbicularis oculi muscles. In the others, responses were seen only in the orbicularis oculi. Because of the small number of responses that were collected, it was not possible to analyse the amplitude and latency data statistically. However, the latency of onset of sternocleidomastoid activation tended to be longer than that seen in the normal group (progressive nuclear palsy patients (n = 3), mean onset latency 87 ms, range 80–131).

The lack of a startle response was unlikely to have been caused by deafness in the patients (Lees, 1987). Auditory evoked potentials are normal in progressive supranuclear palsy (Tolosa and Zeese, 1979), and in our patients hearing was clinically normal. Also, in a separate series of experiments, we confirmed that in normals, the startle was not affected under the present conditions by reducing the intensity of the stimulus from 90 dB to 60 dB, so long as the stimulus was unexpected.

In contrast to the lack of a startle response (even in the orbicularis oculi muscles in some of the patients), the electrical blink reflexes to stimulation of the supraorbital nerve were present and of normal latency.

## Discussion

The major finding of the present study was that the startle response was often absent in progressive supranuclear palsy, and even when present, was severely limited in its distribution. The findings are consistent with the known pathology of the disease, and the postulated site of origin of the startle. Quantitative immunocytohistochemical evaluation of brainstem cholinergic cells in postmortem tissue using polyclonal antibodies against human choline acetyltransferase reveals loss of cholinergic neurones in the lower pontine reticular formation, mainly within the cells of the nucleus pontis centralis caudalis and the nucleus papillioformis (Juncos et al., 1991; Malessa et al., 1991). These nuclei occupy a portion of the human brainstem which is coextensive with corresponding nuclei defined in animals such as the nucleus reticularis pontis centralis caudalis and the nucleus reticularis tegmenti (Olszewski and Baxter, 1954; Taber, 1961). The nucleus reticularis pontis centralis caudalis is directly involved in the startle reflex of the rat (Davis et al., 1982), and we suggest that the absent or poor auditory startle responses in patients with progressive supranuclear palsy could be explained by a pathological lesion of this part of the pontine reticular formation.

In patients with Parkinson's disease, the form of the auditory startle, and the muscles recruited were the same as in normal subjects. This indicates that the neuronal circuits mediating the auditory startle are intact in Parkinson's disease. Nevertheless one abnormality was noted: the timing of EMG responses recruited in the auditory startle was delayed in comparison with the control group. One explanation for this might be a lack of facilitation onto the startle system, perhaps due to abnormal basal ganglia output in Parkinson's disease to brainstem centres.

Little information is available at the present time about the descending projections from the globus pallidus and substantia nigra pars reticulata onto brainstem nuclei. However, clinical observations suggest that the functions of these pathways are not greatly influenced by L-dopa. For example, at least in the elderly or in advanced disease, L-dopa has little effect on postural control, despite its dramatic influence on other aspects of movement (Lakke, 1985; Weiner and Lang, 1989). This would be consistent with the present results which showed no detectable effect of L-dopa treatment on the latency of the startle response in any of the five patients who were studied both *on* and *off* their medication, despite a clear improvement in their motor performance in clinical tasks.

In conclusion, in patients with progressive supranuclear palsy, the startle response is abnormal or absent. We suggest that this may be related to the pathological lesion of the pontine reticular formation, mainly in the nucleus reticularis pontis caudalis which is thought to be an important relay in the startle response. In contrast, the startle response in patients with Parkinson's disease is of normal size and form, but is delayed, findings that may be related to withdrawal of facilitatory input to brainstem centres from the basal ganglia.

# References

Brown P, Rothwell JC, Thompson PD, Day BL, Marsden CD (1991) New observations on the normal auditory startle reflex in man. Brain 114: 1891–1902

Colebatch JG, Barrett G, Lees AJ (1990) Exaggerated startle reflexes in an elderly woman. Mov Dis 5: 167–169

Davis M, Gendelman DS, Tischler MD, Gendelman PM (1982) A primary acoustic startle circuit: lesion and stimulation studies. J Neurosci 2: 791–805

Grillner S, Lund S (1968) The origin of a descending pathway with monosynaptic action on flexor motoneurones. Acta Physiol Scand 74: 274–284

Groves PM, Wilson CJ, Boyle RD (1974) Brainstem pathways, cortical modulation, and habituation of the acoustic startle response. Behav Biol 10: 391–418

Hammond GR (1973) Lesions of the pontine and medullary reticular formation and prestimulus inhibition of the acoustic startle reaction in rats. Physiol Behav 10: 239–243

Hori A, Yasuhara A, Naito H, Yasuhara M (1986) Blink relfex elicited by auditory stimulation in the rabbit. J Neurol Sci 76: 49–59

Juncos JL, Hirsch EC, Malessa S, Duyckaerts C, Hersh LB, Agid Y (1991) Mesencephalic cholinergic nuclei in progressive supranuclear palsy. Neurology 41: 25–30

Lakke JP (1985) Axial apraxia in Parkinson's disease. J Neurol Sci 69: 37–46

Lees AJ (1987) The Steele-Richardson-Olszewski syndrome (progressive supranuclear palsy). In: Marsden CD, Fahn S (eds) Movement disorders, vol 2. Butterworths, London, pp 272–287

Malessa S, Hirsch EC, Cervera P, Javoy-Agid F, Duyckaerts C, Hauw JJ, et al (1991) Progressive supranuclear palsy: loss of choline acetyltransferase-like immunoreactive neurones in the pontine reticular formation. Neurology 41: 1593–1597

Olszewski J, Baxter D (1954) Cytoarchitecture of the human brain stem. Lippincott, Philadelphia

Steele JC, Richardson JC, Olszewski J (1964) Progressive supranuclear palsy: a heterogenous degeneration involving the brainstem, basal ganglia and cerebellum with vertical gaze and pseudobulbar palsy, nuchal dystonia and dementia. Arch Neurol 10: 333–359

Taber E (1961) The cytoarchitecture of the brain stem of the cat. I. Brain stem nuclei of cat. J Comp Neurol 116: 27–69

Tolosa ES, Zeese JA (1979) Brainstem auditory evoked responses in progressive supranuclear palsy. Ann Neurol 6: 369

Vidialhet M, Rothwell JC, Thompson PD, Lees AJ, Marsden CD (1992) The auditory startle response in Steele-Richardson-Olszewski syndrome and Parkinson's disease. Brain 115: 1181–1192

Wiener WJ, Lang AE (1989) Parkinson's disease. In: Movement disorders: a comprehensive survey. Futura, New York, pp 23–115

Wu MF, Suzuki SS, Siegel JM (1988) Anatomical distribution and response patterns of reticular neurons active in relation to acoustic startle. Brain Res 457: 399–406

Authors' address: Dr. J. C. Rothwell, Institute of Neurology, MRC Human Movement and Balance Unit, Queen Square, London, WC1N 3BG, United Kingdom.

# Differential diagnosis of PSP

**R. C. Duvoisin**

Department of Neurology, University of Medicine and Dentistry of New Jersey,
New Brunswick, New Jersey, U.S.A.

**Summary.** The clinical diagnosis of PSP depends primarily on the history and the physical findings. Clinicans should be alerted to the possibility of this condition in assessing patients presenting with atypical parkinsonism and other complex extrapyramidal syndromes in late middle age or later. The differential diagnosis includes MSA (both OPCA and SND), PD, CBD and cerebrovascular disease. PD is probably the most common erroneous diagnosis. Unfortunately, pathognomonic signs do not usually appear until several years after symptom onset. No specific laboratory test is yet available. Neuroimaging studies show characteristic anatomic alterations only late in the course of the illness and must be correlated with the clinical findings.

## A historical perspective

Progressive supranuclear palsy (PSP) is the most prominent member of a group of several chronic progressive neurodegenerative disorders in which extrapyramidal features, chiefly bradykinesia and rigidity, dominate the clinical picture. In some ways, its differential diagnosis repeats the history of its initial recognition and separation from other neurodegenerative disorders barely 3 decades ago. That history continues with the more recent recognition on combined clinical and pathological grounds of additional disorders sharing many similar features and also the discovery of new manifestations of PSP. It is consequently useful to place the nosology of PSP in at least a brief historical context.

Richardson, Olszewski, and Steele (1963) described in the mid 1960's a small series of patients who presented a distinctive clinical picture characterized by "defects of ocular gaze, spasticity of the facial musculature with dysarthria and sometimes dysphonia, extensor rigidity of the neck with head retraction and dementia". On postmortem study (Steele, Rishardson, and Olszewski, 1964) they found neuronal degeneration associated with neurofibrillary tangles chiefly limited to the pontine and midbrain tectum, the substantia nigra, subthalamic nucleus and pallidum. The pathology was reminiscent of post-encephalitic parkinsonism and the amyotrophic lateral sclerosis-parkinson-dementia syndrome of Guam but was distinctive in its topography and cytopathology. They named it *progressive supranuclear palsy* in reference to the supranuclear ophthalmoplegia which remains to the present its most distinctive diagnostic feature. It is most familiar to clinicians today by the acronym "PSP". Its distinctive clinical and pathological features have gained the disorder general recognition as a unique morbid entity *sui generis*.

These authors correctly recognized that it was not an uncommon disorder. Within a decade the condition had become familiar to neurologists throughout the world and had been observed in all geographic regions and ethnic groups. With growing experience, the original clinical picture was enlarged and modified. Steele et al. had emphasized the absence of parkinsonian features such as tremor, stooped posture or parkinsonian hand postures. They thought that associated movements were preserved and that eyeblinks were "more frequent". However, parkinsonism subsequently proved to be a common clinical feature, the usual clinical presentation being that of an atypical parkinson syndrome. The original cases were all males but it later became clear that the disorder occurs equally among men and women. Dementia appeared to be prominent in the early cases but dementia has often been mild and infrequently a presenting symptom.

Several previously reported cases appear in retrospect to have had similar sympoms and pathology and very probably had the same disorder. These were, however, isolated cases and it remains the accomplishment of Steele and his colleagues to have recognized that PSP represented a particular morbid entity distinct from postencephalitic parkinsonism and other extrapyramidal disorders. Posey (1904) and Spiller (1905) had described patients unable to look up or down and a photograph of Spiller's case looks very much like a PSP patient even earlier described an unusual parkinson patient with hyperextension of the trunk and head which in retrospect must have had PSP. These early reports indicate that PSP is not a new disorder.

The question arises why was PSP not recognized as a distinct morbid entity much prior to 1963? One reason is undoubtedly its resemblance, both clinically and pathologically, to post-encephalitic parkinsonism (PEP) occurring as a sequel of encephalitis lethargica. On the clinical side, it must easily have been lost amongst the myriad manifestations of the latter condition, which, it should be emphasized, was the most common form of Parkinsonism in the 1920's and 1930's (Dimsdale, 1946) and still accounted for 10% of the parkinson population in the mid '60's (Duvoisin and Yahr, 1965). Moreover, as in PSP, rigidity and bradykinesia predominated and tremor was minimal. There were also a variety of ocular palsies in the postencephalitics. In addition, many authorities regarded parkinsonism as a broad syndrome of diverse nonspecific etiologies. The pathological findings would also have been confused with those of postencephalitic parkinsonism. Indeed, Chavany et al. (1951) ascribed their case to encephalitis and Olszewski (1966) had been struck by the resemblance of the pathology to that of encephalitis lethargica.

The introduction of levodopa therapy in the treatment of parkinsonism in clinical practice crica 1970 stimulated great interest in the various parkinson syndromes. Differences in prognosis and in therapeutic response rendered attention to the differential diagnosis of parkinsonism a matter of practical importance. PSP soon became very familiar to neurologists specializing in the new field of movement disorders as a particular parkinsonian syndrome characterized by a chronic progressive course, extrapyramidal rigidity, axial dystonia, dementia, a pseudobulbar palsy and a supranuclear ophthalmoplegia affecting vertical gaze preferentially. It soon became recognized as the principal "parkinson-plus" syndrome encountered in neurological practice.

## Prevalence and demography

From 3.9% (Jackson et al.) to 7% (Agid et al.) of patients attending clinics devoted to parkinsonism carry the diagnosis of PSP. In the author's clinic it was 4.9% (Duvoisin et al., 1987). However, the diagnosis is delayed in most cases for several years until the distinctive disorder of ocular motility develops. If one attempts to include the patients who are not yet diagnosed, the true proportion of parkinson patients who have PSP is much larger, perhaps as large as 12% as Duvoisin et al. (1987) have suggested. Thus PSP is today the second most common cause of parkinsonism after PD itself and is a more common cause than multiple system atrophy (MSA) or arteriosclerotic parkinsonism. Today, the possibility of PSP must regularly be considered in the differential diagnosis of parkinsonism.

PSP tends to affect an older population than does PD and MSA. Onset under age 60 is unusual and under 50 quite rare. There is as yet little evidence of familial concentration. A positive family history would be more typical of OPCA and would be consistent with PD and corticobasal degeneration (CBD) but should raise doubts as to the possibility of PSP.

## General considerations

The diagnosis of PSP depends primarily on an informed analysis of the clinical features: the history, the clinical setting, the pattern of progression and the physical findings. In the initial phases, diagnosis may vex the most experienced clinican. Later on, as the disease progresses, it may be suspected but confirmation may require futher observation until pathognomonic signs become evident. The difficulty of establishing the diagnosis in early and intermediate cases should not be underestimated. In contrast, in the fully developed cases, the clinical picture is quite distinctive and the diagnosis is readily made.

Typically, the PSP patient has been symtpomatic and under treatment for a substantial period of time before the diagnosis is made. In several series of patients collected by leading students of the disorder, the interval from symptom onset to diagnosis varied from 3.6 to 4.9 years. In individual cases the delay in diagnosis can be 6 to 7 years or more. During the first several years of the disease the PSP patient is usually treated as a case of parkinsonism. It is often recognized that the parkinsonism is atypical for PD, but differentiation from MSA, especially from the SND variant, from multi-infarct syndromes such as Binswanger's disease and from corticobasal degeneration (CBD) remains difficult by clinical criteria alone.

As recent clinico-pathologic studies have shown, the clinical criteria for the diagnosis of PD, a more common and familiar condition, permit a surprisingly high rate of diagnostic error. Atypical cases thought not to have PD and suspected of having PSP or SND may nevertheless turn out on postmortem study to have the typical Lewy body pathology of PD. In their post-mortem study of 100 cases of clinically diagnosed PD, Hughes et al. (1992) found that in fact only 79 had PD. Thus the false-positive error rate for PD is about 20%. The condition most often erroneously diagnosed as PD in that series of cases was PSP. It accounted for 12 (12%) of the cases.

Unfortunately, there are as yet no specific biochemical or pathophysiologic laboratory abnormalities to aid the clinician in diagnosing PSP. The characteristic gross anatomical alterations of midbrain atrophy can be demonstrated on computed x-ray tomography or magnetic resonance imaging, but usually these anatomical signs are present only late in the course and serve chiefly to confirm the clinical findings. They are not helpful earlier in the course of the illness. Thus in most cases, one must rely principally on clinical crtieria which can consistently identify PSP only relatively late in its clinical course.

The diagnosis cannot be made with reasonable certainty on clinical grounds until the pathognomonic disturbance in ocular motility can be satisfactorily documented. They may not appear for several years and in some cases may not appear at all. In such cases, the diagnosis can only be suspected and must await post-mortern confirmation. It is thus clear that PSP must be grossly underdiagnosed.

Despite the limitations of clinical criteria for the diagnosis, it is important to attempt to establish the correct diagnosis early in the course of the illness to allow for reliable prognosis, to spare the patient and family needless consultations and tests and to avoid inappropriate treatment. In addition, by carefully differentiating PSP from other conditions the clinician can make an important contribution to advancing our understanding of this and similar disorders.

Certain features may be considered essential for the clinical diagnosis of PSP. Others frequently occur and may be considered confirmatory. Still other manifestations may occur but also occur in many other disorders so commonly that they have little diagnostic value. Finally, some clinical phenomena which are not bone fide manifestations of PSP but occur in other conditions which may otherwise resemble it must be recognized. They are incompatible with the diagnosis of PSP and should suggest another diagnosis. This heirarchical ordering of clinical features to be considered in the diagnosis of PSP is displayed in Table 1.

**Table 1.** Diagnostic value of clinical manifestations of PSP

---

*1. Cardinal features*
   Onset over age 40
   Chronic progressive course
   Bilateral supranuclear ophthalmoparesis
       At a minimum: hesitation in voluntary downgaze, impaired vertical
           opticokinetic nystagmus and impaired suppression of vertical vestibulo-oculo
           reflex.
   Rigidity with axial predominance
   Bradykinesia

*2. Common manifestations*
   Onset with gait impairment and frequent falls
   Poor response to levodopa therapy
   Severe bradyphrenia with frontal lobe features (grasping, perseveration and
       utilization behavior)
   Axial dystonia with cervical hyperextension
   Dysarthria and dysphagia
   Ocular fixation instability with macro square wave jerks
   Apraxia of eyelid opening and/or closing
   Extremely low eyeblink frequency (10/min)

*3. Occasional manifestations*
   Tremor, postural, action of resting
   Pyramidal tract signs
   Focal or segmental dystonia
   Amyotrophy
   Depression
   Schizophreniform psychosis

*4. Manifestations inconsistent with a diagnosis of PSP*
   Early or prominent cerebellar signs unexplained polyneuropathy
   Aphasia
   Agnosia
   Sensory defects — primary or cortical

---

The cardinal feature are essential for the clinical diagnosis of PSP. The common manifestations may suggest the diagnosis. The occasional manifestations are consistent with the diagnosis but have no diagnostic value. Manifestations inconsistent with a diagnosis of PSP include signs of peripheral neuropathy, cerebellar involvement or cerebral cortical dysfunction. Their presence suggests that an alternate diagnosis should be considered

## The history

In eliciting the history, the physician should consider both the patient's account of symptoms and the observations of the spouse, family members or even of friends. Others may have noted significant changes in posture, gait, general demeanour and mobility of which the patient is unaware. Even patients with quite obvious bradykinesia sufficient to impair function and cause occupational disability often stoutly deny any problem. This is not merely denial of illness, which may be present, but also because extrapyramidal function and dysfunction is largely outside conscious awareness.

Typically, as with other chronic neurodegenerative disorders, the onset of PSP is insidious. Motor manifestations are often preceded for a long prodromal period by nonspecific symptoms such as fatigue, lethargy, weakness, depression, headache, arthralgias and dizziness. The patient may be found unable to carry on his/her usual occupation or activities. A change of personality is a common early symptom. The patient appears to lack energy and initiative, becomes slower in social interactions and gradually withdraws to an inactive sedentary life.

Family members often report that the patient no longer maintains normal eye contact in conversation. The patient is not aware of this change but it is one of the earliest physical changes observed and a significant clue in the history provided by family members. The family may also have noted that the patient fails to notice steps or other obstacles on the ground and consequently trips over them. They may also have noted a failure to focus on a task at hand such as eating due to failure to look down. These are additional clues to the possibility of PSP.

Visual symptoms including diplopia, difficulty focusing, blurring of vision, impaired depth perception and difficulty reading may then appear. These are common initial manifestations of PSP. They may also occur as early syptoms in some hereditary ataxias but not in PD. Questioning may elicit the observation that the patient cannot visually track a line of print and find the beginning of the next line. More subtle impairments of vision are also common. They may be exposed by difficulties in tasks requiring good spatial perception such as driving an automobile. The patients have trouble orienting their vehicle to the road and in visually judging distance. They are confused in traffic, especially at intersections and accidents may result. Such symptoms in spatial orientation are unusual as initial manifestations of other extrapyramidal disorders.

"Dizziness" is another common but non-specific early complaint. The term is usually used to indicate subjective awareness of unsteadiness standing and walking rather than faintness or lightheadedness as in impending syncope or vertigo. Some patients, however, appear to experieince a true illusion of movement suggestive of vertigo. Unsteadiness walking and fear of falling are typical early symptoms in many PSP patients. They may also occur in PD and OPCA but rarely as early manifesations.

Sudden unexplained falls are also common initial complaints and tend to suggest PSP. They are neither drop attacks nor transient syncopes or siezures but due to loss of postural "righting" reflexes. The patient suddenly falls, usually backwards, for no apparent reason. A brief retropulsion may precede a fall. Their sudden unexpected occurrence has suggested the term "paroxysmal dysequilibrium". Descriptions of these falls as initial manifestations should alert the clinician to the possibility of PSP.

### The neurological examination

As in other neurological disorders, the examination should begin the moment the physician first meets the patient. The general demeanor, the

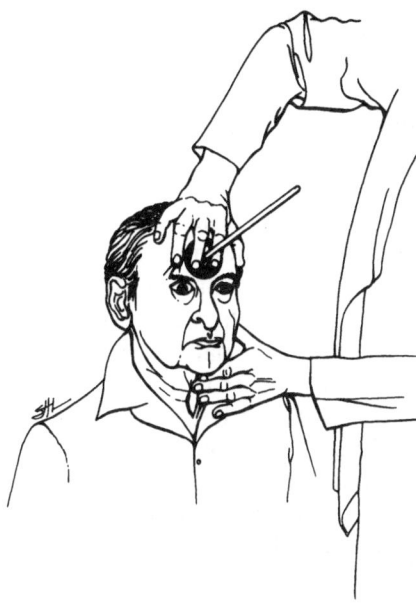

**Fig. 1.** Testing visual suppression of vestibular ocular reflex (VOR) with Gower's reflex hammer

frowning facial expression with lid lag suggesting surprise or astonishment (Fig. 1), poor eye contact, impaired coordination of head and eye movement, posture and general bradykinesia may all be observed at a glance as one welcomes the patient. These features should immediately suggest the possibility of a parkinson-plus syndrome and particularly PSP. The patient's first words of initial greeting may reveal defects of articulation and speech modulation of diagnostic value. Continuing observation while obtaining the history may detect a very low frequency of eye blink and perhaps evidence of eye-lid openning apraxia. These findings should increase the suspicion of PSP. The manner in which the patient arises from a chair, the posture, and the character of the gait as the patient walks to the examining table provides further valuable information to the observant consultant.

Even if the diagnosis seems obvious from the history and from visual inspection, a formal neurological examination should be carried out. It may be focused on features suggested by the history and the preliminary observation but certain items should be regularly checked in all extrapyramidal disorders. The author's routine is to begin with the cranial nerves and follow a standard routine which adheres to the classic neurologic examination, emphasizing some selected aspects such as examintion of ocular motility, gait and postural stability.

*Olfaction* should be routinely tested. Vials of coffess, tea and extract of vanilla are useful for the purpose. Hyposmia is common in PD but is not a feature of PSP. PD patients with hyposmia are usually aware of this defect.

Thus, if not explained by nasal pathology, the finding of hyposmia argues against a diagnosis of PSP.

*The optic fundus.* Neither optic nerve atrophy nor retinal degeneration is a feature of PSP. Optic nerve atrophy is occasionally seen in OPCA and retinal degeneration is a feature of some dominantly inherited ataxias. Macular degeneration is common in elderly subjects and may thus occur incidentally in PSP. Visual acuity, using correction if necessary, should be normal in PSP even in patients compaining of difficulty reading.

*Examining ocular motility.* Examination of ocular motility is of primary importance in assessing all patients with extrapyramidal disorders; and especially important in assessing a patient suspected of having PSP. One may begin by testing stability of ocular fixation. The examiner carefully watches the patient's eyes while the latter attempts to maintain steady gaze on a distant stationary object for a minimum of 10 seconds. In PSP, fixation is impaired by the intrusion of brief transient jerks moving the ocular globes abruptly 0.5 to 3.0 degrees from the point of fixation. These macro square-wave jerks are readily seen on simple visual inspection and are often quite prominent.

It is useful to assess ocular movements with regard to a heirarchy of control from higher cerebral levels to the coordinating mechanisms in the brainstem to the oculomotor nuclei. Thus one may begin by observing willed eye movements in response to command, then pursuit or following movements and finally reflex eye movements. In general, vertical eye movements are the first to be affected in PSP. The earliest change is a hesitancy of downgaze in response to a command. Variability is common. Hesitancy may be present on one examination but not on another a few minutes later. Repeated examination should be done in case of doubt. The patient may use synkinetic movements in an effort to covercome the hesitation. Commonly, the PSP patient will blink the eyes before attempting to look down.

Optokinetic nystagmus (OKN) may be induced in both horizontal and vertical planes by having the patient focus on a moving striped target. The commonly used length of striped cloth is useful for this purpose. Absence of OKN is a common finding in PSP. In early cases, it may be absent only in the vertical plane, and in very early cases only on downard movement of the target.

Failure of suppression of the vestibular-ocular reflex (VOR) by fixation on a target moving with the head is common in PSP. It can easily be tested using the Gower's reflex hammer. The tire of the hammer is held firmly against the forehead by the examiner as the head is moved gently from side to side and up and down while the patient is asked to maintain gaze fixed on the tip of the handle (Fig. 2). Normal subjects have no difficulty maintaining gaze fixed on the moving target. However, PSP patients often fail to suppress the VOR and are unable to maintain fixation on the moving handle. The eyes fail to remain locked on the target and lag behind. They may catch up repeatedly in a series of jerks. The defect may occasionally be found only in the vertical plane, sometimes only on downard movment of the head. Failure of VOR suppression is also a common finding in OPCA

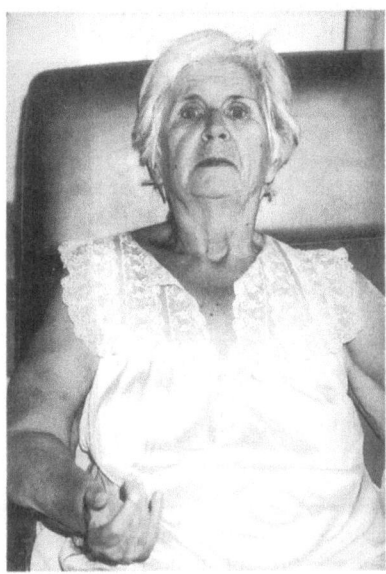

**Fig. 2**. Patient with PSP showing "surprised" facial expression

and in the Parkinson-Dementia Complex of Guam (Lepore et al., 1987) but is not found in PD.

As the disease progresses, hypometric saccades and saccadic slowing develops on upward and lateral gaze as well as down gaze. Later, when limitation of volitional gaze appears, a normal range of eye movements can still be elicited with following movements. With further progression of the disease following movements are also impaired. Reflex eye movements may still be preserved. This is easily demonstrated by asking the patient to fix on a visual target while manipulating the head. For example, in a patient unable to look down or follow a target moving downwards, the eyes may maintain fixation straight ahead while the head is tilted backward by the examiner. This "doll's eye" phenomenon is a classic finding in PSP. In advanced stages of the disease, even reflex eye movements may disappear. Bell's phenomenon may, however, be preserved into very late stages of the disease even in the presence of severe ophthalmoplegia. However, eventually, it may no longer be elicitable.

*Eyelid motility*. Spontaneous eyeblink is profoundly depressed in PSP patients, more markedly than in PD or OPCA. Extreme infrequency of eyeblink may be the first clue that an atypical parkinson patient has PSP. A mean rate of 3.5 blinks per minute was counted by Golbe et al. (1989) in PSP patients. Karson (1983) had found a mean rate of 11 per minute in untreated PD patients and 20 per minute in normal controls. This severe loss of spontaneous eyeblink often results in exposure keratitis and a complaint of burning eyes.

R. C. Duvoisin

**Table 2.** Comparison of eyelid and ocular motility disorders
in PD, PSP, and OPCA

| Abnormality | PD | PSP | OPCA |
|---|---|---|---|
| Nystagmus | 0 | + | ++ |
| Jerky pursuit | + | +++ | ++ |
| Fixation instability | 0 | ++++ | ++ |
| Ocular dysmetria | 0 | 0 | + |
| hypometric saccades | + | ++ | ++ |
| Fautly VOR suppression | + | +++[a] | +++[b] |
| Loss of OKN's | 0 | ++++[a] | ++[b] |
| Slow saccades | + | ++++ | ++ |
| Ocular lateropulsion | + | 0 | 0 |
| Levator inhibition | 0 | +++ | 0 |
| Supranuclear ophthalmoplegia | | | |
|    Verical | 0 | ++++ | + |
|    Horizontal | 0 | +++ | +++ |

+ Uncommon, ++ common or moderate, +++ frequent
or marked, ++++ present in nearly all cases or severe.
[a] chiefly in vertical movements; [b] chiefly in horizonal
movements (modified from Duvoisin, 1987)

Retraction of the eyelid (Cowper's sign), inhibition of levator palpebrae
function which is often referred to as "apraxia" of eye opening, supranuclear paresis of eyelid closing (eyelid closing "apraxia") and spasm of
the orbicularis oculi muscle (blepharospams) occuring singly or in various
combinations are frequent findings in PSP patients. In pure levator inhibition
the eye remains closed in the absence of any activity in the orbicularis oculi
causing blepharospasm (Lepore and Duvoisin, 1985). Blepharospasm may
erroneously be thought to be present if the orbicularis oculi muscle is not
carerully observed. Both may be present. In levator inhibition the brow is
often elevated (Charcot's sign) and the frontalis muscle actively contracted
in a compensatory effort to open the eye. The patient may resort to raising
the lid with a finger. As Golbe et al. (1989) showed in their videotapes,
many patients also employ various synkinetic tricks such as openning the
mouth to induce openning of the eye.

*Facial expression.* Richardson (1963) had referred to the face as "spastic"
in PSP. It may also be considered dystonic in the sense of a sustained
abnormal posture. Retraction of the upper eyelid combined with facial
hypomimia and frontalis contraction produce an expression suggestive of
surprise or astonishment as Jankovic (1983) has vividly illustrated (Fig. 1).
Sustained frontalis contraction may also be seen in arteriosclerotic parkinsonism and contributes to the "perplexed" look Spillane (1968) documented
in various dementias in his photographic atlas of neurology.

*Oropharynx.* Dysarthria and dysphagia disproportionate to the severity
of the parkinsonism is common in PSP and OPCA and may occur early in
the course of these disorders. Movements of the mouth, tongue and palate

are slow. Slowness of the tongue can be readily appreciated by asking the patient to move the tongue rapidly from side to side and to articulate labial and dental consonants. Despite severe immobility of the tongue, however, atrophy and fasciculations are rare and mandibular hyperreflexia is not found.

The gag reflex may be weak or absent. Saliva is often pooled in the fauces. Slowness in initiating deglutition is easily detected. Cinefluororadiographic studies may be useful in defining the disturbance in swallowing and may also reveal defective esophageal motility.

*Speech.* The voice and speech in PSP patients may be normal early in the course of the disease but later on the voice often becomes soft and the speech monotone as in PD with loss of normal prosody. The speech disturbance of PSP, however, differs from that in PD in being slow rather than tachyphemic, slurred and sometimes scanning in quality. The loss of facial expression and gestures of the hands which normally accompany conversational speech is more striking in PSP than in PD. Marked delay initiating verbal responses is seen in more advanced cases. Dysarthria can become so marked that speech is unintelligible.

*Posture and attitude.* Changes in posture standing and walking in PSP differ from those commonly seen in PD. The characteristic simian posture with flexion at the knees, hip and neck with mild thoracic kyphosis so typical of PD are conspicuously lacking in PSP. This fact may be a useful early clue in a patient presenting an atypical parkinsonism. PSP patients tend to stand erect in the early stages. Some stand leaning backward a few degrees from the vertical thereby impairing postural stability.

Hyperextension of the neck persisting even in the supine position is highly typical of PSP and helps to distinguish the disorder from PD, OPCA and other disorders. It is not usually present in the initial stages of the disease and is not invariably present in all PSP cases even in the advanced stages of the disease.

The mild thoracic scoliosis common in PD patients reflecting the asymmetry of its motor manifestations is not common in PSP. In general, the motor manifestations of PSP are more generalized. It rarely presents as hemiparkinsonism.

The characteristic hand posture of PD with flexed metacarpophalangeal joints, hyperextended proximal and flexed distal interphalangeal joints may be seen in PSP but is not so prominent as in PD. Clawing of the toes with tonic hyperextension of the first toe (the "striatal toe") often accompanied by varus posturing of the feet occurrs in PSP as in PD and other extrapyramidal disorders.

*The musculature.* The muscular rigidity common to many extrapyramidal syndromes and readily appreciated on manipulating the limbs at major joints tends to occur with marked axial preponderance in PSP. The examiner may find marked cervical rigidity yet normal tone at the wrist and elbow. Such a pattern would not be expected in PD or other parkinsonian syndromes. Contractures at the shoulder joints are also common in PSP patients. The occurrence of a frozen shoulder after minor trauma should alert the clinican to the diagnosis of PSP. Muscle atrophy, weakness and fasciculations consistent with anterior horn cell disease have been described in PSP but are rarely a prominent feature. Distal muscle weakness and atrophy reflecting an axonal neuropathy are also not uncommon in OPCA and the related hereditary ataxias.

*Tremor.* Tremor is relatively uncommon in PSP and is rarely a presenting symptom. When present it is mild, limited to the upper extremities and usually an action and/or postural tremor. However, a typical parkinsonian resting tremor with alternating pronation and supination does occur in an occasional patient. Masucci and Kutzke (1989) noted in 3 patients an alternating tremor with a frequency of 3 Hz, which is slower than the usual 5–7 Hz of the typical "pill-rolling" tremnor of PD.

*Reflexes.* No specific alteration of reflexes occurs in PSP. Peripheral neuropathy otherwise unexplained is not a primary feature of PSP and hyporeflexia is unexpected. The plantar reflexes are usually flexor. The Babinski sign may be elicited in some patients, but care must be taken not to confuse the tonic extension of the first toe common to many extrapyramidal diseases. As Hunt (1917) pointed out, positioning the leg at rest gently flexed at the knee with the patient supine helps distinguish the two. However, the disctinction is sometimes difficult and occasionally both may be present. The striatal toe may be present at rest and is usually accompanied by clawing of the other toes and varus posturing of the foot. It is not a response to a cutaneous stimulus but instead is provoked or accentuated by voluntary movement of the leg and on walking.

Palmomental reflexes are usually quite easily elicited in PSP and there is usually a failure of the blink response to a glabella tap to accomodate on repeated tapping. These abnormal reflexes commonly elicited in various extrapyramidal syndromes have no specific diagnostic significance. Hand and foot grasping and increased snout, rooting and sucking reflexes may also be elicited in many PSP patients. These familiar cortical release signs also lack diagnostic specificity.

*Sensory function.* Sensory function is not impaired in PSP unless additional illness is present, e.g., dabetes mellitus. However, a sensorimotor polyneuropathy may be seen in OPCA and the hereditary ataxias. Thus testing of pain, temperature and vibration should be done. In testing appreciation of pain with a pin, it is helpful to test for stimulus-induced myoclons. Myoclonic jerks elicited by pinpricks to the sole of the foot and volar surfaces of the fingers suggests multiple system atrophy (MSA). Cortical sensory defecits occur in CBD in which disorder they may be associated with limb apraxia and the "alien hand" syndrome.

Dysautonomia. Orthostatic hypotension may be a rare occurrence in advanced PSP. Its neuropathological basis has not yet been demonstrated. A Shy-Drager syndrome has not yet been documented in pathologically proven PSP.

*Gait impairment.* Gait is usually afected early in the course of PSP and as noted above impairment of gait is a common early symptom. Symmetrical bilateral involvement of gait with the parkinsonian features of bradykinesia, shortness of stride, shuffling, festination and freezing, loss of associated arm swinging and impaired postural stability may all be observed on examination. Postural stability is assessed by pulling the patient while standing upright suddenly backward by the shoulders. The mildly affected parkinsonian patient may step backward reasonably well but without compensatory swing or the arm of flexion of the trunk. In moderately affected patients this test will elicit retropulsion. Severely affected patients will simply fall backwards so that the examiner should be prepared to catch the patient to prevent a fall. The gait is more likely to be broad based reflecting a degree of truncal ataxia in OPCA but usually — though not always — remains within a normal base in PSP and PD.

*Mental status.* The routine mini mental test commonly employed in neurologic practice and standardized by Folstein et al. (1975) is often abnormal in PSP and other parkinsonian syndromes, cheifly because of slowness in mental response. Bradyphrenia is especially prominent in PSP. Given time to respond, however, the PSP patient thought initially to be demented may be capable of providing correct answers and appropriate interpretation of proverbs. Perseveration, difficulty performing sequential tasks and "utilization" behavior consistent with frontal lobe dysfunction are more prominent in PSP than in PD. Difficulty in verbal communication due to dysarthria may also contribute to an impression of dementia.

The extrapyrmidal motor problems of these patients interferes with the performance of even the simplest paper and pencil tests of mental function and give an erroneous impression of mental impairment and apraxia. PSP patients have in addition particular difficulty with visually guided motor tasks. For example, in giving a handwriting sample, the PSP patient may be seen to write without looking at the task.

Cortical deficits such as aphasia and agnosia are not normally found in PSP patients. The routine tests of language function — naming, repition, and verbal comprehension are usually normal even if performed very slowly.

Inappropirate expression of mood, both laughing and weeping, may occur in PSP. Their occurrence has not been recorded in PD or other parkinsonian syndromes. Hypersexuality on levodopa therpay has also, in the author's experience, been much more marked in PSP patients and may suggest the diagnosis in patients previously thought to have PD.

## Laboratory tests

The usual battery of screening blood chemistries are not affected by PSP. Blood counts and routine urinalysis reveal no specific abnormalities. Nor do routine studies of cerebrospinal fluid reveal any specific abnormalities.

Electroencephalography (EEG) shows nonspecific abnormalities comprising frontal and temporal theta activity. Epileptic seizures occurred in 3 of the 8 historic cases of Steele et al. and in 7 of 55 patients studied by Nygaard et al. (1989). EEG studies in the latter cases showed diffuse slowing with temporal accentuation, frontal rhythmic delta activity, sharp temporal transients and bitemporal slow and sharp wave bursts. The occurrence of seizres or the finding of these features in the EEG can alert the clinican to the possibility of PSP.

Magnetic resonance imaging (MRI) should ideally be done in all patients with newly recognized parkinsonian syndromes. Even normal findings are of interest and may serve as a valuable baseline for future comparison. Computed tomography (CT) and MRI imaging can identify characteristic features in patients with OPCA, SND, CBD, cerebral vascular disease and PSP. Schoenberg et al. (1986) found the measurement of the anterior-posterior diameter of the midbrain especially useful in confirming the diagnosis of PSP. Unfortunately, however, diagnostic anatomical changes on imaging studies are not generally present early in the disease and thus serve chiefly to confirm a diagniosis already indicated by the clinical findings.

Blin et al. (1990) have demonstrated on positron emission tomography (PET) frontal lobe hypometabolism in some PSP patients and loss of striatal D-2 dopamine receptors. Thus PET may have some diagnostic utility but its expense and limited availability restrict it primarily to research protocols. Single Photon Emission Computed tomography (SPECT) may reveal parietal hypoperfusion in CBD and frontal hypoperfusion in PSP (Timmons et al., 1984) but its diagnostic value has not yet been validated. SPECT with the dopamine receptor ligand may also reflect loss of striatal D-2 dopamine receptor (Brücke et al., 1993) and offers promise of clinical usefulness but its value in routine clinical work also remains to be established.

### Differential diagnosis

Typically, PSP arises as a diagnostic possibility in a patient presenting an akinetic-rigid syndrome characterized by a chronic progressive course, axial dystonia, dementia, a pseudobulbar palsy and a supranuclear ophthalmoplegia affecting vertical gaze preferentially. Often, the patient may have been misdiagnosed as PD but early failure of levodopa response subsequently arouses suspicion that PD is not the correct diagnosis. The differential diagnosis includes, in addition to PD, multiple system atrophy (MSA) — incuding both olivopontocerebellar atrophy (OPCA) and strionigral degeneration (SND), arteriosclerotic parkinsonism due to cerebrovascular disease, and corticobasal degeneration (CBD). All these conditions generally respond poorly if at all to levodopa therapy.

Many manifestations are common to all these conditions but vary in pattern of evolution over time and bodily distribution. For example, onset with gait impairment and postural instability is far more typical of PSP and MSA than of PD. The differential diagnosis includes gait apraxia due to normal-pressure hydrocephalus, OPCA, cerebral vascular disease (especially Binswanger's disease) and PD. The distinction from the "lower body parkinsonim" associated with cerebral vascular disease may be difficult as Dubinsky and Jankovic (1987) showed. However, as Critchley noted long ago (1926) the patient with "arteriosclerotic parkinsonism" tends to vigorously swing the arms which, however, are constantly flexed at the elbows.

The distribution of muscular rigidity may be helpful. Rigidity tends to be more sharply limited to axial musculature in PSP than in PD or MSA. The marked nuchal rigidity with hyperextension of the neck is distinctive of PSP, however, it has on rare occasions been seen in far advanced PD patients (Sage et al., 1990). Symmetry of signs and symptoms is also more typical of PSP and MSA than of PD. Hughes et al. (1992) found that asymmetry of signs and symtpoms was a useful clinical sign of PD. Asymmetry of clinical features is especially pronounced in CBD (Riley et al., 1990).

Probably disturbances in ocular and eyelid motility provide the greatest assistance in differentiating PSP from PD and MSA. A brief summary of the salient differences observed is presented in Table 1. Moderate decrease of eyeblink and cogwheeling of smooth pursuit movements and even diplopia may be seen in PD, OPCA and other parkinson syndromes. In PD, these occur chiefly in advanced stages of the disease and frank paresis of gaze is rare, occurring only in unusual circumstance, for example, transiently during a febrile episode. Ocular lateropulsion, a rare symptom of PD, has not been described in PSP. Apraxias of eyelid opening and closing, common in PSP, are rare in PD.

Supranuclear defects of ocular motility often occur in OPCA. Generally, however, horizontal movements are more severely affected in that condition and overall they are generally milder than in PSP. Fixation instability evident in macro-square wave jerks are more frequent and more prominent in PSP than in the hereditary ataxias and other cerebellar degenerations. The OPCA patient often uses head thrusting to bring the eyes onto a desired visual target; this has not been observed in PSP. The slow eye movements seen in some cases of hereditary ataxia, as noted by Wadia (1984), are smooth and not improved on reflex movement. The traditional cerebellar eye signs of gaze evoked nystagmus and ocular dysmetria are commonly seen in OPCA and especially the cerebellar degenerations and help distinguish those entities.

About half the recorded cases of CBD (Riley et al., 1990) have shown supranuclear defects of gaze and eyelid movement and in some cases these strikingly resemble the ophthalmoparesis of PSP. Supranuclear ophthalmoparesis with differential involvement of voluntary as opposed to pursuit movements and reflex eye movements and apraxias of eye openning and closing may occur in both conditions. However, impersistence of gaze and the preservation of spontaneous saccades in a patient who cannot induce ocular movements in response to a command is more consistent with CBD. Other features such as the striking apraxia, the "alien hand" syndrome and the marked asymmetry of manifestations should point to the latter condition.

## References

Agid Y, Javoy-Agid F, Ruberg M, et al (1986) Progressive supranuclear palsy: anatomical and biochemical considerations. Adv Neurol 45: 191–206

Baron JC, Maziere B, Loc'h C, et al (1985) Progressive supranuclear palsy: loss of striatal dopamine receptors demonstrated in vivo by positron tomography. Lancet i: 1163–1164

Blin J, Baron JC, Pillon B, Cambon H, Cambier J, Agid Y (1990) Positron emission tomography in progressive supranuclear palsy: brain hypometabolic pattern and clinicometabloic correlations. Arch Neurol 47: 747–752

Brücke T, Wenge S, Asenbaum S, Fertl E, Pfafflmeyer N, Müller C, Podreka I, Angelburger (1993) Dopamine $D_2$ receptor imaging and measurement with SPECT. Adv Neurol 60: 494–500

Chavany JA, Van Bogaert L, Godlewski S (1951) Sur un syndrome de rigidite a predominance axiale avec perturbation des automatismes oculo-palpebraux d'origine encephalitique. Presse Med 50: 958–962

Doty RL, Golbe LI, McKeown DA, Stern MB, Lehrach CM, Crawford D (1993) Olfactory testing differentiates between progressive supranuclear palsy and idiopathic Parkinson's disease. Neurology 43: 962–965

Dubinsky RM, Jankovic J (1987) Progressive supranuclear palsy: relationship to multinfarct state and cerenral amyloid angiopathy. Neurology 37: 570–576

Duvoisin RC (1987) The olivopontocerebellar trophies. In: Marsden CD, Fahn S (eds) Movement disorders, vol 2. Butterworths, London, pp 249–269

Duvoisin RC, Yahr MD (1965) Postencephalitic parkinsonism. Arch Neurol 35: 487–495

Duvoisin RC, Golbe LI, Lepore FE (1987) Progressive supranuclear palsy. Can J Neurosci 14: 547–554

Folstein MF, Folstein SE, McHugh PR (1975) "Mini-mental state": a practical method for grading the cognitive state of patients for the clinician. J Psychiatr Res 12: 188–198

Golbe LI (1989) SPECT imaging in corticobasal degeneration. Neurology 28: 298–299

Golbe LI, Davis PH, Schoenberg BS, Duvoisin RC (1988) Natural history and prevalence of progressive supranuclear palsy. Neurology 38: 1031–1034

Golbe LI, Davis PH, Lepore FE (1989) Eyelid movement abnormalities in progressive supranuclear palsy. Mov Disord 4: 297–302

Hunt JR (1917) Progressive atrophy of the globus pallidus. Brain 40: 58–148

Jackson JA, Jankovic J, Ford J (1983) Progressive supranuclear palsy: clinical features in 16 patients. Ann Neurol 13: 273–278

Jankovic J (1984) Progressive supranuclear palsy: clinical and pharmacological update. Neurol Clin 2: 473–486

Karson CN (1983) Spontaneous eye-blink rates and dopaminergic systems. Brain 106: 646–653

Lepore FE, Duvoisin RC (1985) "Apraxia" of eyelid openning: an involuntary levator inhibition. Neurology 38: 423–427

Lepore FE, Steele JC, Tilson G, Calne DB, Duvoisin RC, Lavine L, McDarby JV (1988) Supranuclear disturbances of ocular motility in Lytioco-Bodig. Neurology 38: 1849–1853

Maher ER, Lees AJ (1986) The clinical features and natural history of the Steele-Richardson-Olszewski syndrome (progressive supranuclear palsy). Neurology 36: 1005–1008

Nygaard TG, Duvoisin C, Manocha M, Chokroverty S (1989) Seizures in progressive supranuclear palsy. Neurology 39: 122–140

Olsjewski J (1966) Heterogenous system degeneration: progressive supranuclear palsy — ocular, facial and bulbar. J Neurosurg 24: 250–254

Posey WC (1904) Paralysis of upward movement of the eyes. Ann Ophthalmol 13: 523–529

Richardson JC, Steele JC, Olzewski J (1963) Supranuclear opthalmoplegia, pludobulbar palsy, nuchal dystopnia and dementia: a clinical report on eight cases of "heterogenous system degeneration". Trans Am Neurol Assoc 88: 25–27

Riley DE, Lang AE, Lewis A, Resch L, Ashby P, Hornykiewicz O, Black S (1990) Cortico-basal ganlionic degeneration. Neurology 40: 1203–1212

Sage JI, Miller DC, Golbe LI, Walters AS, Duvoisin RC (1990) Clincally atypical expression of pathologically typical Lewy-body parkinsonism. Clin Neuropharmacol 13: 36–47

Schonfeld SM, Golbe LI, Safer J, Sage JI, Duvoisin RC (1987) Computed tomographic findings in progressive supranuclear palsy: correlation with clinical grade. Mov Disord 2: 263–278

Spiller WG (1905) The importance in clinical diagnosis of paralysis of associative movements of the eyeballs (Blick-Lahmung) especially of upwards and downward associated movements. J Nerv Ment Dis 32: 417–488; 4997–530

Spillane JD (1968) An atlas of clinical neurology. London, pp 16–17

Su PC, Goldensohn ES (1973) Progressive supranuclear palsy: electroencephalographic studies. Arch Neurol 29: 183–186

Timmons JH, Bonikowski FW, Harshorne MF (1984) Iodoamphetamine-123 brain imaging demonstrating cortical deactivation in 2 patient with progressive supranuclear palsy. Clin Nucl Med 14: 841–842

Wadia NH (1984) A variety of olivopontocerebellar atrophy distinguished by slow eye movements and peripheral neuropathy. In: Duvoisin A, Plaitzkis A (eds) The olivopontocerebellar atrophies. Raven Press, New York, pp 149–177

Author's address: Prof. Dr. R. C. Duvoisin, Department of Neurology, University of Medicine and Dentistry of New Jersey, One Robert Wood Johnson Place, New Brunswick, NJ 08903-0019, U.S.A.

# Cognitive disturbances in progressive supranuclear palsy

## I. Litvan

Neuroepidemiology Branch, National Institutes of Neurological Disorders and Stroke, National Institutes of Health, Bethesda, Maryland, U.S.A.

**Summary.** The cognitive disturbances in progressive supranuclear palsy (PSP) gave rise to the term "subcortical dementia." PSP patients demonstrate prominent recall deficits and moderate forgetfulness although their short-term and implicit perceptual memory processes are intact. PSP patients have both slowed motor responses and dramatically slowed information processing speed. Executive dysfunction appears early in the course of the disease and is relatively severe. The combination of severely slowed information processing and marked executive dysfunction are characteristic of PSP and differentiates it from other dementias.

In their landmark description of progressive supranuclear palsy (PSP) as a clinicopathological entity, Steele et al. (1964) reported that cognitive disturbances were present in seven out of their nine patients. Ten years later, Albert et al. (1974) characterized these changes to be part of a "subcortical dementia." They analyzed 5 of their own PSP cases and also reviewed the published literature; they found a common cluster of symptoms, including the presence of forgetfulness, slowness of thought process, emotional or personality changes, and impaired ability to manipulate acquired knowledge. Albert et al. analysis was qualitative, but in the authors' view, clearly differentiated PSP patients from patients with "cortical dementia" who presented with aphasia, apraxia, and/or agnosia. They also suggested that the symptoms found in PSP were similar to those that had previously been described in patients with frontal lobe lesions.

### Memory in PSP

Can forgetfulness in PSP be quantified? Are there additional components of memory affected in PSP? We found that short-term memory (i.e., the limited capacity, data-driven stores from which information is transferred to more permanent conceptually-driven stores) is unaffected in PSP. We studied short-term memory using the digit span and the Sternberg paradigm (1966), and found that PSP patients performed normally on both (Litvan et al., 1989). The Sternberg paradigm measures the speed of inter-item scan in short term-memory. Patients are shown a set of 1 to 6 digits on a computer

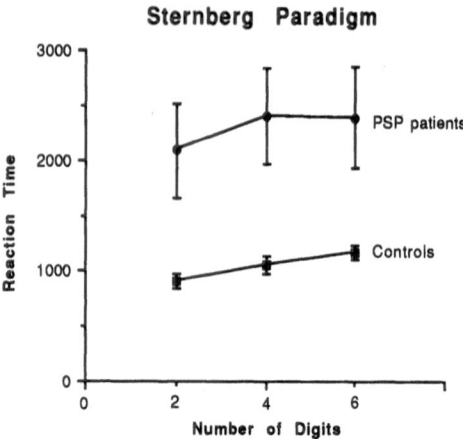

**Fig. 1.** Sternberg Memory Paradigm: there was no significant difference between PSP patients and controls in the inter-item scanning speed, although patients' response times were significantly slower (p < 0.0001). [Reprinted with permission from Arch Neurol]

screen for 200 msec per digit, the screen goes blank, and then a single probe digit is presented. Subjects are required to decide whether or not the digit presented was a member of the previously memorized set. The larger the set, the longer it takes to decide whether a probe digit was a member of that set. Reaction time is measured for different set sizes, and the resultant slope serves as a measure of the speed of inter-item scanning in short-term memory. Our results (Fig. 1) showed that patients and controls slopes were parallel, indicating that the PSP patients' inter-item scanning speed in short-term memory is normal. Compared to controls, patients' response times were consistently slower, despite the similarity in scanning time. We interpreted these findings as representing slowed response execution time.

On the other hand, several aspects of long term memory appear impaired in PSP patients. We measured verbal learning with the Rey-Auditory Verbal Learning Test in which patients need to learn a list of 15 words that are repeatedly presented for 5 consecutive trials (Litvan et al., 1989). Patients had more difficulty than controls retrieving words across trials, although some learning was evident (Fig. 2). On trial 6, which represents recall of the words after a 30-minute filled delay, not only did PSP patients recall fewer words than controls, they also had a steeper forgetting slope. A recognition memory test that followed the delayed recall revealed that PSP patients recognized fewer words from the Rey Auditory Verbal Learning Test than controls but their performance on a recognition test was qualitatively superior to their free recall.

We also examined the hypothesis that PSP patients demonstrate rapid forgetting with the Brown-Peterson paradigm (Brown, 1958). In this paradigm, subjects are presented with three words and after interference-filled intervals of different durations (3–36 sec), they are asked to recall the

**Rey-Auditory Verbal Learning Test**

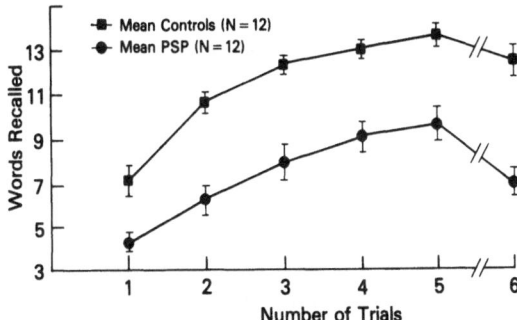

**Fig. 2.** There was a significant difference between 12 PSP patients and equal number of educated matched controls in the Rey Auditory Verbal Learning Test (trials 1 to 5) ($p < 0.005$). Patients had a steeper decay of information from trial 5 to trial 6. [Reprinted with permission from Arch Neurol]

previously presented words. Not only did PSP patients generally recall fewer words than controls but the longer the interference duration, the poorer their recall became relative to the controls (Fig. 3). In a forced-choice recognition test that included old and new words, patients were able to reject new words as well as controls but had difficulty recognizing old words. These findings suggest that PSP patients demonstrate exaggerated forgetting as compared to controls.

Given the involvement of basal ganglia structures in PSP, we were also interested in measuring procedural learning (Grafman et al., 1990b). For this purpose, we administered the Serial Reaction Time Test. In this task, an asterisk appears on a computer screen at one of four locations arranged horizontally and separated by 3.4 cm (Nissen and Bullemer, 1987). Subjects were instructed to keep their fingers on four keys located below each position where an asterisk appeared on the monitor. Reaction time to each stimulus was recorded in milliseconds. The asterisk remained on the monitor until the subject pressed the correct key. Subjects completed 7 blocks of 100 trials each. Blocks 1, 2 and 7 were 100 random presentations. Blocks 3 to 6 contained 10 repeated sequences of 10 trials each. The difference in reaction times between block 6 (the last repeated block) and block 7 (a random block) was considered a reflection of procedural learning. If reaction times improved simply as a result of faster Reaction Times and not from learning the repeated sequence, then there should be no decrement in reaction times from block 6 to block 7. PSP patients, as opposed to Alzheimer's disease patients (AD), failed to show a decrement in reaction times from block 6 to block 7. However, they continued to show a decrease in Reaction Times suggesting an improved familiarity with the task demands rather than a specific effect of procedural learning (Fig. 4). We can contrast these results with those obtained on another test of implicit memory (Grafman et al., 1990b). On this test of perceptual priming, subjects try to read words shown on a computer monitor. The exposure dura-

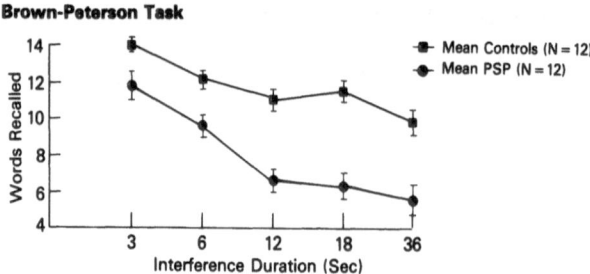

**Fig. 3.** There was a significant difference between PSP patients and controls for all interference filled intervals of the Brown Peterson Paradigm (p < 0.0001). [Reprinted with permission from Arch Neurol]

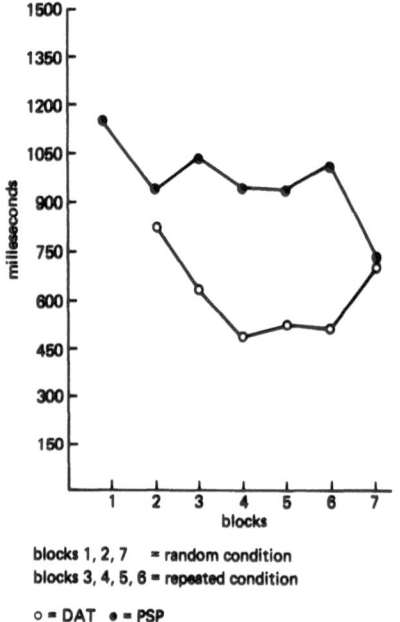

blocks 1, 2, 7  = random condition
blocks 3, 4, 5, 6 = repeated condition

o = DAT   • = PSP

**Fig. 4.** Procedural learning task, comparison between PSP and AD patients. Reaction time was measured in milliseconds. The difference between block 6 (repeated sequence) and block 7 (random sequence) suggest lack of procedural learning in patients with PSP

tion of these words is reduced until the subject recognizes only 30% of the words. Then another list of words is presented at the exposure duration. This list is composed of 15 new words and 15 words from the original set. Priming is shown when words from the original list are recognized at a higher rate than the new words. PSP patients showed significantly greater priming than AD patients (Fig. 5). Thus, PSP patients performed poorly on a visuomotor procedural learning task compared to AD patients, but superior to AD patients on a perceptual priming task.

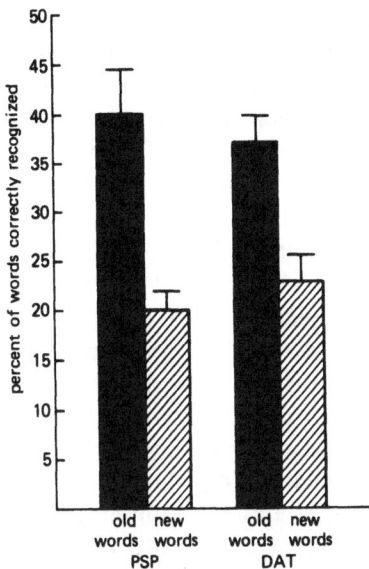

**Fig. 5.** Perceptual priming, comparison between PSP and AD patients. PSP patients showed significant priming effect of old over newly recognized words

Pillon et al. (1986, 1991, 1992) also compared the performance of PSP and AD patients on several measures of memory. They found that in general, explicit episodic and remote memory measures are more impaired in AD than in PSP.

*In summary*: Long-term memory (learning, retrieval, forgetting) is clearly affected in PSP but less than in AD patients. Rapid forgetting appears to be one of several impaired components. While procedural learning is impaired, short-term memory and perceptual priming are preserved.

### Executive and other cognitive functions

Are executive functions impaired in PSP patients? Executive functions include reasoning, problem-solving, concept-formation, planning and social cognition. These functions have been studied by several groups of investigators and there is agreement that they are definitely abnormal in PSP (Cambier et al., 1985; Grafman et al., 1990a; Maher et al., 1985; Pillon et al., 1986; Rafal and Grimm, 1991). Patients have problems in forming concepts: i.e., they are unable to interpret proverbs (Albert et al., 1974; Cambier et al., 1985; Grafman et al., 1990a). They have difficulties in shifting established conceptual sets as can be seen in their performance on the Wisconsin Card Sorting and Trail Making Tests (Pillon et al., 1986, 1991; Grafman et al., 1990a). In the Wisconsin Card Sorting Test, subjects are given cards that they have to sort according to three possible concepts "color, form or number." The examiner gives the patients feedback if they made a correct response. After 10 consecutive correct responses, the concept changes. This test is scored for the number of concepts attained and type of error, and is considered to be a measure of concept formation and shifting. PSP patients made significantly more sorting errors than controls and the majority of these were perseverative errors. PSP patients perform more poorly on the trail making B Test which also requires conceptual shifting, in this case between numbers and letters. They also had difficulty in ordering events in a correct sequence. This has been tested with the "Picture Arrangement" subtest of the WAIS-R (Cambier et al., 1985; Grafman et al., 1990a; Maher et al., 1985). In this test, subjects are presented with cartoon pictures that need to be arranged in the correct order to be able to tell a story. Patients with PSP demonstrate problems with analogic thinking. This was seen, for example in their performance on the "Similarities" subtest of the WAIS-R, where they have to find the commonality between two items belonging to the same category (Cambier et al., 1985; Grafman et al., 1990; Maher et al., 1985; Pillon et al., 1985). PSP patients also have difficulty with initiation and fluency (Cambier et al., 1985; Grafman et al., 1990; Maher et al., 1985; Pillon et al., 1985, 1991). In these tests, as well as in others where verbal or drawing production is required, perseveration was frequently observed.

One of the main clinical characteristics of PSP patients is the slowness in their response. Does this slowness reflect a motor problem or cognitive slowing? Dubois et al. (1988) attempted to evaluate cognitive slowing in a series of experiments in which they subtracted reaction times of simple tasks from complex ones while the motor demands of the tasks were held constant. In these experiments, PSP, but not Parkinson's disease (PD) patients, were observed to have "central cognitive slowing." Johnson et al. (1991, 1992), using event-related brain potentials (ERPs), were able to confirm that PSP patients have remarkably slowed information processing. Using an Oddball task, they found that the early N1 component was normal in amplitude and latency but that the P2 and P300 components had increased latencies and decreased amplitudes. These potentials showed a normal scalp distribution, suggesting that all component cognitive processes were functioning but that the process of stimulus identification in these patients was slowed and degraded. PSP patients had significantly increased latencies for the P2 component (Fig. 6). The difference in latency for the P300 component between patients and controls was even more exaggerated than the P2 latency differences. The difference in amplitude between controls and patients for the P300 was also dramatic. These remarkably delayed latencies have not been reported in any other type of dementia (Johnson, 1992). Since increased latencies began with the earlier P2 component, it suggested that PSP patients may have altered sensory processing which affects later conceptually driven cognitive functions.

*In summary*: not only do PSP patients demonstrate slowed motor responses but they have increased latencies in central processes that include both data-driven and conceptually-driven analyses.

Attention is also impaired in PSP. On a task in which patients need to tap their finger whenever they hear a target letter presented among

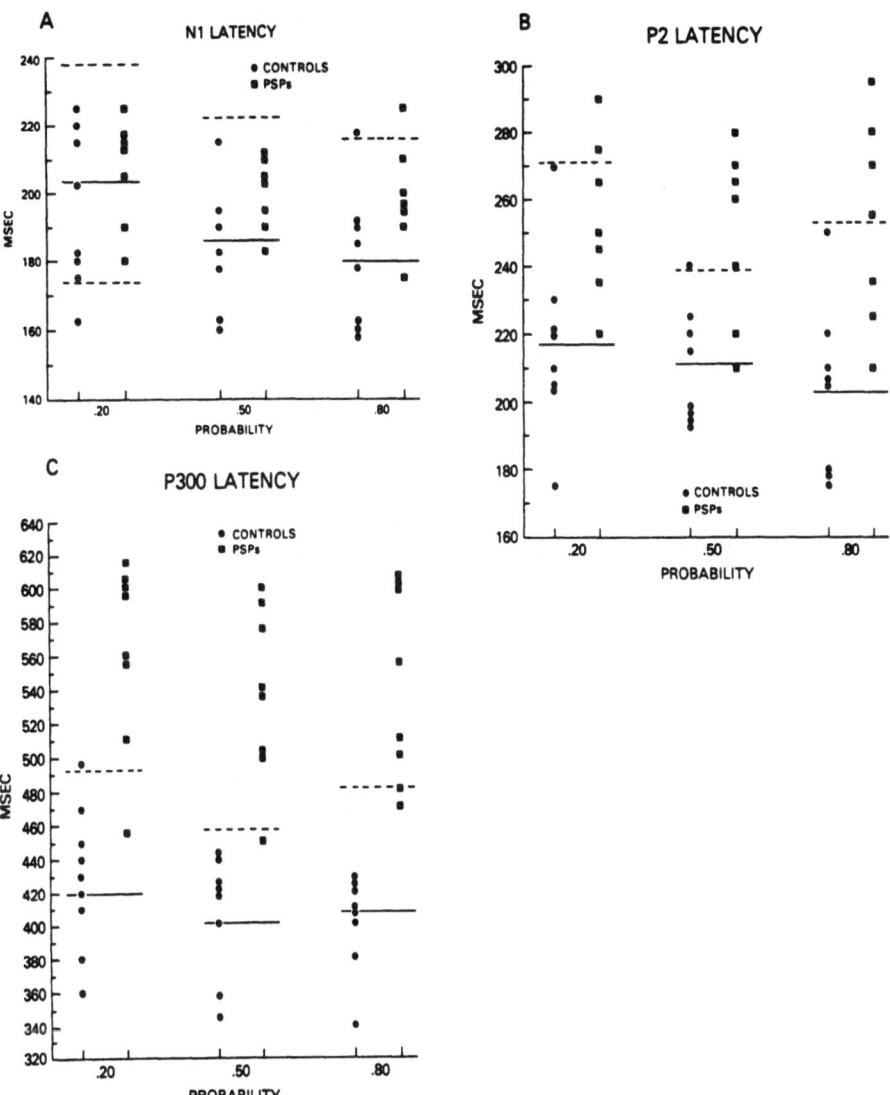

**Fig. 6.** Case-by-case latencies of the occipital (*N1*), frontal (*P2*) and P300 event related brain potentials using an Oddball task at 3 different stimulus probabilities. The solid horizontal line indicates the mean and the dashed line ± SD. P2 and P300 components were significantly slower in PSP patients than in controls

different letters, PSP patients demonstrated mildly impaired vigilance by identifying fewer targets over time. PSP patient visuospatial attention has been evaluated using a task designed by Posner (1980). Several investigators (Kertzman et al., 1990; Rafal et al., 1988, 1992) found that PSP patients were impaired in orienting their attention when compared to controls. This impairment was related to the duration of their disease (Kertzman et al., 1990).

Progression and severity of cognitive deficits in PSP have recently been addressed by Pillon et al. (1991, 1992) who tested 44 AD, 45 PSP, 35 Huntington's disease (HD) and 164 PD patients. These investigators defined dementia as a global intellectual performance two standard deviations below the mean control values. This definition was chosen because in a population with movement disorders it is difficult to use the DSM-III-R criteria of dementia which links cognitive impairment with patients inability to pursue their social and work related obligations. Their results indicated that 93% of the AD patients were demented, 66% of HD, 58% of PSP patients, but only 18% of the PD patients. They also found that there was consistent progression of cognitive deficits in PSP and that executive functions were the most severely affected. AD patients seemed to have a different neuropsychologic profile than patients with the "subcortical dementias." Pillon et al. argued that it was clinically important to preserve the distinction between cortical and subcortical dementia in spite of the lack of clear anatomopathologic differentiation.

What are the neuroanatomic concomitants of the cognitive deficits seen in PSP? Three of the five corticosubcortical circuits described by Alexander (1986, 1990) are affected in PSP. The dorsolateral-frontal circuit related to cognition, the lateral-orbitofrontal circuit related to social affect, and the anterior cingulate circuit related to attention. Frontal lobe neuropathologic changes are minimal in PSP but these patients have significant frontal hypometabolism as seen on positron emission tomography (PET) (Hauw et al., 1990; D'Antona et al., 1985; Blin et al., 1992). Several investigators have proposed that the cognitive deficits in PSP are due to deafferentation of frontal cortex (Agid et al., 1986; D'Antona et al., 1985; Pillon et al., 1986, 1992).

### Conclusion

Executive dysfunction and slowed information processing appear early in the course of PSP and are relatively severe. Both of these deficits are characteristic of PSP and differentiate it from other dementias. Attention and memory, although impaired, are less severely affected. These cognitive deficits are attributed to frontal deafferentation secondary to subcortical lesions.

Future research is needed to identify the cause(s) and effects of slowed information processing, evaluate how executive functions fail in PSP, as well as to assess the relative contribution of impaired attention to their cognitive disorder.

### Acknowledgements

I am grateful to the PSP patients who kindly participated in our studies and to Dr. Grafman and D. Schoenberg for their helpful comments of a previous draft.

# References

Agid Y, Javoy-Agid F, Ruberg M, Pillon B, Dubois B, Duyckaerts C, Hauw JJ, Baron JC, Scatton B (1986) Progressive supranuclear palsy: anatomo-clinical and biochemical considerations. In: Yahr MD, Bergmann KJ (eds) Advances in neurology, vol 45. Raven Press, New York, pp 191–206

Albert ML, Feldman RG, Willis AL (1974) The subcortical dementia of progressive supranuclear palsy. J Neurol Neurosurg Psychiatry 37: 121–130

Alexander GE, DeLong MR, Strick PL (1986) Parallel organization of functionally segregated circuits linking basal ganglia and cortex. Annu Rev Neurosci 9: 357–381

Alexander GE, Crutcher MD, Delong MR (1990) Basal ganglia-thalamocortical circuits: parallel substrates for motor, oculomotor, "prefrontal" and "limbic" functions. In: Uylungs HBM, Van Eden CG, De Bruin MN, Corner MA, Feenstra MGP (eds) The prefrontal cortex: its structure, function and pathology, vol 85. Elsevier, pp 119–143

Blin J, Ruberg M, Baron JC (1992) In: Litvan I, Agid Y (eds) Progressive supranuclear palsy: clinical and research approaches. Oxford University Press, New York, pp 155–168

Brown J (1958) Some test of decay theory of immediate memory. Q J Exp Psychol 39: 15–22

Cambier J, Masson M, Viader F, Limodin J, Strube A (1985) Le syndrome frontal de la paralysie supranucleaire progressive. Rev Neurol (Paris) 141: 528–536

D'Antona R, Baron JC, Samson Y, Serdaru M, Viader F, Agid Y, Cambier J (1985) Subcortical dementia: frontal cortex hypometabolism detected by positron tomography in patients with progressive supranuclear palsy. Brain 108: 785–799

Dubois B, Pillon B, Legault F, Agid Y, Lhermitte F (1988) Slowing of cognitive processing in progressive supranuclear palsy. Arch Neurol 45: 1194–1199

Grafman J, Litvan I, Gomez C, Chase TN (1990a) Frontal lobe function in progressive supranuclear palsy. Arch Neurol 47: 553–558

Grafman J, Weingartner H, Newhouse PA, Thompson K, Lalonde F, Litvan I, Molchan S, Sunderland T (1990b) Implicit learning in patients with Alzheimer's disease. Pharmacopsychiatry 23: 94–101

Hauw JJ, Verny M, Delaere P, Cervera P, He Y, Duyckaerts C (1990) Constant neurofibrillary changes in the neocortex in progressive supranuclear palsy. Basic differences with Alzheimer's disease and aging. Neurosci Lett 119: 182–186

Johnson R Jr, Litvan I, Grafman J (1991) Progressive supranuclear palsy: altered sensory processing leads to degraded cognition. Neurology 41: 1257–1262

Johnson R Jr (1992) Event-related brain potentials. In: Litvan I, Agid Y (eds) Progressive supranuclear palsy: clinical and research approaches. Oxford University Press, New York, pp 122–154

Kertzman C, Robinson DL, Litvan I (1990) Effects of physostigmine on spatial attention in patients with progressive supranuclear palsy. Arch Neurol 47: 1346–1350

Litvan I, Grafman J, Gomez C, Chase TN (1989) Memory impairment in patients with progressive supranuclear palsy. Arch Neurol 46: 765–767

Maher ER, Smith EM, Lees AJ (1985) Cognitive deficits in the Steele-Richardson-Olszewski syndrome (progressive supranuclear palsy). J Neurol Neurosurg Psychiatry 48: 1234–1239

Nissen MJ, Bullemer P (1987) Attentional requirements of learning: evidence from performance measures. Cogn Psychol 19: 1–32

Pillon B, Dubois B, Lhermitte F, Agid Y (1986) Heterogeneity of cognitive impairment in progressive supranuclear palsy, Parkinson's disease and Alzheimer's disease. Neurology 36: 1179–1185

Pillon B, Dubois B, Ploska A, Agid Y (1991) Severity and specificity of cognitive impairment in Alzheimer's, Huntington's, Parkinson's disease and progressive supranuclear palsy. Neurology 41: 634–643

Pillon B, Dubois B (1992) Cognitive and behavioral impairments. In: Litvan I, Agid Y
    (eds) Progressive supranuclear palsy: clinical and research approaches. Oxford
    University Press, New York, pp 223–239
Posner MI (1980) Orienting of attention. Q J Exp Psychol 32: 3–25
Rafal RD, Grimm RJ (1981) Progressive supranuclear palsy: functional analysis of the
    response to methysergide and antiparkinsonian agents. Neurology 31: 1507–1518
Rafal RD, Posner MI, Friedman JH, Inhoff A, Bernstein E (1988) Orienting of visual
    attention in progressive supranuclear palsy. Brain 111: 267–280
Rafal RD (1992) Visually guided behavior. In: Litvan I, Agid Y (eds) Progressive
    supranuclear palsy: clinical and research approaches. Oxford University Press, New
    York, pp 204–222
Steele JC, Richardson JC, Olszewski J (1964) Progressive supranuclear palsy: a hetero-
    geneous degeneration involving the brain stem, basal ganglia, and cerebellum, with
    vertical gaze and pseudobulbar palsy, nuclear dystonia, and dementia. Arch Neurol
    10: 333–359
Sternberg S (1966) High speed scanning in human memory. Science 1953: 652–654

Author's address: Dr. I. Litvan, Neuroepidemiology Branch, National Institute of
Neurological Disorders and Stroke, National Institutes of Health, Federal Building,
Rm 714, Bethesda, MD 20892, U.S.A.

# Progressive supranuclear palsy and corticobasal ganglionic degeneration: differentiation by clinical features and neuroimaging techniques

**S. Giménez-Roldán, D. Mateo, C. Benito, F. Grandas,** and
**Y. Pérez-Gilabert***

Departments of Neurology and Neuroradiology, Hospital General Gregorio Marañón,
Madrid, Spain

**Summary.** To assess the extent of overlap between clinically diagnosed patients with progressive supranuclear palsy (PSP) and corticobasal ganglionic degeneration (CBGD) we compared clinical scores for rigidity, bradykinesia, supranuclear gaze abnormalities, hemineglect and limb apraxia, postural instability, neck rigidity, and limb dystonia in 15 patients with a degenerative rigid-akinetic syndrome at presentation and at follow-up 3 to 120 months later. Only the presence of hemineglect, usually in combination with limb apraxia, was a reliable and early clinical factor for discriminating between these two conditions. These symptoms were present at admission in all 4 CBGD patients but not in any of the 11 PSP patients either at presentation or later during serial examinations. Though supranu-clear ophthalmoplegia, neck rigidity, and postural instability were already observed in most CBGD patients at presentation, their scores remained low compared to those for PSP patients over the longterm. CT-scans and MRI were helpful in supporting clinically-based diagnoses made at presentation in that the vast majority of the PSP patients exhibited various degrees of midbrain atrophy and 50 percent of the CBGD patients exhibited asymmetric pericentral cortical atrophy.

## Introduction

Progressive supranuclear palsy (PSP) and corticobasal ganglionic degener-ation (CBGD) are multisystem degenerative disorders of the brain presenting in late life and causing rigidity and akinesia (Steele et al., 1964; Rebeitz et al., 1968), often in conjunction with other neurological abnormalities. Even though such "plus" symptoms may be helpful in making clinical diagnoses

* Present address: Hospital Eugenio Espejo, Quito, Ecuador; her work was supported by a grant provided by Consejería de Salud, Comunidad Autónoma de Madrid

(Jankovic, 1989), similarities between these two conditions were soon recognized (Rebeitz et al., 1968; Scully et al., 1985). Indeed, some authors claim that these two disorders are clinically indistinguishable (Gibbs et al., 1989). If true, this may be frustrating for practising neurologists, adds to the uncertainties of prognosis, prevent identification of specific risk factors, and give rise to difficulties in selecting patients for therapeutic trials.

Differences between PSP and CBGD are based not only on the unique histological features of degenerating neurons particular to each of these disorders, namely globose neurofibrillary tangles in PSP and corticobasal inclusions in CBGD (Gibb, 1989), but also in the distinctive atrophy distribution pattern observed on gross inspection of the brain. Thus, shrinkage of the midbrain is often a conspicuous feature on macroscopic examination of the brain in PSP (Jellinger et al., 1980), while asymmetric atrophy is often observed in the frontoparietal cortex of CBGD patients (Rebeitz et al., 1968; Scully et al., 1985; Gibb et al., 1989; Greene et al., 1990). Clinical-anatomical correlates of relatively circumscribed atrophic changes may include supranuclear gaze palsy, particularly impairment of downward eye movements, which is consistently observed in most PSP patients (Brusa et al., 1979), and the ill-defined alien hand phenomena (Doody and Jankovic, 1992), often associated with apraxia and motor neglect in CBGD patients. With few excepcions (Akashi et al., 1989; Sasaki et al., 1991), the neocortex appears histologically preserved in PSP (Steele et al., 1964), thus explaining the absence of aphasia or perceptive motor abnormalities in patients suffering from this condition. PSP patients can therefore be expected not to develop such "cortical" disturbances as hemineglect or unilateral limb apraxia or neuroimaging evidence of asymmetric cortical degeneration during the course of their illness.

The object of the present study was two-fold. First, to evaluate to what extent there is overlap between the abnormal neurological findings observed in PSP and CBGD, either early at presentation or during progression of the disease in follow-up studies. In particularly, we were interested in ascertaining whether asymmetric motor involvement was a common feature in PSP patients and whether these patients develop limb apraxia or allied hand phenomena at some stage in their illness. And second, to assess the value of CT-scans and MRI in identifying early distinctive features, which might be helpful in supporting clinically based diagnoses.

## Material and methods

We reviewed the records of 15 patients admitted during the past 10 years for investigation of a progressive rigid-akinetic syndrome. To be included as a subject, patients had to fullfil the following requirements: 1) Exhibit an unequivocal disturbance of supranuclear eye movements and/or unilateral motor apraxia and neglect, either present at admission or documented later at some time during follow-up; 2) Failure of levodopa to give any substantial and enduring benefit, on administration of conventional doses for at least three months; 3) Undergo a CT-scan as part of the routine initial work-up, eventually supplemented in some patients by a MRI study; and 4) After discharge, undergo follow-up at regular intervals at the outpatient Movement Disorder Clinic for a minimum of three months. Patients were excluded if they had an identifiable cause for the disorder, a history of postencephalitic parkinsonism, a family history of olivopontocerebellar degeneration or overt vascular disease of the brain.

A severity score (0–3) was designed to assess seven different neurological signs. The presence and severity of five of these signs (limb rigidity, bradykinesia, neck stiffness, postural instability, and disturbed speech) were routinely recorded using the UCLA Rating Scale for Parkinsonism (McDowell et al., 1970), on each patient's visit to the clinic. Asymmetric involvement was considered to exist when there was a difference of at least one point in the scores for rigidity or akinesia for the limbs on each side of the body. Upper limb dystonia was considered to exist when the patient exhibited characteristic postural abnormalities of the limbs either at action or while standing or walking, such as ulnar deviation and hyperextended fingers or a tendency to place a hyperpronated hand behind the back. Hemidystonia was considered to exist when both the upper and lower limbs on one side of the body showed conspicuous postural abnormalities (Marsden et al., 1985). Fahn and Marsden's dystonia severity evaluation scale was used, and provoking and severity factor scores 3 and 4 were unified as a score 3 for the sake of uniformity (Marsden and Schachter, 1981). Ideomotor apraxia was defined as an inability to carry out an activity, such as gestures or the use of common objects on verbal command with either the left or right hand that was unexplained by motor weakness or extreme rigidity (Benson, 1978). Unilateral motor neglect was defined as evident lack of use of one limb in the absence of motor weakness or incapacitating rigidity (Binder et al., 1992). We restricted the term "alien hand" to recognition that actions of one's limbs or hands are autonomous, patients therefore failing to recognize the actions as their own, independently of any feelings of foreign control (Levine, 1992). Cognitive assessment was rated as normal (0), mildly impaired (1) for scores below 23 on the Mini-Mental State Examination (Folstein et al., 1975), moderately impaired (2) when there was a demonstrable abnormality in social or occupational abilities, and severe (3) when patient fulfilled the criteria for definite dementia (American Psychiatric Association, 1987).

The diagnostic criteria for the clinical diagnosis of PSP were those of Lees (1987). Thus, besides down-gaze abnormalities and parkinsonism, patients had to present at least one of the following cardinal symptoms: axial dystonia or rigidity, pseudobulbar palsy, frontal lobe signs such as perseveration or utilization behaviour, and postural instability with backwards falls. Clinical diagnosis of CBGD relied on the presence of unilateral limb apraxia, motor hemineglect, or alien hand sign in addition to parkinsonism, irrespective of the presence of other neurological abnormalities, including disturbances in eye movements (Weiner and Lang, 1989).

All scans were obtained at admission with a model 9800 General Electric scanner without contrast enhancement using 5-mm cuts. Magnetic resonance imaging was performed with a T5 Philips model, 0.5-Tesla supraconducting magnet using T2-weighted (TR/TE 2000 to 2500/30 to 6 milliseconds) sequences in the transverse orientation and T1-weighted images (TR/TE 600/30 milliseconds) generated in the sagittal plane.

## Results

A clinically based diagnosis of PSP was achieved in 11 patients, and 4 others were considered to be suffering from CBGD. Table 1 gives the demographic details on all 15 patients. Symptom duration was half as long in PSP as in CBGD, possibly because the course of the former disorder was more rapid or disabling (1.9 ± 1.9 years versus 4.2 ± 1.8 years). All 15 patients were in

**Table 1.** Demographics of 11 progressive supranuclear palsy (PSP) patients and 4 corticobasal ganglionic degeneration (CBGP) patients

|  | P.S.P. (n = 11) | D.G.C.B (n = 4) |
|---|---|---|
| Sex (M/F) | 6/6 | 2/2 |
| Mean age at diagnosis | 65.4 ± 4.8 | 69.5 ± 5.4 |
| Range | (57–71) | (64–77) |
| Mean disease duration at admission (years) | 1.97 ± 1.94 | 4.2 ± 1.8 |
| Range | (0.7–4.0) | (3.0–7.0) |
| Mean follow-up (months) | 32 ± 33 | 15.5 ± 16.6 |
| Range | (3–120) | (3–84) |

**Table 2.** Presenting symptoms in 11 PSP and 4 CBGD patients

| Symptom | PSP | | CBGD | |
|---|---|---|---|---|
|  | No. | % | No. | % |
| Unexplained falls | 7 | 63.6 | 2 | 50.0 |
| Gait disorder | 5 | 45.4 | 1 | 25.0 |
| Cognitive dysfunction | 3 | 27.2 | 1 | 25.0 |
| Speech difficulties | 3 | 27.2 | 0 | — |
| Visual disturbances | 2 | 18.1 | 1 | 25.0 |
| Behavioural disorder | 2 | 18.1 | 1 | 25.0 |
| Generalized slowness | 2 | 18.1 | 0 | — |
| Unilateral limb apraxia | 0 | — | 3 | 75.0 |
| Other complaints |  |  |  |  |
|    Retrocollis | 1 | 9.0 | 0 | — |
|    Resting tremor | 1 | 9.0 | 0 | — |
|    Emotional lability | 1 | 9.0 | 0 | — |
|    Lingual dysesthesias | 1 | 9.0 | 0 | — |

their sixties, though PSP patients were on average four years younger than CBGD patients (65.4 ± 4.8 years versus 69.5 ± 5.4 years) at presentation.

Table 2 lists the presenting symptoms in both disease groups. As expected, frequent, unexplained falls were often an initial complaint in PSP patients (64%), but this same complaint was also claimed by two of the four CBGD patients. Symptoms related to parietal lobe dysfunction were an early complaint in three CBGD patients. At the time of admission these symptoms were clearly described by all CBGD patients and their relatives, often as a combination of unilateral motor neglect while eating, writing or washing and inability to use simple instruments with one hand, causing them to shift combs or shavers to the other hand. Gait difficulty was a conspicuous complaint in all but one PSP patient at admission, as was a speech disorder, a symptom presented by all PSP patients on admission. In comparison, walking and language disturbances were uncommon early complaints in CBGD patients and were mentioned only by a single patient.

**Table 3.** Neurological findings at presentation and at follow-up in 11 PSP and 4 CBGD patients

| Neurological findings | PSP (n = 11) | | | | CBGD (n = 4) | | | |
| | Presentation | | Follow-up | | Presentation | | Follow-up | |
| | No | % | No | % | No | % | No | % |
| --- | --- | --- | --- | --- | --- | --- | --- | --- |
| Supranuclear ophthalmoplegia | 9 | 81.8 | 11 | 100 | 3 | 75 | 3 | 75 |
| Neck rigidty | 8 | 72.7 | 10 | 90.9 | 4 | 100 | 4 | 100 |
| Postural instability | 10 | 90.9 | 11 | 100 | 2 | 50 | 2 | 50 |
| Rigidity | 9 | 81.8 | 9 | 81.8 | 2 | 50 | 2 | 50 |
| Bradykinesia | 10 | 90.9 | 11 | 100 | 4 | 100 | 4 | 100 |
| Dysarthria | 10 | 90.9 | 11 | 100 | 4 | 100 | 4 | 100 |
| Cognitive impairment | 5 | 45.4 | 7 | 63.6 | 5 | 75 | 3 | 75 |
| Increased deep reflexes | 8 | 72.7 | 9 | 81.8 | 1 | 25 | 1 | 25 |
| Extensor responses | 4 | 36.3 | 6 | 54.5 | 0 | — | 0 | — |
| Spasmodic laughing | 2 | 18.1 | 3 | 27.2 | 0 | — | 0 | — |
| Neglect and apraxia | 0 | — | 0 | — | 4 | 100 | 4 | 100 |
| Unilateral limb dystonia | 0 | | 0 | — | 3 | 75 | 4 | 100 |
| Dysmetria | 0 | — | 1 | 9 | 0 | — | 0 | — |
| Resting tremor | 1 | 9 | 1 | 9 | 0 | — | 0 | — |

Table 3 shows the neurological findings at presentation and at the last outpatient visit during follow-up. Mean follow-up of the entire group was $27.7 \pm 29$ months, ranging from 3 to 120 months. Two PSP patients failed to show gaze abnormalities initially but eventually developed them 2 and 3.5 years later, respectively. No neck rigidity was present at admission in three PSP patients and indeed was still absent in one of these patients three years later. Neck rigidity was initially mild in three other patients, though it later became severe enough to prevent passive forward movement of the neck. Eventually, 10 of the 11 PSP patients developed classic cervical dystonia with hyperextended neck. Thus, the full clinical triad of parkinsonism, supranuclear opthalmoplegia, and neck dystonia was absent at admission in 3 of the 11 patients eventually diagnosed to be suffering from PSP. In contrast, both eye movement abnormalities, and neck rigidity were already present both at admission in three of the four CBGD patients.

No PSP patient showed any evidence of hemineglect or overt limb apraxia at any time during progression of the disease as determined by careful follow-up. However, asymmetric involvement of parkinsonian rigidity and bradykinesia in the limbs was observed in 6 of the 11 PSP patients. Asymmetric rigidity of the limbs was marked in one patient who scored a difference of two points between the right and left limbs on the UCLA scale, while another patient presented with unilateral resting tremor. Limb dystonia, usually mild in intensity, was present in all four CBGD patients but in none of the PSP patients. Figure 1 despicts the mean severity scores of five different neurological signs (supranuclear ophthalmoplegia, cervical dystonia, postural instability, rigidity and bradykinesia) in PSP and CBGD

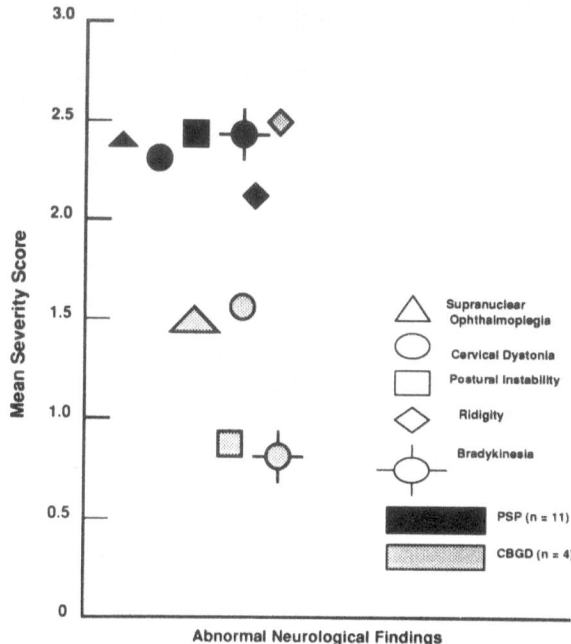

**Fig. 1.** Mean severity scores for five different abnormal neurological findings as assessed at the last follow-up visit in 11 PSP patients and 4 CBGD patients

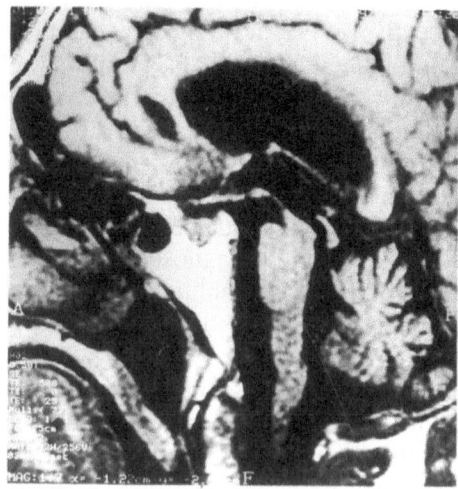

**Fig. 2.** MRI for a PSP patient at admission. There is marked atrophy of the midbrain tegmentum and ventral pons as well as aqueductal dilatation. The cerebral cortex appears normal for patient's age (SE 500/25)

patients at follow-up. The differences in limb rigidity between the two disorders, at presentation or at follow-up, were small. Supranuclear ophthalmoplegia, cervical rigidity, and postural instability though often presented in CBGD patients, remained mild in severity during follow-up, as opposed to the PSP patients in which, as a rule, these abnormalities became severe as the disease progressed.

CT-scans at admission showed various degrees of midbrain atrophy in 10 of the 11 PSP patients. MRI was available for seven patients (four PSP patients and three CBGD patients) and essentially confirmed the CT-scan findings. Atrophy was restricted to the quadrigeminal plate in three instances whereas there was global shrinkage of the midbrain in seven other patients (Fig. 2). Mild to moderate ventricular enlargement was observed in two patients, and mild widening of the subarachnoid sulci at the cerebral cortex was noted in four others. In no case was the cortical atrophy either focal or asymmetric. Furthermore, in no CBGD patient was there involvement of the brainstem. Two of the four patients exhibited marked asymmetric atrophy of the cortex, mainly circumscribed to the frontal-parietal areas (Fig. 3). Repeated MRI studies in one patient 11 months later showed progression of the asymmetric atrophy (Fig. 4).

## Discussion

In the present series, the final allocation of patients into specific diagnostic categories was established tentatively purely on clinical grounds. The authors are well aware that the conditions dealt with here represent discrete clinical-pathological entities. Even so, patients with supranuclear gaze palsy and parkinsonism may at postmortem be found to have alternative pathological findings that do not correspond to PSP, such as progressive subcortical gliosis (Will et al., 1988), diffuse Lewy body disease (Fearnley et al., 1991; de Bruin et al., 1992), or olivopontocerebellar degeneration (Al-Din et al., 1990). Also, patients with clinical findings consistent with a diagnosis of PSP may prove to be unclassifiable on a pathological basis (Paulus and Selim, 1990; Calabrese and Hadfield, 1991).

Despite these drawbacks, our findings corroborate the existence of some degree of overlap between the neurological abnormalities found in PSP and CBGD patients, as suggested by others (Rebeitz et al., 1968; Gibb et al., 1989). Clinical similarities between PSP and CBGD are not restricted to rigid-akinetic disturbances. Thus, frequent, unexpected falls early at disease onset are not only often reported by PSP patients (Brusa et al., 1979) but were also a prominent complaint in two of the four CBGD patients in this study as well. Early unsteadiness with a propensity to frequent falling spells was reported in CBGD by Scully et al. in 1985. Though our results confirm these findings, the mean score for postural instability was higher in our PSP patients as compared to the CBGD patients at follow-up (2.5 versus 0.8, respectively) suggesting a self-limiting progression of postural instability in CBGD.

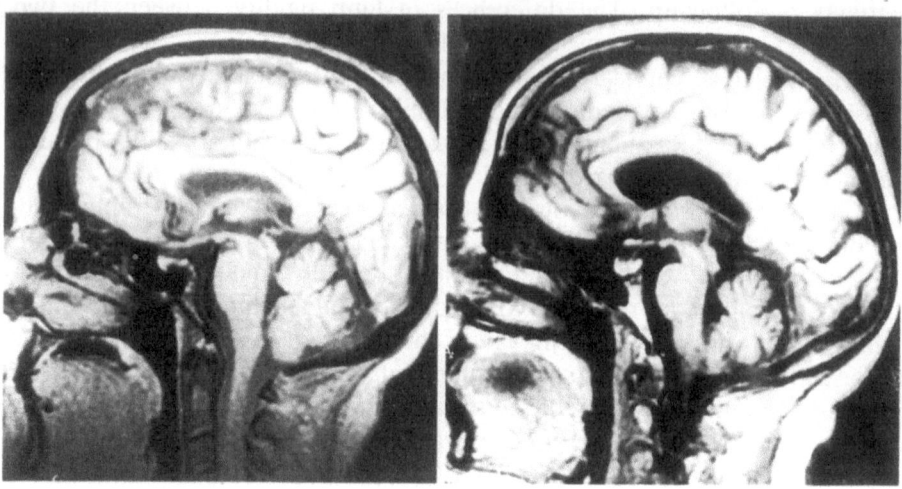

**Fig. 3.** MRI for two CBGD patients showing frontal cortex atrophy rostral to the sulcus centralis (left), and particularly marked around the perisulcal area (right). The occipital cortex appears relatively preserved (SE 500/25)

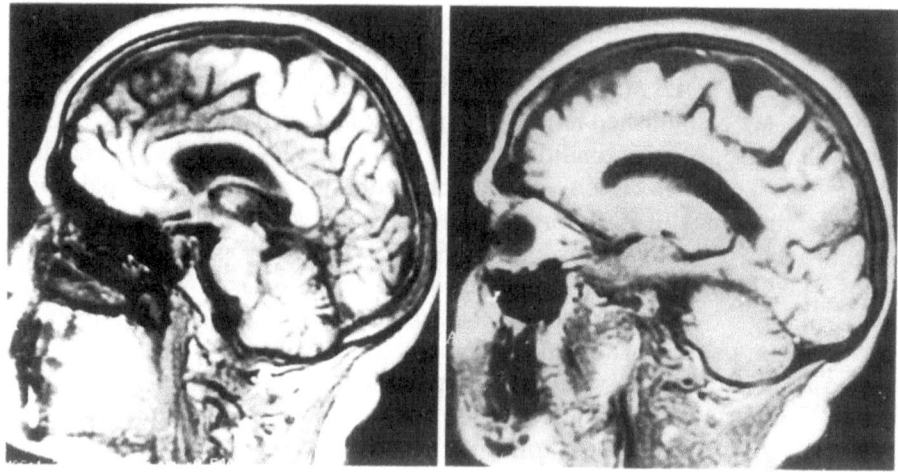

**Fig. 4.** Serial MRI studies from a CBGD patient. Images on the right showing disease progression were taken 11 months later (SE 500/25)

Supranuclear ophthalmoplegia may represent another source of difficulty. Three of our four CBGD patients developed eye movement abnormalities at some point during the course of the disease. Conversely, such disturbances were absent at presentation in 2 of the 11 PSP patients, a fact well documented in the literature (Dubas et al., 1983). Though the severity score of ocular motor abnormalities remained lower in the CBGD patients than in the PSP patients and preservation of pursuit extraocular movements is said to be a feature of CBGD until late in the course of the disease (Green et al., 1991), we feel that gaze abnormalities are not a reliable indicator of help in differentiating CBGD from PSP.

Mean disease duration in our PSP patients at the final examination was 4.5 years. It seems reasonable to assume that, at that time, the clinical spectrum of the disease was already fully developed (Golbe et al., 1988) and indeed in the end all the patients fulfilled current diagnostic criteria for PSP (Lees, 1987; Golbe et al., 1988). Though rigidity and akinesia in the limbs commonly exhibited some degree of asymmetry, and was present in 54.5 percent of our PSP patients, we were unable to detect any evidence of motor hemineglect, complex unintentional movements, feelings of non-belongingness or illusion of limbs movements in any of our PSP patients.

These findings sharply differentiated the PSP patients from the CBGD patients, in whom unilateral apraxia and hemineglect were early symptoms in three of four patients and were fully developed in all four patients by the time of admission. Most patients acknowledged experiencing feelings of foreingness of the neglected limb, though mainly in response to direct questionning. Unintentional movements of the hand and arm, such as posturing or levitation during standing or walking, were often observed, though none of our patients exhibited the purposeful, self-destructive be-haviours described by Doody and Jankovic (1992). We do not claim that the presence of unilateral motor neglect and apraxia is mandatory in order to sustain a clinically oriented diagnosis of CBGD. For instance, Greene et al. (1991) failed to detect apraxia in 6 of 18 clinically diagnosed patients after a mean course of 3.3 years. However, in the absence of such symptoms, clinical differentiation from PSP appears to be less consistent.

Mild dystonic posturing of the hand and upper limb was a common associated feature in our CBGD patients, and was often manifested at presentation. Unilateral limb dystonia was not recorded over the course of the disease in any of our PSP patients, though it has been described by others (Léger et al., 1987). Clinicopathological correlation of this form of symptomatic dystonia is as yet unknown. Case 2 in the series of Rebeiz et al. (1968) showed no histopathological abnormalities in the caudate and putamen, even though the patient exhibited a dystonic posture of the left hand with the wrist flexed and the metacarpophalangeal joints stiffly extended. Gliosis and myelin fiber damage, but no cell loss, was found in the posterolateral thalamus of that case. Though a lesion in the thalamus might have caused these symptoms (Obeso and Giménez-Roldán, 1988), dystonia related to unilateral parietal lobe dysfunction remains a possible explanation in that similar hand and forearm dystonic postures following parietal lobe infarction have been documented by CT-scan (Obeso and Giménez-Roldán, 1988). In the absence of limb apraxia or neglect, however, the value of limb dystonia in distinguishing PSP from CBGD remains unclear.

Computed tomography findings in patients with PSP include atrophy of the midbrain tegmentum and superior colliculi, aqueductal dilatation, pontine atrophy, third ventricular and quadrigeminal plate cistern dilatation, as well as a deep interpeduncular cistern (Schonfeld et al., 1987). Interestingly, in our series, despite the fact that there was a clear correlation of radiological findings with clinical disability, even fully ambulant patients displayed some degree of midbrain atrophy. Though we did not attempt any measurements of scan data, some degree of midbrain atrophy was evident in 10 of the 11 PSP patients at presentation, including two in whom ophthalmoplegia was not yet apparent at that stage. Cortical atrophy was not a prominent radiological feature in any of the patients.

Information on the value of neuroimaging in support of clinically-based diagnoses of CBGD is scarce and contradictory. Greene et al. (1991) reviewed 18 patients and failed to disclose any specific abnormality in any of the brain CT-scans or MRI studies. Riley et al. (1990) reported asymmetric cerebral atrophy on CT in 8 of 15 patients, while MRI yielded no additional information in 11 patients, except to demonstrate ipsilateral midbrain and pontine hemiatrophy in one patient. In our patients, asymmetric perisulcal atrophy occurred at presentation in two of the four CBGD. Serial investigation in one of the cases added evidence of the progressive nature of the disorder. Discrepancies probably reflected underlying disease severity, hence patients in an advanced stage of the disease correlated better with such gross neuropathological findings as abnormally narrow gyri and correspondingly wide sulci, a feature that is more pronounced in the superior frontoparietal region where it may occurs asymmetrically (Rebeiz et al., 1968). Absence of detectable lesions on CT or MR scans is more likely to occur early in the course of the disease. The glucose consumption pattern as studied by positron emission tomography and [$^{18}$F]Fluorodeoxyglucose may detect asymmetrical decreases in the temporal and sensorimotor cortex an subcortical structures at this stage (Blin et al., 1992). The pattern of brain energy metabolism in PSP as measured by positron emission tomography also differs from that in CBGD (Leenders et al., 1988; Karbe et al., 1992).

In summary, we believe that motor hemineglect and limb apraxia, which are often combined ipsilaterally, represent the most consistent marker for a clinically-based diagnosis of CBGD in patients with progressive, degenerative rigid-akinetic syndromes. Though asymmetric involvement of the limbs in rigidity and akinesia is not uncommon in PSP, patients do not develop limb apraxia, hemineglect or alien hand phenomena. Supranuclear ocular abnormalities, neck rigidity, frequent falls at disease onset, and postural instability often appear in both disorders, though to different extents. Neuroimaging studies are often helpful in supporting a clinically-based diagnosis at presentation.

## References

Akashi T, Arima K, Maruyama N, Anda S, Imase T (1989) Severe cerebral atrophy in progressive supranuclear palsy: a case report. Clin Neuropathol 8: 195–199
Al-Din ASN, Al-Kurdi A, Al-Salem MK (1990) Autosomal recessive ataxias, slow eye movements, dementia and extrapyramidal disturbances. J Neurol Sci 96: 191–205

American Psychiatric Association (1987) Committee on nomenclature and statistics. Diagnostic and statistical manual of mental disorders, 3rd edn. American Psychiatric Association, Washington DC, pp 205–224

Benson DF (1978) Neurological correlates of aphasia and apraxia. In: Matthews WB, Glaser GH (eds) Recent advances in clinical neurology. Churchill Livingstone, Edinburgh, pp 163–175

Blin J, Vidailhet MJ, Pillon B, Dubois B, Feve JR, Agid Y (1992) Corticobasal degeneration: decreased and asymmetrical glucose consumption as studied with PET. Mov Disord 7: 348–354

Brusa A, Mancardi GL, Bugiani O (1979) Progressive supranuclear palsy 1979: an overview. It J Neurol Sci 1: 205–222

Calabrese VP, Hadfield MG (1991) Parkinsonism and extraocular motor abnormalities with unusual neuropathological findings. Mov Disord 6: 257–260

De Bruin VMS, Lees AJ, Daniel SE (1992) Diffuse Lewy body disease presenting with supranuclear gaze palsy, parkinsonism, and dementia: a case report. Mov Disord 7: 355–358

Doody RS, Jankovic J (1992) The alien hand and related signs. J Neurol Neurosurg Psychiatry 55: 806–810

Dubas F, Grey F, Escourolle R (1983) Maladie de Steele-Richardson-Olszewski sans ophtalmoplégie. Six cas anatomo-cliniques. Rev Neurol (Paris) 139: 407–416

Fearnley JM, Revesz T, Brooks DJ, Frackowiak RSJ, Lees AJ (1991) Diffuse Lewy body disease presenting with a supranuclear gaze palsy. J Neurol Neurosurg Psychiatry 54: 159–161

Folstein MF, Folstein SE, McHugh PR (1975) The "mini-mental state", a practical method for grading the cognitive state of patients for the clinician. J Psychiatr Res 12: 189–198

Gibb WRG, Luthert PJ, Marsden CD (1989) Corticobasal degeneration. Brain 112: 1171–1192

Gibb WRG (1989) The pathology of parkinsonian disorders. In: Quinn NP, Jenner PG (eds) Disorders of movement: clinical, pharmacological and physiological aspects. Academic Press, London, pp 33–55

Greene P, Przedborski S, Kostic V, Giladi N, Eidelberg D, Fahn S (1991) Cortical-basal ganglionic degeneration: survey of 18 clinically diagnosed patients. Neurology 41 [Suppl] 1: 344

Greene PE, Fahn S, Lang AE, Watts RL, Eidelberg D, Powers JM (1990) Case 1, 1990: progressive unilateral rigidity, bradykinesia, tremulousness, and apraxia, leading to fixed postural deformity of the involved limb. Mov Disord 5: 341–351

Golbe LI, Davis PH, Schoenberg BS, Duvoisin RC (1988) Prevalence and natural history of progressive supranuclear palsy. Neurology 38: 1031–1034

Jankovic J (1989) Parkinsonism-Plus syndromes. Mov Disord [Suppl] 1 (4): S95–S119

Jellinger K, Riederer P, Tomonaga M (1980) Progressive supranuclear palsy: clinico-pathological and biochemical studies. J Neural Transm [Suppl] 16: 111–128

Karbe H, Grond M, Huber M, Herholz K, Kenler J, Heiss WD (1992) Subcortical damage and cortical dysfunction in progressive supranuclear palsy emonstrated by positron emission tomography. J Neurol 239: 98–102

Leenders KL, Frackowiak RSJ, Lees AJ (1988) Steele-Richardson-Olszewski syndrome. Brain energy metabolism, blood flow and fluorodopa uptake measured by positron emission tomography. Brain 111: 615–630

Léger JM, Girault JA, Bolgert F (1987) Deux cas de dystonie isolée d'un membre supérieur inaugurant une maladie de Steele-Richardson-Olszewski. Rev Neurol (Paris) 143: 140–142

Levine DN (1992) The alien hand. In: Joseph AB, Young RR (eds) Movement disorders in neurology and neuropsychiatry. Blackwell, Oxford, pp 691–695

Lees AJ (1987) The Steele-Richardson-Olsezewski syndrome (progressive supranuclear palsy). In: Marsden CD, Fahn S (eds) Movement disorders, vol 2. Butterworths, London, pp 272–287

Marsden CD, Schachter M (1981) Assessment of extrapyramidal disorders. Br J Clin Pharmacol 11: 129–151

Marsden CD, Obeso J, Zarranz JJ, Lang AE (1985) The anatomical basis of symptomatic hemidystonia. Brain 198: 463–483

McDowell F, Lee JE, Swift T, Sweet RD, Ogsbury JS, Kessler JT (1970) Treatment of Parkinson's syndrome with L-dihydroxyphenylalanine (levodopa). Ann Intern Med 72: 29–35

Obeso J, Giménez-Roldán S (1988) Clinicopathological correlations in symptomatic dystonia. In: Fahn S, Marsden CD, Calne DB (eds) Advances in neurology, vol 50. Dystonia 2. Raven Press, New York, pp 113–122

Paulus W, Selim M (1990) Corticonigral degeneration with neuronal achromasia and basal neurofibrillary tangles. Acta Neuropathol (Berl) 81: 89–94

Riley DE, Lang AE, Lewis A, Resch L, Ashby P, Hornykiewicz O, Block S (1990) Cortical-basal ganglionic degeneration. Neurology 40: 1203–1212

Rebeitz JJ, Kolodny EH, Richardson EP (1968) Corticodentatonigral degeneration with neuronal achromasia. Arch Neurol 18: 20–33

Sasaki S, Maruyama S, Toyoda Ch (1991) A case of progressive supranuclear palsy with widespread senile plaques. J Neurol 238: 345–348

Schonfeld SM, Golbe LI, Sage JI, Safer JN, Duvoisin RC (1987) Computed tomographic findings in progressive supranuclear palsy: correlation with clinical grade. Mov Disord 2: 263–278

Scully RE, Mark EJ, McNeely BV (1985) Case records of the Massachusetts General Hospital. N Engl J Med 313: 739–748

Steele JC, Richardson JC, Olszewski J (1964) A heterogeneous degeneration involving the brainstem, basal ganglia, cerebellum with vertical gaze and pseudobulbar palsy, nuchal dystonia, and dementia. Arch Neurol 10: 333–359

Weiner WJ, Lang AE (1989) Other akinetic-rigid and related syndromes. In: Weiner WJ, Lang AE (eds) Movement disorders. A comprehensive survey. Futura, Mount Kisko NY, pp 179–181

Will RG, Lees AJ, Gibb W, Barnard RO (1988) A case of progressive subcortical gliosis presenting clinically as Steele-Richardson-Olzewski syndrome. J Neurol Neurosurg Psychiatry 51: 1224–1227

Authors' address: Dr. S. Giménez-Roldán, Department of Neurology, Hospital General Gregorio Marañón, Doctor Esquerdo, 42, E-28007-Madrid, Spain.

**Neuroimage analysis, cerebral blood flow and metabolism**

# Magnetic resonance imaging in progressive supranuclear palsy and other parkinsonian disorders

**M. Savoiardo**[1], **F. Girotti**[2], **L. Strada**[1], and **E. Ciceri**[1]

Departments of [1] Neuroradiology, and [2] Neurology, Istituto Nazionale Neurologico "C. Besta", Milano, Italy

**Summary.** High field intensity MRI may demonstrate signal abnormalities consistent with deposits of iron or other paramagnetic substances in several extrapyramidal disorders. Hallervorden-Spatz disease was the only disorder widely known to have iron deposits in the pallidum, that are now easily demonstrated in vivo by MRI. However, lower field intensity MRI may also demonstrate characteristic findings.

In progressive supranuclear palsy, definite atrophy of the midbrain and of the region around the third ventricle is seen in slightly more than half of the cases. Minimal signal abnormalities are sometimes seen in the periaqueductal region, but MRI studies remain of little help in establishing the diagnosis of the disease.

Asymmetric atrophy in the parietal regions is seen in corticobasal degeneration, as expected from pathological studies. Minimal alterations may be seen in the substantia nigra in Parkinson's disease.

The most interesting MRI findings are observed in multiple system atrophies. Variable abnormal signal intensities, depending on the field intensity, are visible in the putamen in striatonigral degeneration and in Shy-Drager syndrome; in this latter condition the abnormalities are due to its striatonigral degeneration component. Atrophy of the pons, middle cerebellar peduncles, and cerebellum, and signal abnormalities in a characteristic distribution are visible in olivopontocerebellar atrophy.

A combination of these posterior fossa abnormalities and putaminal alterations may confirm the involvement of the cerebellar and extrapyramidal systems in multiple system atrophies.

The interest in Magnetic Resonance Imaging (MRI) of parkinsonian disorders arose about six years ago, when it was recognized that high field intensity MRI could demonstrate iron in the brain (Drayer et al., 1986a) and that some parkinsonian syndromes presented abnormal distribution of iron or other paramagnetic substances in the basal ganglia (Drayer et al., 1986b; Pastakia et al., 1986).

Iron is not present in the brain at birth, but accumulates during life with an uneven distribution; it is present in greater amounts in the basal ganglia, particularly in the pallidum, in the substantia nigra, red nucleus and dentate nucleus. Iron or other paramagnetic elements cause a shortening of T2 relaxation time; therefore, at high field intensity MRI, the areas where iron accumulates present a low signal intensity in T2-weighted images (Drayer et al., 1986b; Gomori et al., 1985). The distribution of this low signal intensity correlates exactly with the intensity of the blue coloration of the brain at the Perls' staining for iron (Drayer et al., 1986a).

The extrapyramidal disorder which for years has been known to present abnormal accumulation of iron in the pallidum is Hallervorden-Spatz disease. A remarkable loss of signal intensity in the pallidum in T2-weighted images was expected, therefore, in this disease; indeed, the first reports of MRI findings in Hallervorden-Spatz disease confirmed this expectation, but also demonstrated the presence of a small area of high signal intensity in the anteromedial part of this nucleus (Rutledge et al., 1987; Sethi et al., 1988). This was called by Sethi et al. the "eye-of-the-tiger sign" (Sethi et al., 1988). The significance of this high signal intensity area was unclear from the previous pathological reports; only Dooling et al. had mentioned that destructive changes and gliosis were more evident in the internal segment of the pallidum (Dooling et al., 1974). Recently, we could compare the MRI findings of our eight clinical cases of Hallervorden-Spatz disease, which all presented the "eye-of-the-tiger sign", with the pathological specimens of two proven cases of Hallervorden-Spatz disease examined by Halliday in Winnipeg (Canada). The area of high signal intensity corresponds to an area of "loose" tissue, with vacuolization and less iron than found in the rest of the pallidum, where the tissue is more "dense" and contains greater amounts of iron (Savoiardo et al., 1993).

An additional observation we could make was that when we examined our cases of Hallervorden-Spatz disease with intermediate field intensity MRI (i.e. 0.5 Tesla (T) rather than 1.5 T) we could easily demonstrate the area of high signal intensity, while the loss of signal intensity due to iron was poorly detectable; with gradient echo images, however, iron was detected or better demonstrated also at 0.5 T (Fig. 1).

Therefore, when discussing the magnetic susceptibility effects of iron or other paramagnetic substances, it is important to make clear whether intermediate or high field intensity MRI is used, and whether spin echo or gradient echo techniques, which are more sensitive to the magnetic susceptibility effects, are employed.

We shall now discuss the MRI findings of progressive supranuclear palsy, which is the main topic of this discussion, and then of other parkinsonian disorders, mainly of multiple system atrophies.

### Progressive supranuclear palsy or Steele-Richardson-Olszewski syndrome

From pathological reports, progressive supranuclear palsy (PSP) is known to be associated with atrophy of the midbrain (Barr, 1979). Neuroradiological studies, therefore, may support the diagnosis by demonstrating this

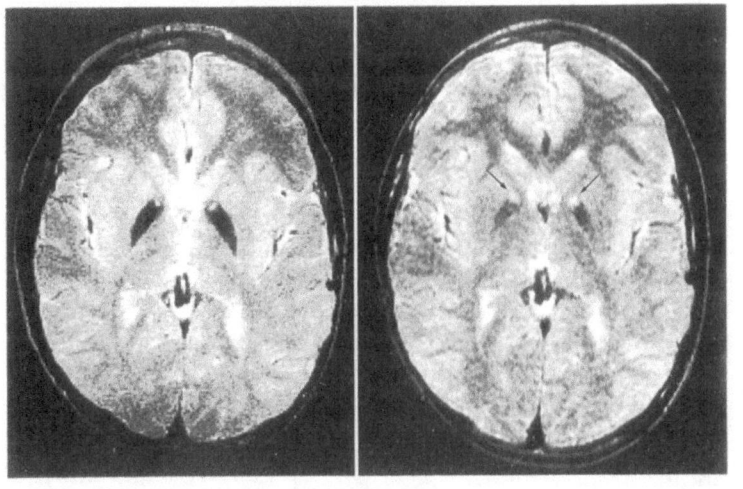

**A**                                                                          **B**

**Fig. 1.** Hallervorden-Spatz disease. T2-weighted images obtained at different field intensities (**A** 1.5 T; **B** 0.5 T) demonstrate the dependence of the magnetic susceptibility effects due to iron in the pallidum on the field strength; the loss of signal intensity caused by iron is much more evident at 1.5 T (**A**). The area of high signal intensity in the anteromedial part of the pallidum (eye-of-the-tiger sign) is more evident at 0.5 T (arrows, **B**). [From AJNR (Savoiardo et al., 1993), with permission]

atrophy; this demonstration was obtained first by pneumoencephalography (Bentson and Keesey, 1974) and then by computed tomography (CT) (Masucci et al., 1985; Schonfeld et al., 1987). The first MRI report included PSP among the multiple system atrophies (MSAs) and suggested that abnormal distribution of iron in the lentiform nucleus (with predominant loss of signal intensity in T2-weighted images at high field intensity in the putamen rather than in the pallidum) was the characteristic feature of these diseases (Drayer et al., 1986b). We could not confirm these findings in PSP (Savoiardo et al., 1989). We subsequently expanded our observations to 20 cases and the relevant findings will be reported here.

The 20 patients with clinical diagnosis of PSP were followed by a group of neurologists with particular experience in extrapyramidal disorders. There were 11 males and 9 females; age ranged from 56 to 79 years (mean 64.6 years). Length of the disease ranged from 2 to 10 years (mean 3.8 years).

The MRI studies were performed with a 0.5 T equipment in 10 cases and with 1.5 T machine in 13 cases. Three patients were therefore examined at both field strengths.

The MRI findings were evaluated by three experienced neuroradiologists, but measurements of the size of the brainstem or other structures were not used.

The expected finding of atrophy of the midbrain was unquestionable in 11 cases; in the other 9 patients the size of the midbrain was borderline or appeared normal. When atrophy was present, it was recognizable both in

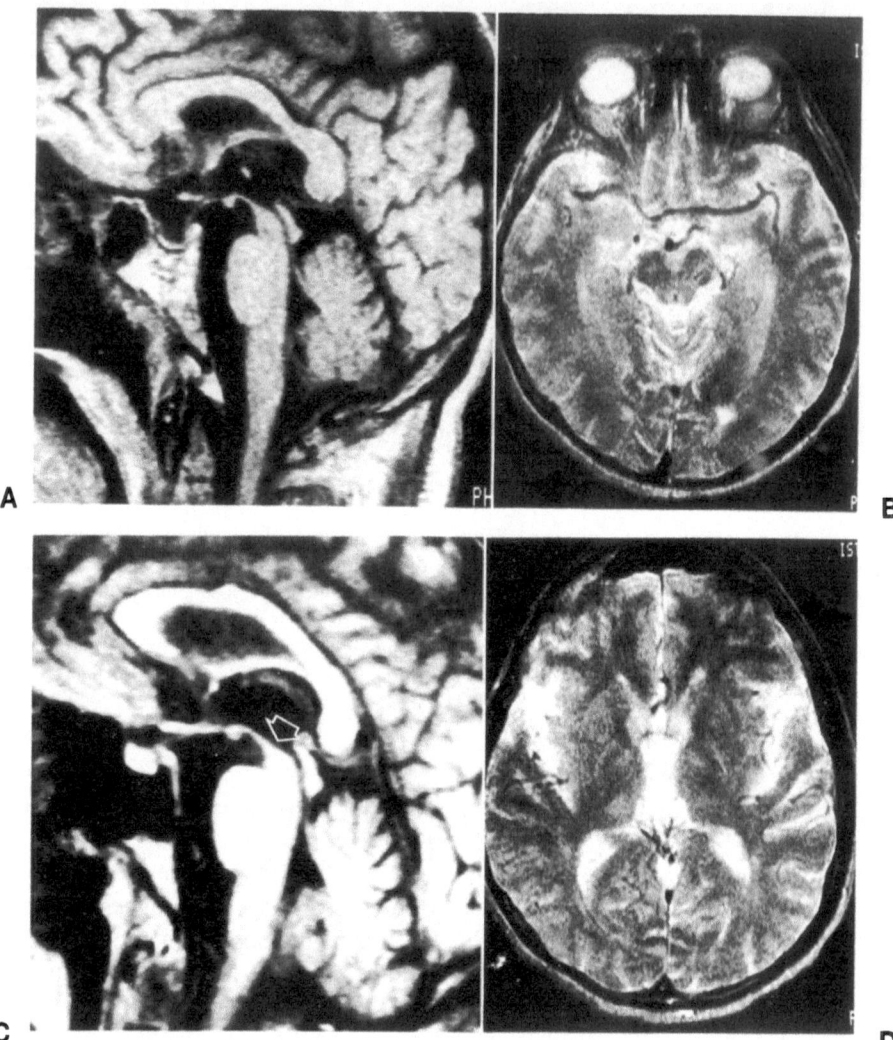

**Fig. 2.** Progressive supranuclear palsy. Sagittal T1-weighted images demonstrate definite midbrain atrophy in about half of the cases (**A**); atrophy of the dorsal midbrain is also recognizable in axial sections (**B** T2-weighted image). Hyperextension of the head may be present (**C**); atrophy involving the brain tissue around the third ventricle also causes an inferior convexity of the floor of the third ventricle (open arrow, **C** sagittal T1-weighted section), and a disproportionate enlargement of the third ventricle compared to the lateral ventricles (**D** axial T2-weighted image)

transverse and in sagittal sections. Sagittal sections were particularly helpful in demonstrating thinning of the quadrigeminal plate, more marked in its superior part (Fig. 2A and B).

Another atrophic feature consisted in dilatation of the third ventricle, which was disproportionately wider than the lateral ventricles in 11 patients; 10 of them coincided with the patients who also had definite midbrain atrophy. The widening of the third ventricle and the midbrain atrophy often determined a concave aspect of the posterior part of the floor of the third ventricle on the midline sagittal section (Fig. 2C and D).

Occasionally one could guess the diagnosis of PSP merely observing the first sagittal sections, when attention was called by a hyperextended position of the head and the midbrain atrophy was then recognized (Fig. 2C).

Regarding signal abnormalities, the findings were also subtle and inconstant. The most common signal abnormality observed in our series of PSP was a slight increase in signal intensity in intermediate or proton density images in the pariaqueductal region (Fig. 3A and B), where gliosis is found in pathologic specimens. No definite abnormalities could be recognized in this region in T2-weighted images, which sometimes demonstrated a prominent loss of signal intensity in the region of the substantia nigra, with smudging of its margin toward the red nucleus at high field intensity MRI. Of the 13 cases examined with high field intensity MRI, 11 presented normal distribution of iron in the lentiform nucleus, i.e. low signal intensity in T2-weighted images in the pallidum and normal signal or minor loss of signal intensity in the putamen (Fig. 3C). In 2 cases, the loss of signal intensity of the putamen equalled or was perhaps superior to that seen in the pallidum; however, one was a patient of 79 years of age, and it has been reported that in old age the accumulation of iron in the putamen may equal that in the pallidum. Therefore, only one patient of age 66 is left with abnormal distribution of signal intensity in the basal ganglia, a finding that remains distinctive of MSAs.

We had the chance of examining with 1.5 T MRI a 1 cm thick coronal section through the basal ganglia in a case of pathologically proven PSP; the T2-weighted images did not exhibit abnormalities of signal intensity consistent with abnormal iron deposition, a finding which was confirmed by the normal Perls' staining.

A rare signal abnormality observed in 3 of the 13 cases examined at 1.5 T was low signal intensity in T2-weighted images in the superior colliculi which did not appear atrophic (Savoiardo et al., 1989).

Finally, several patients had diffuse supratentorial atrophy with widening of cisterns and sulci, and some of them had rare areas of signal abnormalities in the white matter of the cerebral hemispheres, mostly in the periventricular region. However, these changes were considered not in excess to what is usually seen in a group of subjects of the same age.

In conclusion, all the significant abnormalities that correlate with the known pathological findings of PSP were seen in the region of the midbrain and around the third ventricle. However, definite although subtle abnormalities were seen in only about half of the patients with clinical diagnosis of PSP and, therefore, MRI remains of little help in establishing the diagnosis of this disease.

## Corticobasal degeneration

Corticobasal degeneration (CBD) is a recently rediscovered disease, for which the importance of imaging studies has not yet been determined. From

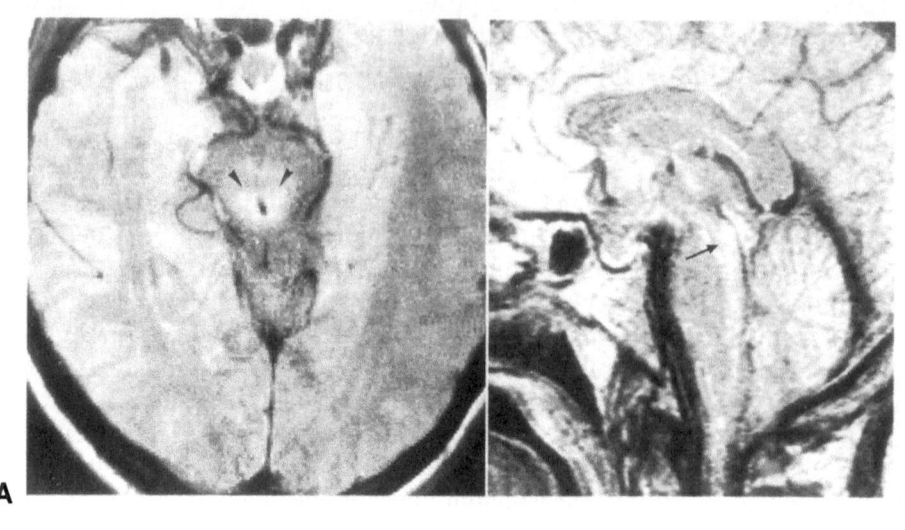

A                                                                                          B

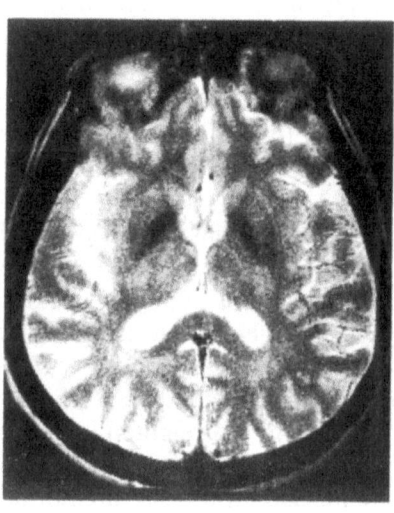

C

**Fig. 3.** Progressive supranuclear palsy. Minimal signal hyperintensity may be observed only in intermediate or proton density images in the periaqueductal region (**A** axial section, arrowheads; **B** sagittal section, arrow, in two different patients). The signal intensity in the lentiform nucleus is normal; low signal intensity consistent with abnormal putaminal deposits of iron is not observed (**C** T2-weighted axial section in a 66-year-old man with PSP, 1.5 T)

pathological studies diffuse histological changes in the substantia nigra, midbrain tegmentum, subthalamic and lentiform nuclei are known to occur in association with parietal or frontoparietal atrophy on the side opposite to the alien hand, which is a characteristic clinical feature (Gibb et al., 1989; Rebeiz et al., 1968).

There are clinical similarities with PSP, in terms of axial rigidity, supranuclear gaze palsy, and bradykinesia (Gibb et al., 1989).

With MRI we examined six patients with clinical diagnosis of CBD, 4 males and 2 females. Their age ranged from 63 to 76 years (mean 67.5 years). In only one patient was the midbrain considered to present borderline atrophy. In all the other cases the midbrain presented normal size. Diffuse supratentorial atrophy was observed in all the cases but predominance in one parietal region was mentioned only in two cases on the first reading. However, after being alerted of possible asymmetries, we reviewed the cases and could clearly identify a greater degree of parietal atrophy on the side opposite to the dystonic alien hand in all six cases, without knowing which side should have been affected. Awareness of the disease is therefore essential for noticing the focal atrophy (Fig. 4). No signal abnormalities were seen except for a minimal increase in signal intensity in the affected parietal cortex in one case (Fig. 4C), and for a few lacunes in the basal ganglia in two other cases.

## Parkinson's disease

MRI studies on Parkinson's disease have been focussed on the changes in signal intensity in the substantia nigra. High field MRI may demonstrate low signal intensity in this region with smudging of its posterior border toward the red nucleus (Braffman et al., 1988; Duguid et al., 1986). The anterior extension of the low signal intensity usually includes the medial part of the cerebral peduncle. Therefore, the assessment that the hypointensity is the expression of iron deposition in the pars reticulata and that the reduction of the normal signal intensity between substantia nigra and red nucleus is due to atrophy of the pars compacta seems hazardous. To our knowledge, there are no studies that correlate the MRI findings with the histological demonstration of iron deposition in this region.

Another study on Parkinson's disease emphasized the "restoration" of normal signal intensity in T2-weighted images in the dorsolateral part of the substantia nigra (Rutlcdge et al., 1987) (Fig. 5).

These two signs may coexist, but their importance in supporting the diagnosis of Parkinson's disease or other parkinsonian disorders is modest. In patients with parkinsonian disorders poorly responding to therapy, it seems more important to look for changes in the putamen (see the following section on MSAs) that may indicate a poor prognosis with evolution toward a multiple system atrophy.

## Multiple system atrophies

When we started to study patients with MSAs, we decided to try to separate patients who had prominent signs of autonomic failure, accompanied by cerebellar and parkinsonian signs of variable severity, from patients who presented predominant extrapyramidal signs not responding to therapy, and from patients who presented mainly a cerebellar disorder with no or mild

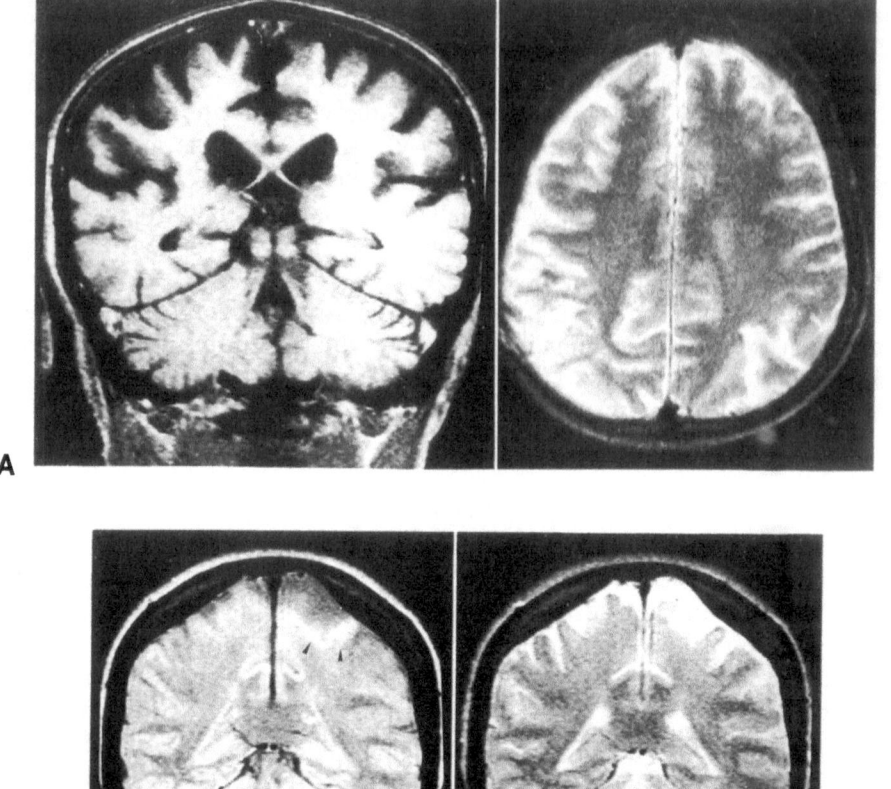

**Fig. 4.** Corticobasal degeneration. Coronal T1- (**A**) and axial T2-weighted images (**B**) in the same patient demonstrate right parietal atrophy; the patient had a left alien hand. Minimal high signal intensity in the left parietal atrophic cortex was seen only in one patient in proton density images (**C** arrowheads); no abnormal signal intensity is recognizable in the T2-weighted image (**D**). This patient had a right alien hand sign

extrapyramidal and autonomic disturbances. In other terms, we tried to separate patients with *Shy-Drager syndrome (SDS)* from patients with *striatonigral degeneration (SND)* and with *olivopontocerebellar atrophy (OPCA)*. This separation was aimed at finding the characteristic MRI features of these different entities, in order to subsequently verify whether patients who presented with marked involvement of all the three systems, i.e. extrapyramidal, autonomic, and cerebellar, could also be recognized by imaging studies as having multiple systems involvement.

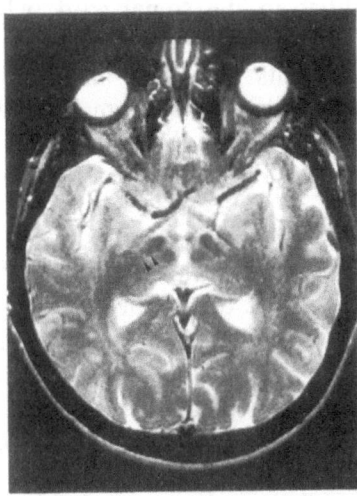

**Fig. 5.** Parkinson disease. In this patient with left hemiparkinson a slight asymmetry of signal suggests restoration of signal intensity in the lateral part of the substantia nigra on the right side (arrowheads, T2-weighted image, 1.5 T)

After collecting several cases of SDS, SND and OPCA, we realized that there was identity of MRI findings in patients with SDS and SND, while patients with OPCA presented abnormalities confined to the posterior fossa, that corresponded exactly to the distribution of gross and microscopic abnormalities described in the pathology of this disease. There were also cases, indeed, with marked clinical involvement of the three systems, which presented, on MRI, combination of abnormalities in posterior fossa, as seen in OPCA, and in the putamen, as seen in SDS and SND.

We now have seen more than 60 cases with MSA and, although an occasional patient clinically diagnosed as an MSA case may present normal or questionable MRI study, we found a quite reliable correspondence between clinical and MRI findings, particularly in OPCA. In our Institute, a tentative clinical diagnosis of OPCA in a sporadic case is now accepted only if it is supported by the results of MRI studies.

*Striatonigral degeneration and Shy-Drager syndrome*

SDS is considered a synonym of MSA by many authors (Oppenheimer, 1984; Quinn, 1989). Here we considered as SDS cases the patients who had prominent dysautonomic signs, with minimal cerebellar and extrapyramidal involvement. They were separated from patients with marked involvement of all the three systems, who were more clearly MSA patients.

We have already said that the MRI findings observed in SND and SDS are identical. In fact, SND is an associated or constitutive finding in a high proportion of cases of SDS (Oppenheimer, 1984). MRI, so far, has not demonstrated abnormalities in the spinal cord, where the abnormalities responsible for the autonomic failure should be sought for in the intermediolateral columns. Patients with pure autonomic failure studied with MRI of the brain did not exhibit any abnormalities (Brown et al., 1987).

The abnormalities seen in brain MRI of SND and SDS at 1.5 T are different from those seen in 0.5 T studies, but in both are confined to the putamen.

In 1.5 T studies, there is low signal intensity in T2-weighted images in the putamen, which is more marked than in the pallidum, with an inversion, therefore, of the normal signal distribution within the lentiform nucleus (Fig. 6). This putaminal low signal intensity is consistent with increased deposits of iron, but other paramagnetic substances, such as manganese, neuromelanin and "hematin" pigments (Borit et al., 1975; Dexter et al., 1991; Pastakia et al., 1986), can also be responsible for or contribute to this finding.

In the most lateral and posterior part of the putamen a thin rim of hyperintensity bordering the hypointensity is sometimes visible in T2-weighted images (Fig. 6).

The substantia nigra and the red nucleus sometimes presented a lower signal intensity than in normal subjects, but the low signal intensity in the putamen remained the most reliable and easily observable abnormality.

In 0.5 T studies, the low signal intensity in T2-weighted images due to iron or other paramagnetic substances is barely evident or absent. On the contrary, the MRI hallmark of these diseases at 0.5 T is putaminal hyperintensity in proton density and T2-weighted images, even in cases in which hyperintensity is not recognizable at 1.5 T. In cases with a thin band of hyperintensity on 1.5 T studies, 0.5 T examinations show a more evident and larger putaminal hyperintensity in T2-weighted images (Savoiardo et al., 1989, 1990) (Fig. 7A and B).

In our series, low signal intensity in the putamen at 0.5 T was well seen only in 3 of 10 SND cases (Fig. 7C), while was not seen in the 9 SDS cases. Although we did not find any difference between SND and SDS in 1.5 T studies, it is possible that the detection of iron at 0.5 T in some SND cases may indicate a greater amount of iron and a greater putaminal involvement in SND compared with SDS cases. This may correlate with the more severe extrapyramidal involvement of the patients with SND than of the patients with SDS.

The different findings obtained by 1.5 and 0.5 T studies are not contradictory; in 0.5 T examinations, the high signal intensity consistent with increased amounts of water, likely related to gliosis, neuronal loss, and spongiosis, is not masked by the presence of paramagnetic substances, which cause a signal loss in T2-weighted images proportional to the square of the magnetic field; therefore, at 1.5 T, these substances exert a magnetic susceptibility effect 9 times greater than at 0.5 T (Gomori et al., 1985; Savoiardo et al., 1989, 1990).

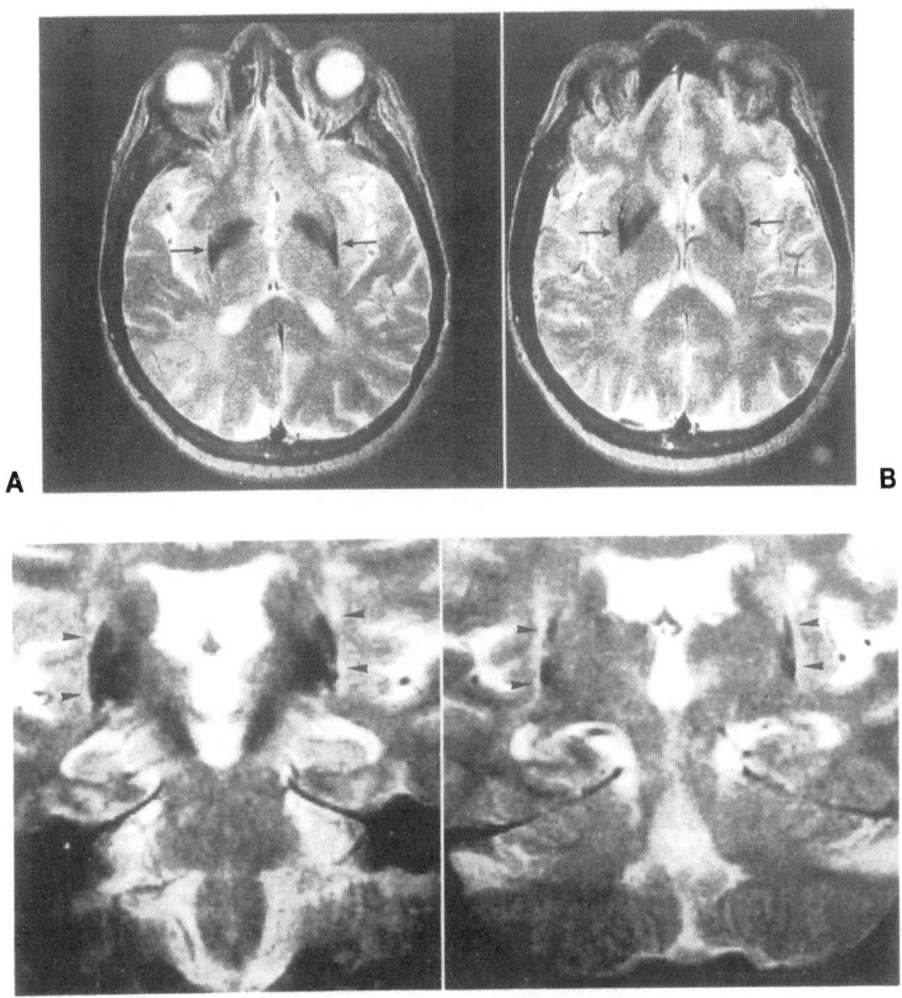

**Fig. 6.** Multiple system atrophies. At 1.5 T, involvement of the putamen is manifested by very low signal intensity in T2-weighted images. The loss of signal intensity of the putamen (arrows) prevails over that of the pallidum with reversal of the normal distribution (**A** and **B** axial sections; compare with Fig. 3C). This patient had severe involvement of extrapyramidal, autonomic and cerebellar systems and presented MRI abnormalities also in posterior fossa. Similar findings are present in a patient with diagnosis of Shy-Drager syndrome, who did not present clinical nor MRI findings of OPCA. Also note thin band of lateral hyperintensity (arrowheads; **C** and **D** coronal sections). [From J Comput Assist Tomogr (Savoiardo et al., 1989), with permission]

### Olivopontocerebellar atrophy

OPCA may be diagnosed by the pathologist immediately at the inspection of the brain, when he observes atrophy of the pons, middle cerebellar peduncles, and cerebellum, usually more marked in the hemispheres than in the vermis. The diagnosis is then confirmed by the histological examination

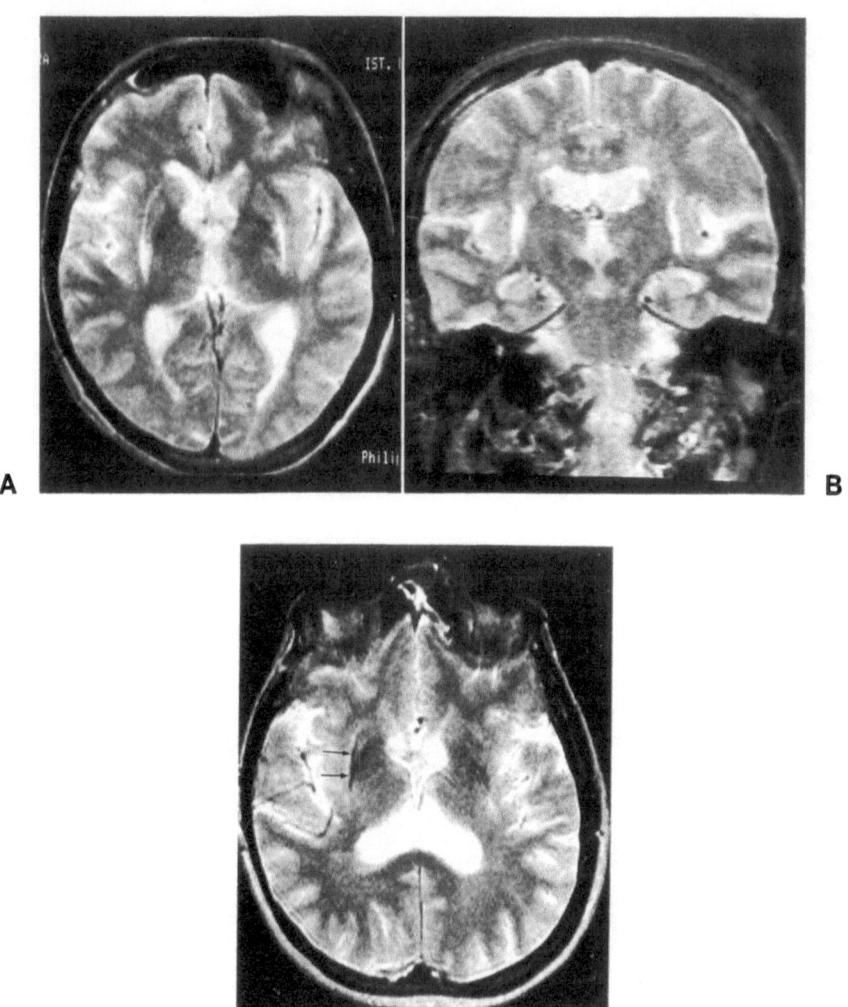

**Fig. 7.** Multiple system atrophies. In 0.5 T studies, high signal intensity consistent with increased amounts of water is observed (**A** axial, and **B** coronal T2-weighted sections in a case of Shy-Drager syndrome; same patient as in Fig. 6C and D. [From J Comput Assist Tomogr (Savoiardo et al., 1989), with permission.] In a case of striatonigral degeneration, low signal intensity particularly in the right putamen is present (**C** axial T2-weighted section, arrows)

of these structures. There is loss of cells in the pontine nuclei and degeneration of their fibers, which course as transverse pontine fibers in the anterior and posterior part of the basis pontis crossing on the midline in the raphe, running then in the middle cerebellar peduncles to reach the cerebellum. In the cerebellum there is a variable degree of degeneration of Purkinje cells; their fibers to the dentate nuclei also degenerate and the cerebellum becomes gliotic. However, since the cells of the dentate nuclei do not degenerate, their fibers which run in the superior cerebellar peduncles to reach the red nuclei and the thalami remain intact. A retrograde cell loss of the inferior olives, secondary to cortical cerebellar lesions, is also seen (Oppenheimer, 1984). With MRI studies, the same findings can be observed. The distribution of atrophy is particularly well seen in T1-weighted images in sagittal and coronal sections. In midline sagittal sections, the normal bulge of the pons above the profile of the medulla oblongata diminishes, due to flattening of its inferior part. In more advanced cases, the whole pons is atrophic. The fourth ventricle enlarges and the cerebellar vermis becomes atrophic, with widened sulci (Fig. 8A and B). The coronal sections, however, demonstrate that atrophy involves equally or in a more severe degree the cerebellar hemispheres (Fig. 8C). These sections also demonstrate the atrophy of the middle cerebellar peduncles. Normally, a coronal section through the central or posterior part of the pons shows the rounded contour of the middle cerebellar peduncles; in OPCA, this contour becomes pointed (Fig. 8D). Obviously, these atrophic changes are also recognizable in axial and sagittal sections. No signal abnormalities are recognizable in T1-weighted images (Savoiardo et al., 1990).

In intermediate or proton density and T2-weighted images, slight signal hyperintensity is seen in the degenerated areas. Therefore, signal hyperintensity is seen in the transverse pontine fibers, in the middle cerebellar peduncles, and in the whole cerebellum (Fig. 9). Very rarely mild signal abnormalities are also recognizable in the inferior olives. The fibers that do not degenerate maintain normal signal intensity. Therefore, the pyramidal tracts, the superior and inferior cerebellar peduncles become well identifiable because they stand out by their normal signal intensity against the abnormal background (Savoiardo et al., 1990) (Fig. 9A, C and D). The most convincing demonstration of the distribution of signal abnormalities is obtained by comparing the proton density and T2-weighted images with the histological sections stained for myelin (Fig. 10, compare with Fig. 9A and C).

In axial sections, sometimes, the signal abnormalities of the cerebellum are poorly appreciated, particularly if they are very mild, because of adjustment of the window setting. To obviate this problem, it is useful to obtain coronal sections; in coronal sections the difference between the signal intensity of the cerebellum and that of the normal cerebral hemispheres becomes quite obvious (Savoiardo et al., 1990) (Fig. 9B and D).

These atrophic changes and signal abnormalities correspond so exactly to the pathological cases of OPCA that we consider them pathognomonic. A clinical diagnosis of OPCA in a sporadic case, therefore, should not be accepted if it is not supported by the MRI findings we described, observed in more than 40 cases, almost all sporadic. Of course, there are borderline cases early in the course of the disease in which the clinical suspicion of OPCA does not correspond to definite MRI abnormalities; in a couple of such cases, however, repeat MRI 2 or 3 years later demonstrated a clear progression of atrophy and appearance of signal abnormalities, thus allowing the confirmation of the diagnosis (Fig. 8A and B).

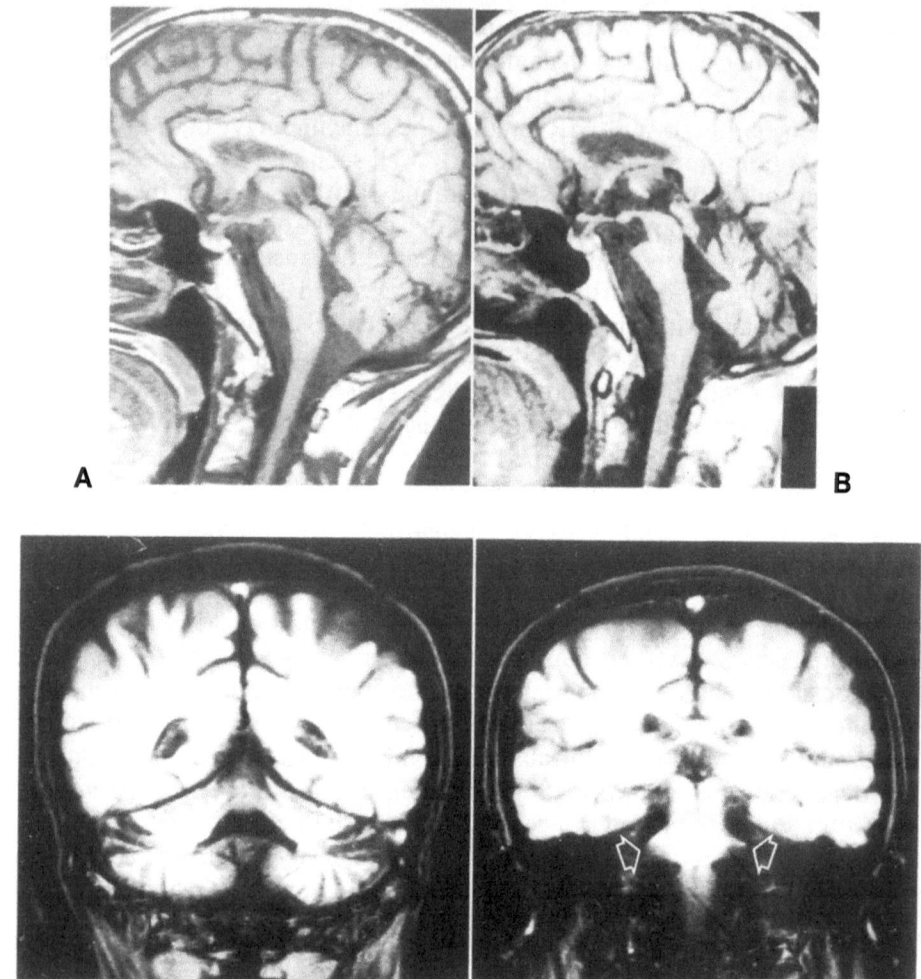

**Fig. 8.** Olivopontocerebellar atrophy. Early in the course of the disease, clinical and MRI findings may remain uncertain; note minimal flattening of the inferior part of the profile of the pons in **A**. Three years later, with clinically advanced disease, the atrophy of the pons and cerebellum is unquestionable (**B**) (both A and B, sagittal T1-weighted sections). T1-weighted coronal sections on the cerebellum (**C**) and pons (**D**) in a different patient demonstrate atrophy of the cerebellar hemispheres and of the middle cerebellar peduncles (open arrows)

## MRI demonstration of multiple system atrophy

In a group of 11 MSA patients with marked cerebellar, autonomic, and extrapyramidal involvement, MRI abnormalities were present both in posterior fossa and in the putamen; there was MRI evidence, therefore, of involvement of multiple systems. However, in patients less severely involved, still classifiable as MSA, some discrepancies between clinical

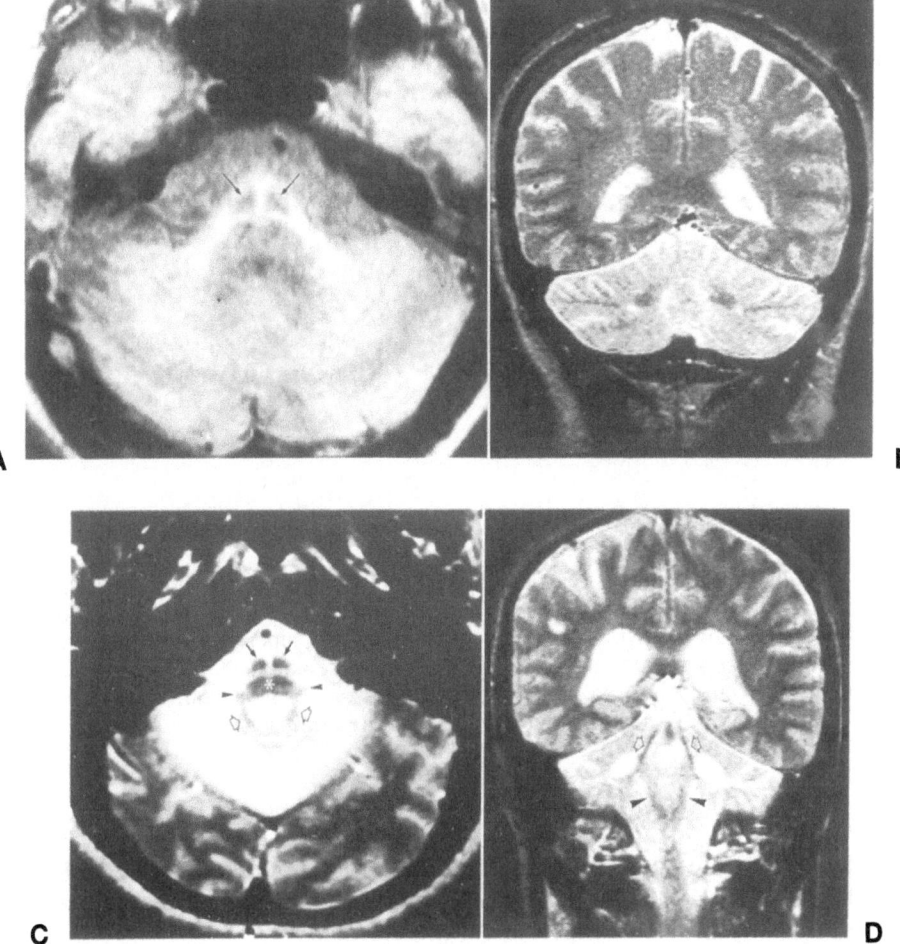

**Fig. 9.** Olivopontocerebellar atrophy. Signal hyperintensity is visible in T2-weighted and mainly in proton density images in the degenerate transverse pontine fibers and middle cerebellar peduncles (**A** axial proton density section; compare with Fig. 10; arrows indicate the normal pyramidal tracts). Cerebellar hyperintensity is particularly well demonstrated in coronal sections, by comparison with the cerebral hemispheres (**B**). The structures that in OPCA do not degenerate stand out by their normal signal intensity; these are: pyramidal tracts (arrows), pontine tegmentum (asterisk), inferior cerebellar peduncles (arrowheads), superior cerebellar peduncles (open arrows), (**C** T2-weighted axial section). The normal inferior (arrowheads) and superior (open arrows) cerebellar peduncles may be seen against the abnormal background in coronal section (**D**). [C and D, from Radiology (Savoiardo et al., 1990), with permission]

involvement of extrapyramidal or cerebellar systems and MRI demonstration of putaminal or posterior fossa involvement were observed.

In one case clinically labelled as OPCA, putaminal abnormalities were present on MRI, in addition to posterior fossa findings of OPCA; in 5 cases considered MSA but with not very severe extrapyramidal involvement,

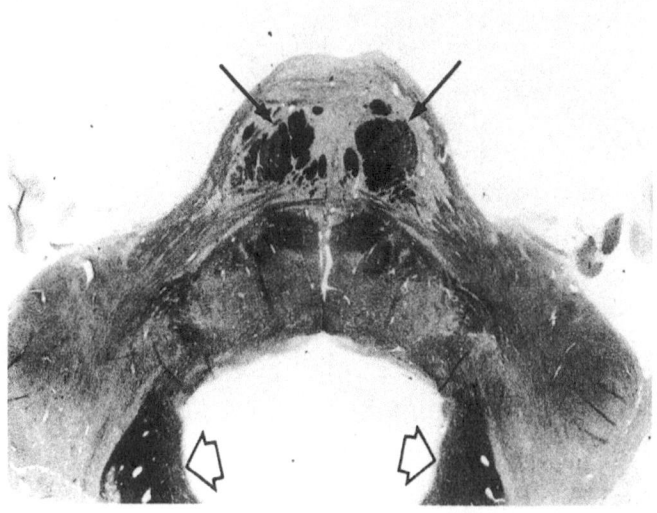

**Fig. 10.** Olivopontocerebellar atrophy. Axial section through the pons in a case of OPCA, stained for myelin (Woelcke-Heidenhain), demonstrates the validity of the MRI findings. Only the pyramidal tracts (arrows) and the superior cerebellar peduncles (open arrows) are normally stained, while the transverse pontine fibers are not stained. Compare with Fig. 9A and C. [Courtesy of Professor Orso Bugiani; from Radiology (Savoiardo et al., 1990), with permission]

MRI demonstrated cerebellar and pontine atrophy but no putaminal abnormalities. It is possible that MRI sensitivity is inferior to clinical sensitivity in the early diagnosis of extrapyramidal involvement.

## Conclusions

In spite of the minor incongruities here described, MRI has become a very powerful tool in demonstrating abnormalities in posterior fossa in patients with OPCA. However, MRI sensitivity is inferior to that of clinical examination in the early diagnosis of several extrapyramidal disorders, but MRI can, indeed, provide evidence of putaminal involvement in the majority of MSA cases. It is also possible that other techniques such as PET studies may be more sensitive than MRI in the early detection of MSAs by demonstrating involvement of multiple systems in cases in which clinical and MRI evaluation indicates the involvement of only the cerebellar or the extrapyramidal system.

# References

Barr AN (1979) Progressive supranuclear palsy. In: Vinken PJ, Bruyn GW, Klawans HL (eds) Handbook of clinical neurology, vol 49. North-Holland, Amsterdam, pp 233–256

Bentson JR, Keesey JC (1974) Pneumoencephalography of progressive supranuclear palsy. Radiology 113: 89–94

Borit A, Rubinstein LJ, Urich H (1975) The striatonigral degenerations: putaminal pigments and nosology. Brain 98: 101–112

Braffman BH, Grossman RI, Goldberg HI, et al (1988) MR of Parkinson's disease using spin-echo and gradient-echo sequences. AJNR 9: 1093–1099

Brown NT, Polinsky RJ, Di Chiro G, Pastakia B, Wener L, Simmons JT (1987) MRI in autonomic failure. J Neurol Neurosurg Psychiatry 50: 913–914

Dexter DT, Carayon A, Javoy-Agid F, et al (1991) Alterations in the levels of iron, ferritin and other trace metals in Parkinson's disease and other neurodegenerative diseases affecting the basal ganglia. Brain 114: 1953–1975

Drayer BP, Burger P, Darwin R, Riederer S, Herfkens R, Johnson GA (1986a) Magnetic resonance imaging of brain iron. AJNR 7: 373–380

Drayer BP, Olanow W, Burger P, Johnson GA, Herfkens R, Riederer S (1986b) Parkinson plus syndrome: diagnosis using high field MR imaging of brain iron. Radiology 159: 493–498

Dooling EC, Schoene WC, Richardson EP Jr (1974) Hallervorden-Spatz syndrome. Arch Neurol 30: 70–83

Duguid JR, De La Paz R, De Groot J (1986) Magnetic resonance imaging of the midbrain in Parkinson's disease. Ann Neurol 20: 744–747

Gibb RG, Luthert PJ, Marsden CD (1989) Corticobasal degeneration. Brain 112: 1171–1192

Gomori JM, Grossman RI, Goldberg HI, Zimmerman RA, Bilaniuk LT (1985) Intracranial hematomas: imaging by high-field MR. Radiology 157: 87–93

Masucci EF, Borts FT, Smirniotopoulos JG, Kurtzke JF, Schellinger D (1985) Thin-section CT of midbrain abnormalities in progressive supranuclear palsy. AJNR 6: 767–772

Oppenheimer DR (1984) Diseases of the basal ganglia, cerebellum and motor neurons. In: Hume Adams J, Corsellis JAN, Duchen LW (eds) Greenfield's neuropathology, 4th edn. Wiley, New York, pp 699–747

Pastakia B, Polinsky R, Di Chiro G, Simmons JT, Brown R, Wener L (1986) Multiple system atrophy (Shy-Drager syndrome): MR imaging. Radiology 159: 499–502

Quinn N (1989) Multiple system atrophy — The nature of the beast. J Neurol Neurosurg Psychiatry 1989 [Special Suppl]: 78–89

Rebeiz JJ, Kolodny EH, Richardson EP (1968) Corticodentatonigral degeneration with neuronal achromasia. Arch Neurol 18: 20–33

Rutledge JN, Hilal SK, Silver AJ, Defendini R, Fahn S (1987) Study of movement disorders and brain iron by MR. AJNR 8: 397–411

Savoiardo M, Strada L, Girotti F, et al (1989) MR imaging in progressive supranuclear palsy and Shy-Drager syndrome. J Comput Assist Tomogr 13: 555–560

Savoiardo M, Strada L, Girotti F, et al (1990) Olivopontocerebellar atrophy: MR diagnosis and relationship to multisystem atrophy. Radiology 174: 693–696

Savoiardo M, Halliday WC, Nardocci N, et al (1993) Hallervorden-Spatz disease: MR and pathological findings. AJNR 14: 155–162

Schonfeld SM, Golbe LI, Sage JI, Safer JN, Duvoisin RC (1987) Computed tomographic findings in progressive supranuclear palsy: correlation with clinical grade. Mov Disord 2: 263–278

Sethi KD, Adams RJ, Loring DW, El Gammal T (1988) Hallervorden-Spatz syndrome; clinical and magnetic resonance imaging correlations. Ann Neurol 24: 692–694

**Addendum:** Since this work was completed, the following chapter appeared:

Rutledge JN, Schallert T, Hall S (1993) Magnetic resonance imaging in parkinsonisms. In: Narabayashi H, Nagatsu T, Yanagisawa N, Mizuno Y (eds) Advances in neurology, vol 80. Raven Press, New York, pp 529–534

Authors' address: Dr. M. Savoiardo, Department of Neuroradiology, Istituto Nazionale Neurologico "C. Besta", Via Celoria 11, I-20133 Milano, Italy.

# Clinical progressive supranuclear palsy: differential diagnosis by IBZM-SPECT and MRI

**G. Arnold[1], K. Tatsch[2], W. H. Oertel[1], Th. Vogl[3], J. Schwarz[1], E. Kraft[1,3], and C. M. Kirsch[2]**

[1] Department of Neurology, [2] Division of Nuclear Medicine, and [3] Department of Radiology, Klinikum Großhadern, Ludwig-Maximilians-Universität München, München, Federal Republic of Germany

**Summary.** In order to in vivo identify subgroups in eight patients with the clinical diagnosis of progressive supranuclear palsy (PSP), we have performed [123]I-iodobenzamide single photon emission computed tomography (IBZM-SPECT), a nuclear medicine technique, to visualize dopamine D2 receptors in vivo, and high resolution (TE/TR 2900/20–90) magnetic resonance imgaging (MRI) to evaluate morphological CNS changes. All patients exhibited similar clinical features including supranuclear vertical gaze palsy, especially of downward gaze, predominantly axial rigidity especially in the neck, bradykinesia, instability of balance with easy falls, and poor response to dopaminergic drugs. Specific striatal dopamine D2 receptor binding in IBZM-SPECT, as calculated by a basal ganglia to frontal cortex ratio (BG/FC) was reduced in 5 patients, but normal in 3 patients. In MRI, these 3 patients exhibited multiple hyperintense white matter lesions; 2 of them had no midbrain atrophy. In contrast, all 5 patients with reduced IBZM binding lacked multiple white matter lesions in MRI, but 4 of them showed marked midbrain atrophy. This pilot study with IBZM-SPECT for in vivo imaging of striatal dopamine D2 receptors and T2-weighted MRI supports published neuropathological findings that clinical signs of PSP appeared to be due to heterogeneous neuropathology.

## Introduction

The clinical diagnosis of progressive supranuclear palsy (PSP, Steele-Richardson-Olszewski syndrome) relies on the presence of the following clinical characteristic sign: supranuclear vertical downgaze and/or upgaze palsy, which leads to inability to shift gaze and to follow moving objects; and additionally two of the five following cardinal features: instability of balance with easy falls, predominantly axial rigidity and dystonia, especially of the neck, bradykinesia, pseudobulbar palsy, frontal lobe signs; and other signs as cerebellar ataxia, depression, occasional rest tremor, and poor response to dopaminergic agents (Lees, 1987). The gross appearance of the brain at autopsy shows moderate atrophy of brainstem and cerebellum or is normal (for review see Barr, 1986). Typical microscopic findings in PSP include degenerative changes, i.e. neuronal cell loss, globose neurofibrillary tangles, gliosis and granulovacuolar degeneration within the mesencephalic and pontine reticular formation, dentate nucleus, substantia nigra, sub-thalamic nucleus and the periaqueductal grey matter (Steele et al., 1964). In separate reports, however, brainstem atrophy has not always be seen in CT and MRI scans, and furthermore a high frequency of multi-infarct lesions in CT and in MRI has been reported in patients with vertical gaze palsy (Jankovic et al., 1990). A neuropathological report about one patient with the clinical diagnosis of PSP and multiple white matter lesions detected by magnetic resonance imaging (MRI) showed that this patient did not exhibit the characteristic changes of PSP (Dubinsky and Jankovic, 1987), but amyloid angiopathy, suggesting at least another neuropathological entity.

Striatal dopamine D2 receptors are reduced in patients with PSP as shown in neuropathological (Pierot et al., 1988) and in positron emission tomography (PET) studies (Baron et al., 1985).

Based on these independent studies, we have attempted to identify subgroups of patients with the clinical diagnosis PSP by the two imaging techniques [123]I-iodobenzamide single photon emission tomography (IBZM-SPECT) and magnetic resonance imaging (MRI). Iodobenzamide (IBZM), a substituted benzamide and close analogue of the PET ligand raclopride, shows specific dopamine D2 antagonistic activity, high binding affinity in rat striatum tissue preparations and low non-specific binding (Kung et al., 1989). IBZM-SPECT has been employed to study striatal dopamine D2 receptors in humans (Brücke et al., 1991; Kung et al., 1990; Schwarz et al., 1992a,b). Magnetic resonance imaging (MRI) demonstrates morphological changes in patients with neurodegenerative disorders and with multiple vascular lesions. In order to establish homogeneous subgroups of patients with progressive supranuclear palsy, we investigated in a pilot study a probable relation between dopamine D2 receptor binding, multiple white

**Table 1.** Neurological signs of eight patients with clinical diagnosis of PSP

|  | Group 1 | | | | | Group 2 | | |
|---|---|---|---|---|---|---|---|---|
| Patient | 1 | 2 | 3 | 4 | 5 | 6 | 7 | 8 |
| Age | 58 | 64 | 61 | 65 | 62 | 62 | 71 | 66 |
| Vertical gaze palsy | + | + | + | + | + | + | + | + |
| Instability of balance with easy falls | + | + | + | + | + | + | + | + |
| Axial rigidity | − | + | − | + | + | + | + | + |
| Bradykinesia | + | − | + | + | + | + | + | − |
| Pseudobulbar palsy | − | − | − | + | − | − | − | − |
| Frontal lobe signs | + | + | − | − | − | + | + | − |
| Ataxia | + | − | + | − | − | − | + | + |
| Depression | − | − | + | − | − | + | − | − |
| Response to L-Dopa | − | − | − | − | − | − | − | − |

matter lesions and brainstem atrophy by employing two methods, i.e. IBZM-SPECT to detect changes of dopamine D2 receptor densities and MRI to detect morphological changes.

## Patients and methods

Eight patients with clinical symptoms of PSP were studied (see Table 1). The mean age was 63.6 years (range 58 to 71 years). All patients had vertical down and/or up gaze palsy and limitations of voluntary eye movements, instability of balance with easy falls and a progressive akinetic and/or rigid syndrome with marked neck rigidity. None of these patients had clearly benefitted from a subcutaneous challenge with apomorphine (up to 6 mg s.c.) or long term oral L-DOPA therapy (up to 1,000 mg). L-DOPA therapy was stopped at least 4 weeks prior to IBZM-SPECT investigation. None of these patients was on therapy with dopamine agonists excluding an influence on IBZM binding by this substances.

IBZM-SPECT was performed two hours after i.v. injection of 185 MBq IBZM (3-iodo-6-methoxybenzamide, Cygne BV, Netherlands/DuPont de Nemour, FRG). For acquisition a rotating double head gamma camera (Siemens Rota II, high resolution collimator) connected to a commercially available computer system was used. Data was collected for 60 projections (360° rotation) in a 64 × 64 matrix. The acquisition time was 50 s per projection. Transverse (horizontal) images were reconstructed by filtered backprojection (Butterworth filter) with a subsequent computation of coronal slices (slice thickness 6.0 mm). Attenuation correction was performed in selected transverse slices. A basal ganglia to frontal cortex ratio (BG/FC) was calculated using the region of interest technique; for this purpose the two slices with the highest IBZM signals representing the basal ganglia were selected. BG/FC ratio is normal at values greater than 1.44. Control values were obtained from 7 persons of comparable age (mean age 57 years, range 25 to 85), published in a former study (Schwarz et al., 1992a).

Cranial MRI was performed using a Siemens Magnetom scanner operated at 1.5 Tesla in 6 patients. Images were recorded using an optimized heavily T2 weighted spin echo sequence (TR/TE = 2900/20−90 msec) in axial and coronal orientation without gap (individual slice thickness = 3 mm). Two patients were investigated using a Siemens Magnetom scanner operated at 1.0 Tesla using a T2 weighted sequence (TR/TE 2500/25−90 msec) in axial orientation (individual slice thickness = 8 mm).

Midbrain diameter was measured in the first 6 patients according to the criteria described by Doraiswamy et al. (1992): anteroposterior diameter was considered reduced, if it was samller than 24 mm (Doraiswamy et al., 1992). Midbrain diameter was visually assessed by an experienced neuroradiologist (T.V.) in the remaining 2 patients.

## Results

Five patients had marekdly reduced IBZM binding [group 1, Fig. 1b] this group was additionally characterized by the absence of white matter lesions [Fig. 2b]. Three patients had normal IBZM binding [group 2, Fig. 1a] and additionally white matter lesions in MRI [Fig. 2a].

All 5 patients with reduced IBZM binding lacked multiple hyperintense white matter lesions; 4 of these patients of group 1 had marked midbrain atrophy with anteroposterior diameters between 20.2 and 22.2 mm. This diameter was normal in the fifth patient (cf. Table 2). All three patients of

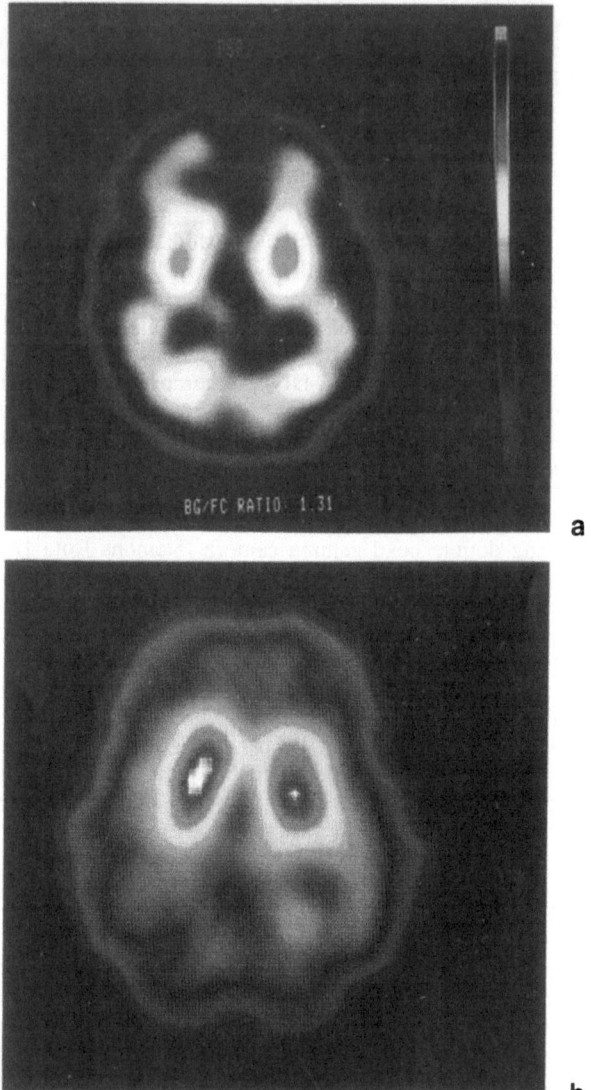

BG/FC RATIO 1.31

**a**

**b**

Fig. 1. a Transverse IBZM-SPECT image in a patient with normal IBZM binding.
b Transverse IBZM-SPECT image in a patient with reduced IBZM binding

group 2 (normal IBZM binding) had lesions within white matter, basal
ganglia and brain stem as well as marked cortical atrophy; the latter findings
are compatible with a multi infarct syndrome. Midbrain diameter was normal
in 2 and reduced in 1 patient (see Table 2). A summary of the results is
presented in Table 3.

## Discussion

This pilot study suggests that the combination of IBZM-SPECT and MRI
can distinguish at least two subgroups of patients with the clinical diagnosis

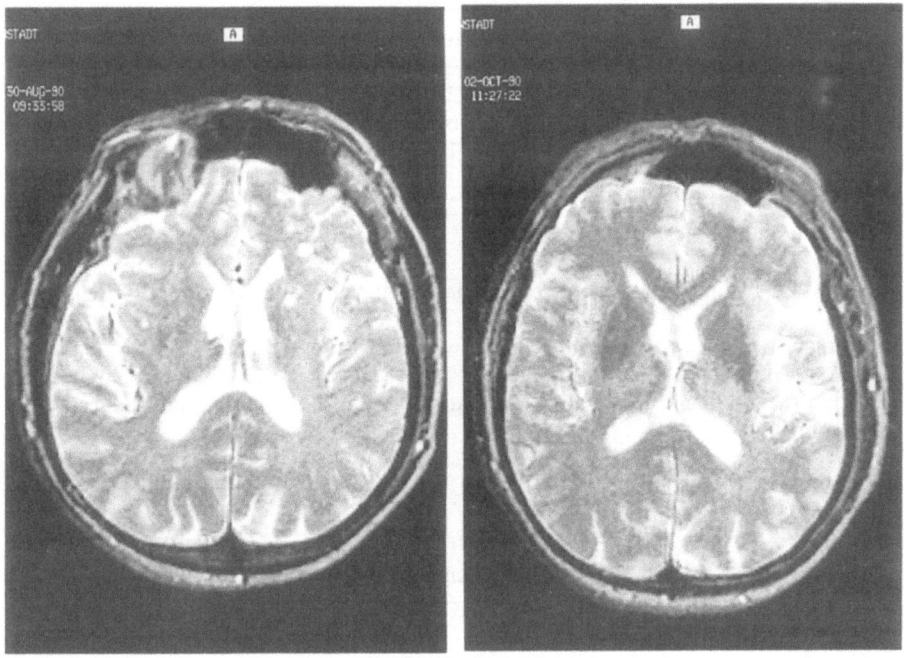

a                                                                          b

**Fig. 2. a** Transverse T2 weighted magnetic resonance image of a patient with multiple hyperintense lesions (but with normal brain stem, and normal IBZM-SPECT binding). **b** Transverse T2 weighted magnetic resonance image of a patient without hyperintense lesions (but with brain stem atrophy, and reduced IBZM-SPECT binding)

PSP. The first subgroup is characterized by reduced IBZM binding, lack of white matter lesions and reduced midbrain diameter in MRI; the second group by normal IBZM-SPECT and multiple white matter lesions. Since the diagnosis of PSP can only be established at autopsy, there is a need for such procedures to support the diagnosis in vivo. CT studies have shown that midbrain and pontine atrophy are characteristic findings in patients with PSP (Haldeman et al., 1981; Schonfeld et al., 1987). In a large series of more than hundred patients, Jankovic et al. (1990) found a multi infarct state in 17% of PSP patients when studied by CT scan, and in 37% when studied with MRI; 61% of the patients had generalized or midbrain atrophy in MRI scans; it was not reported, whether some of these patients had both, a midbrain atrophy and a multi infarct state. MRI is considered to be a more sensitive method to detect morphological changes within the CNS than CT. Also in our pilot study, 37.5% of the patients had white matter lesions compatible with multi infarct state.

IBZM-SPECT may be a useful tool to study the nigrostriatal dopaminergic system postsynaptically in vivo as shown in patients with Parkinson syndromes (Schwarz et al., 1992a) and Wilson's disease (Schwarz et al., 1992b). Decreased IBZM binding has been reported in few patients clinically diagnosed as PSP (Brücke et al., 1991; Schwarz et al., 1992a).

**Table 2.** Results of IBZM-SPECT and T2-MRI

| | Group 1 | | | | | Group 2 | | |
|---|---|---|---|---|---|---|---|---|
| Patient | 1 | 2 | 3 | 4 | 5 | 6 | 7 | 8 |
| Age | 58 | 64 | 61 | 65 | 62 | 62 | 71 | 66 |
| IBZM-SPECT (BG/FC) | 1.22 | 1.31 | 1.33 | 1.34 | 1.37 | 1.44 | 1.47 | 1.48 |
| White matter lesions | − | − | − | − | − | + | + | + |
| Mid brain diameter (T2-weighted MRI) | 24.5° | 22.2⁻ | 21.2⁻ | 20.2⁻ | red⁻ | 22.2⁻ | 25.1° | nor° |

IBZM-SPECT binding is expressed as basal ganglia/frontal cortex (BG/FC) ratio. Values ≥1.44 are normal (group 2)
Mid-brain atrophy was determined in T2-weighted MRI as anteroposterior diameter; values ≥24 mm are normal (°), values <24 mm reduced (⁻). Mid-brain diameter was visually assessed in 2 patients and classified as normal (nor) in one, and reduced (red) in the other

**Table 3.** MRI and IBZM-SPECT findings in patients with clinically diagnosed PSP

| | IBZM-SPECT | |
|---|---|---|
| n = 8<br>MRI | Reduced <1.44<br>Group 1 | Normal ≥1.44<br>Group 2 |
| *No white matter lesions* and | 5 | 0 |
| Normal mid brain | 1 | 0 |
| Mid-brain atrophy | 4 | 0 |
| *White matter lesions* and | 0 | 3 |
| Normal mid brain | 0 | 2 |
| Mid-brain atrophy | 0 | 1 |

When IBZM binding was compared to MRI findings in our study, all five patients with reduced IBZM binding lacked multiple white matter lesions in MRI, but four of them had midbrain atrophy; one patient had no clear midbrain atrophy and no white matter lesions (and markedly reduced IBZM binding); as this patient (patient 1) was only 58 years (the youngest) and he had only mild clinical signs of PSP, one may speculate that IBZM binding can be reduced before morphological brain stem changes are detectable by MRI. Patients with multiple vascular lesions on MRI had normal IBZM binding. One patient (patient 6), however, had just normal IBZM binding, multiple white matter lesions, and midbrain atrophy. At the present state of her disease, we cannot determine, whether her clinical symptoms have their basis in a multi infarct state, or whether this patient has additionally "classical" PSP with striatal degeneration in its early phase.

There is only one neuropathological case report about a patient with clinical diagnosis of PSP and multiple white matter lesions in MRI. This patient did not exhibit the characteristic neuro-pathological changes of PSP, but was diagnosed as amyloid angiopathy (Dubinsky and Jankovic, 1987).

Early diagnosis of PSP is difficult, if clear oculomotor signs are absent, and clinical signs can be misinterpreted as Parkinson's disease or multiple system atrophy. It remains to be shown, whether the combination of T2-MRI and IBZM-SPECT allows to predict the development of classical PSP in patients with a progressive akinetic syndrome, axial rigidity, marked postural instability, and unresponsiveness to dopamimetic therapy, who still lack the pathognomonic oculomotor signs.

In summary, our preliminary findings suggest that the combination of reduced IBZM binding, lack of multiple white matter lesions, and in addition brain stem atrophy on MRI supports the diagnosis of PSP. In contrast, patients with identical clinical symptoms but normal IBZM binding and multiple hyperintense lesions on MRI may have a vascular disorder.

Larger studies and neuropathological confirmation are needed to show whether the groups of patients here described in fact represent distinct disease entities. This in vivo distinction could be important for clinical drug trials in search for an effective therapy in PSP.

## Acknowledgement

This study was supported by Bundesministerium für Forschung and Technologie, Federal Republic of Germany, grant 01KL9001, "Morbus Parkinson and other basal ganglia disorders."

## References

Baron JC, Mazière B, Loc'h C, Cambon H, Sgouropoulos P, Bonnet AM, Agid Y (1985) Loss of striatial [76-Br]bromospiperone binding sites demonstrated by positron emission tomography in progressive supranuclear palsy. J Cereb Blood Flow Met 6: 131–136

Barr AN (1986) Progressive supranuclear palsy. In: Vinken PJ, Bruyn GW, Klawans HL (eds) Handbook of clinical neurology, vol 49. Extrapyramidal disorders. Elsevier, Amsterdam, pp 239–254

Brücke T, Podreka I, Angelberger P, Wenger S, Topitz A, Kufferle B, Müller C, Deeke L (1991) Dopamine D2 receptor imaging with SPECT: studies in different neuropsychiatric disorders. J Cereb Blood Flow Met 11: 220–228

Doraiswamy PM, Na C, Husain MM, Figiel GS, McDonald WM, Ellinwood EH, Boyko OB, Krishnan KRR (1992) Morphometric changes of the human midbrain with normal aging: MRI and stereologic findings. Am J Neuroradiol 13: 383–386

Dubinsky RM, Jankovic J (1987) Progressive supranuclear palsy and a multi infarct state. Neurology 37: 570–576

Haldeman S, Goldman JW, Hyde J, Pribram HFW (1981) Progressive supranuclear palsy, computed tomography, and response to antiparkinsonian drugs. Neurology 31: 442–447

Jankovic J, Friedmann DI, Pirozzolo FJ, McCrary JA (1990) Progressive supranuclear palsy: motor, neurobehavioral, and neuro-ophthalmic findings. In: Streifler MB, Korczyn AD, Melamed E, Youdim MBH (eds) Advances in neurology, vol 53. Parkinson's disease: anatomy, pathology, and therapy. Raven Press, New York, pp 293–304

Kung HF, Pan S, Kung MP, Kasliwal R, Reilly J, Alavi A (1989) In vitro and in vivo evaluation of [123I]IBZM: a potential CNS D-2 dopamine receptor imaging agent. J Nucl Med 30: 88–92

Kung HF, Alavi A, Chang W, Kung M-P, Keyes JW, Velchik MG, Billings J, Pan S, Noto R, Rausch A, Reilley J (1990) In vivo SPECT imaging of dopamine D2 receptors: initial studies with iodine-123-IBZM in humans. J Nucl Med 31: 573–579

Lees AJ (1987) The Steele-Richardson-Olszewski syndrome (progressive supranuclear palsy). In: Marsden CD, Fahn S (eds) Movement disorders, vol 2. Butterworth, London, pp 272–287

Pierot L, Desnos C, Blin J, Raisman R, Scherman D, Javoy-Agid F, Ruberg M, Agid Y (1988) D1 and D2-type dopamine receptors in patients with Parkinson's disease and progressive supranuclear palsy. J Neurol Sci 86: 291–306

Schwarz J, Tatsch K, Arnold G, Gasser T, Trenkwalder C, Kirsch CM, Oertel WH (1992a) 123I-Iodobenzamide-SPECT predicts dopaminergic responsiveness in patients with "de novo" parkinsonism. Neurology 42: 556–561

Schwarz J, Tatsch K, Vogl T, Kirsch CM, Trenkwalder C, Arnold G, Gasser T, Oertel WH (1992b) Marked reduction of dopamine D2 receptors as detected by 123IBZM-SPECT in a Wilson's disease patient with generalized dystonia. Mov Disord 7: 58–61

Schonfeld SM, Golbe LI, Sage JI, Safer JN, Duvoisin RC (1987) Computed tomographic findings in progressive supranuclear palsy: correlation with clinical grade. Mov Disord 2: 263–278

Steele JC, Richardson JC, Olszewski J (1964) Progressive supranuclear palsy: a heterogenous degeneration involving the brain stem, basal ganglia and cerebellum with vertical gaze and supranuclear palsy, nuchal dystonia and dementia. Arch Neurol 10: 333–359

Authors' address: Dr. G. Arnold, Neurologische Klinik, Klinikum Grosshadern, Marchioninistrasse 15, D-81366 München, Federal Republic of Germany

# PET studies in progressive supranuclear palsy

## D. J. Brooks

MRC Cyclotron Unit, Hammersmith Hospital, London, United Kingdom

**Summary.** Functional imaging (PET and SPECT) can be used to non-invasively demonstrate the patterns of metabolic and dopaminergic dysfunction associated with progressive supranuclear palsy. In this chapter the findings of published functional imaging studies are reviewed and the value of PET for distinguishing between the various degenerative causes of parkinsonism is discussed.

## Introduction

Progressive supranuclear palsy (PSP) is conventionally taken to refer to the syndrome first described by Steele, Richardson, and Olszewski (1964). Affected patients have associated rigidity, axial dystonia, bradykinesia, bulbar palsy, and dementia of frontal type. At post-mortem neuronal loss with neurofibrillary tangle (NFT) inclusions and gliosis are characteristically found in brainstem, diencephalic, and cerebellar nuclei, though cerebral cortex may also be mildly affected (Jellinger et al., 1980; Lees, 1986; Maher and Lees, 1986). The NFT's comprise 15 nm straight filaments, rather than the paired helical filaments characteristic of Alzheimer's disease, post-encephalitic parkinsonism, and Lytico-Bodig disease. In the basal ganglia the subthalamus, substantia nigra and globus pallidus are targeted, with lesser involvement of neostriatum. Unlike Parkinson's disease (PD), where ventrolateral nigra compacta shows most severe degeneration, the nigra is uniformly involved in PSP (Fearnley and Lees, 1991; Jellinger et al., 1980). Other targeted brainstem areas include the pretectal area, midbrain and pontine tegmentum, periaqueductal grey matter, corpora quadrigemina, cranial nerve nuclei controlling eye and tongue movement, pontine, cuneate, gracile, and dentate nuclei.

While the pathology of PSP is distinctive, this condition can pose diagnostic problems in life when clinical criteria alone are applied. The supranuclear ophthalmoplegia may be absent (Dubas et al., 1983), or occur late into the disease (Perkin et al., 1978). On presentation the parkinsonism of PSP has been estimated to be levodopa responsive in as many as 50% of cases, though this response is rarely sustained (Jackson et al., 1983). Up to 40% of PSP patients have been reported to have tremor (Jellinger et al., 1980). As

**Table 1.** Tracers used to study PSP

| Biological application | Tracer |
| --- | --- |
| Blood Flow | $C^{15}O_2$, $H_2^{15}O$, $^{123}IMP$ |
| Oxygen metabolism | $^{15}O_2$ |
| Glucose metabolism | $^{18}F$-2-Fluoro-2-deoxyglucose ($^{18}FDG$) |
| Dopa metabolism | $^{18}F$-6-fluorodopa ($^{18}F$-dopa) |
| Dopamine $D_2$ sites | $^{11}C$-raclopride (RAC) |
|  | $^{18}F$-fluoroethylspiperone (FESP) |
|  | $^{76}Br$-bromospiperone (BSP) |

a consequence a number of PSP patients may be initially labelled as having PD, and entered on presentation into therapeutic trials of anti-parkinsonian agents only to be withdrawn later. It has been estimated that between 3%–6% of cases diagnosed as PD in life turn out to have PSP at autopsy (Hughes et al., 1992; Jackson et al., 1983; Rajput et al., 1991). The supranuclear gaze palsy which characterises PSP has also been described in association with a number of other neurodegenerative disorders including diffuse Lewy body disease (Fearnley et al., 1991), corticobasal degeneration (Gibb et al., 1989), progressive subcortical gliosis (Will et al., 1988), olivopontocerebellar atrophy (Koeppen and Hans, 1976), and Creutzfeldt-Jakob disease (Ross-Russell, 1980) as well as rarely in inflammatory disorders such as multiple sclerosis, Whipple's disease, and neurosyphilis. Clearly it would be helpful to have additional means of distinguishing in vivo the Steele-Richardson-Olszewski form of PSP from these other neurological conditions in order to define homogeneous patient populations.

Functional imaging provides a relatively non-invasive means of examining the patterns of regional cerebral dysfunction associated with different multiple system degenerations. Most of the reported work concerning PSP has involved positron rather than single photon emission tomography and this review will concentrate on PET findings. Table 1 lists some of the tracers that have been used to study PSP to date.

## Metabolic studies in PSP

D'Antona et al. (1985) were the first to report PET findings in PSP. They studied six cases with $^{18}FDG$ PET who were diagnosed as having PSP on clinical criteria. Findings were compared with those of eight age-matched controls. Only four of their six patients had voluntary downgaze problems, however, and one of these four had mild parkinsonism so their PSP cohort constituted three clinically probable and three clinically possible cases. Four of the six patients had frontal lobe syndromes. Cortical glucose utilisation (rCMRGlc) was globally reduced (83% of normal) though this reduction did not reach significance. Medial and lateral frontal rCMRGlc were most affected, being reduced to 71% and 75% of normal, respectively. The authors found no significant correlation between frontal rCMRGlc and the presence of frontal release signs, though formal psychometry was not performed. The authors did not report striatal metabolism in their subjects.

Leenders et al. (1988) measured regional cerebral blood flow (rCBF) and oxygen metabolism (rCMRO$_2$) in four cases of clinically probable PSP, one of whom has had the diagnosis subsequently confirmed at autopsy. Their findings were compared with those of five age-matched controls. Like D'Antona et al. (1985) these workers found that cortical function in PSP was globally affected. Striatal rCMRO$_2$ was reduced to 78% of normal and the PSP patients exhibited hypofrontality — see Table 2. Frontal rCMRO$_2$ levels correlated inversely with disease duration, but did not correlate with psychometric scores on the WAIS. No specific tests of frontal lobe function were performed.

Foster et al. (1988) studied 14 clinically probable PSP and 21 age matched controls with $^{18}$FDG PET. These workers also found global cerebral hypometabolism with relative targeting of frontal cortex. Caudate and putamen rCMRGlc were uniformly reduced being 78% and 79% of normal, respectively, and thalamic and brainstem rCMRGlc were impaired. No correlation of rCMRGlc with locomotor or psychometric performance was reported. Three of these workers' PSP cases have subsequently come to autopsy (Foster et al., 1992). Two had the diagnosis of PSP confirmed but the third was found to have progressive subcortical gliosis (PSG). The PSG case showed similar levels of basal ganglia hypometabolism to the PSP cases but had more severely affected frontal function.

Goffinet et al. (1989) compared rCMRGlc in nine cases of probable PSP with 10 age-matched controls. The diagnosis in one of their PSP patients was later confirmed at autopsy. Seven of their PSP patients showed hypofrontality, one case had entirely normal cerebral metabolism, and in one case it was diffusely reduced. Motor and premotor cortex glucose utilisation was reduced to 73% of normal in PSP and was marginally more affected than prefrontal rCMRGlc (77% of normal). Mean striatal, thalamus, and cerebellar rCMRGlc values were significantly reduced but less affected than frontal rCMRGlc — see Table 2. The authors found no correlation between rCMRGlc and cognitive function; no correlations of rCMRGlc with locomotor status were reported.

**Table 2.** Regional cerebral metabolism in PSP (% of normal)

| | No. | Whole cortex | Frontal | Occipital | Striatal | Thalamic | Cerebellar |
|---|---|---|---|---|---|---|---|
| D'Antona | 6 | 83% | 73% | 93% | — | — | — |
| Leenders | 4 | — | 78% | 89% | 78% | — | — |
| Foster | 14 | 81% | 75% | — | 79% | 84% | 91% |
| Goffinet | 9 | 85% | 75% | 86% | 79% | 69% | 76% |
| Blin | 41 | — | 78% | 83% | 81% | 73% | 79% |
| Karbe | 9 | — | 79% | 87% | 76% | 77% | 81% |
| Johnson | 11 | — | 78% | — | 21% | — | — |

Blin et al. (1991) extended the original study of D'Antona et al. (1985), reporting PET findings on a total of 41 PSP cases (25 clinically probable and 16 clinically possible). Two different PET cameras were used during this series, and some patients had glucose while others had oxygen metabolic studies. In order to facilitate comparisons between PSP and controls regional cerebral metabolic data for individual patients were normalised to the mean control rCMRGlc or rCMRO$_2$ values for the particular PET camera used. Cerebral metabolism was significantly and diffusely reduced in PSP. As a group the PSP patients showed hypofrontality. Caudate, putamen, thalamic, and cerebellar metabolism were also significantly impaired — see Table 2. Frontal lobe metabolism correlated with general intelligence scores and performance on specific tests of frontal lobe function, and inversely with disease duration. Locomotor function correlated with caudate and thalamic, but surprisingly not putamen, metabolism. Interestingly there were no significant differences in severity of reduction of regional cerebral metabolism between those PSP groups classified as clinically probable or possible.

Karbe et al. (1992) have reported [18]FDG findings in nine cases of probable PSP. Like other authors these workers found reduced frontal and striatal metabolism in PSP but in their series striatal function was marginally more affected than frontal function. Johnson et al. (1992) have examined regional cerebral blood flow with [123]IMP-SPECT in 11 PSP cases. With this less sensitive, but more available, technique these authors have shown that frontal and striatal hypofunction are also evident.

Table 2 summarises the regional reductions in cerebral metabolism reported in PSP. All studies have shown a diffuse fall in cerebral metabolism in PSP, with most patients having hypofrontality. Striatal, thalamic, and cerebellar metabolism are also generally reduced. Two studies have suggested that superior frontal cortex may be more affected than prefrontal cortex in PSP, though the differences in the relative degrees of involvement have not been statistically significant. Only one study to date has reported a significant correlation between frontal metabolism and cognitive performance, and between striatal metabolism and locomotor function. The sensitivity of PET for detecting regional cerebral hypometabolism in PSP has been addressed by one study (Goffinet et al., 1989). These authors found that significant hypofrontality was present in seven out of their nine PSP patients; inspection of their data suggests that six also had significant striatal hypometabolism and that altogether eight out of the nine showed significant abnormalities of regional cerebral metabolism compared to controls.

The mechanism of the reduction in cortical metabolism with hypofrontality in PSP is still under debate. There is little cortical, but widespread subcortical, pathology in this condition. Noradrenergic, serotonergic, and dopaminergic, projections all degenerate in PSP but loss of these fibres in both animal models and Parkinson's disease does not appear to be associated with hypometabolism (McCulloch et al., 1984; Savaki et al., 1984; Wolfson et al., 1985). Lesions of the nucleus basalis with loss of cholinergic projections results in transient diffuse cortical hypometabolism in monkeys (Kiyosawa et al., 1989) while thalamic infarcts cause sustained diffuse cortical dysfunction (Pappata et al., 1990). As there is loss of both nucleus basalis and thalamic neurones in PSP this may contribute towards the diffuse cortical hypofunction present. The most likely explanation for the hypofrontality in PSP, however, is pallidal degeneration. Internal pallidal neurones are known to project via ventrolateral thalamus to pre-motor and prefrontal areas (Alexander et al., 1986). Toxic and vascular pallidal lesions have been shown to cause selective decreases in frontal metabolism associated with obsessive behaviour (Laplane et al., 1989). As the internal pallidum is targeted in PSP damage to this structure is likely to contribute to the frontal cognitive deficits found in PSP patients.

There has recently been some debate as to whether patients with isolated progressive akinesia may have a variant of PSP. Taniwaki et al. (1992) have studied three such patients and, as in PSP, found reduced frontal and striatal glucose hypometabolism in support of this hypothesis. Their akinesia patients all showed pretectal and pontine atrophy on MRI.

### The dopaminergic system in PSP

At autopsy, a uniform loss of nigro-striatal dopaminergic projections has been reported in PSP, caudate and putamen dopamine content being re-duced to 9%–25% of normal levels (Bokobza et al., 1984; Kish et al., 1985; Ruberg et al., 1985). Striatal $^{18}$F-dopa uptake reflects the capacity of the caudate and putamen to decarboxylate and retain this tracer as $^{18}$F-dopamine, and so provides a functional measure of the integrity of nigro-striatal dopaminergic terminals. Leenders et al. (1988) measured striatal: temporal cortex $^{18}$F-dopa uptake ratios in five PSP patients 120–180 mins after tracer administration and reported that the mean value was signifi-cantly reduced to 87% of normal. The 1.7 cm resolution of their camera was unable to seperate caudate from putamen signal. Striatal $^{18}$F-dopa uptake correlated with disease duration in their PSP patients, but not with locomotor disability. Other workers have also reported reduced striatal $^{18}$F-dopa uptake both in PSP (Bhatt et al., 1990; Taniwaki et al., 1992) and in progressive pure akinesia (Taniwaki et al., 1992).

Brooks et al. (1990) measured putamen and caudate $^{18}$F-dopa influx constants in 10 patients with clinically probable PSP. Figure 1 shows PET images of striatal $^{18}$F-dopa uptake, accumulated 30–90 minutes after tracer administration, in a normal subject and patients with Parkinson's disease and PSP. It can be seen that loss of striatal activity in the PSP case is uniform. Mean putamen and caudate tracer uptake were reduced to 37%, and 48%, of normal, respectively, for the PSP group — see Fig. 2. All ten PSP patients had significantly reduced putamen, and nine had significantly reduced caudate, $^{18}$F-dopa uptake. Neither locomotor function, assessed using the Hoehn and Yahr scale twelve hours after stopping medication, or disease duration of individual PSP patients correlated with putamen or caudate $^{18}$F-dopa uptake.

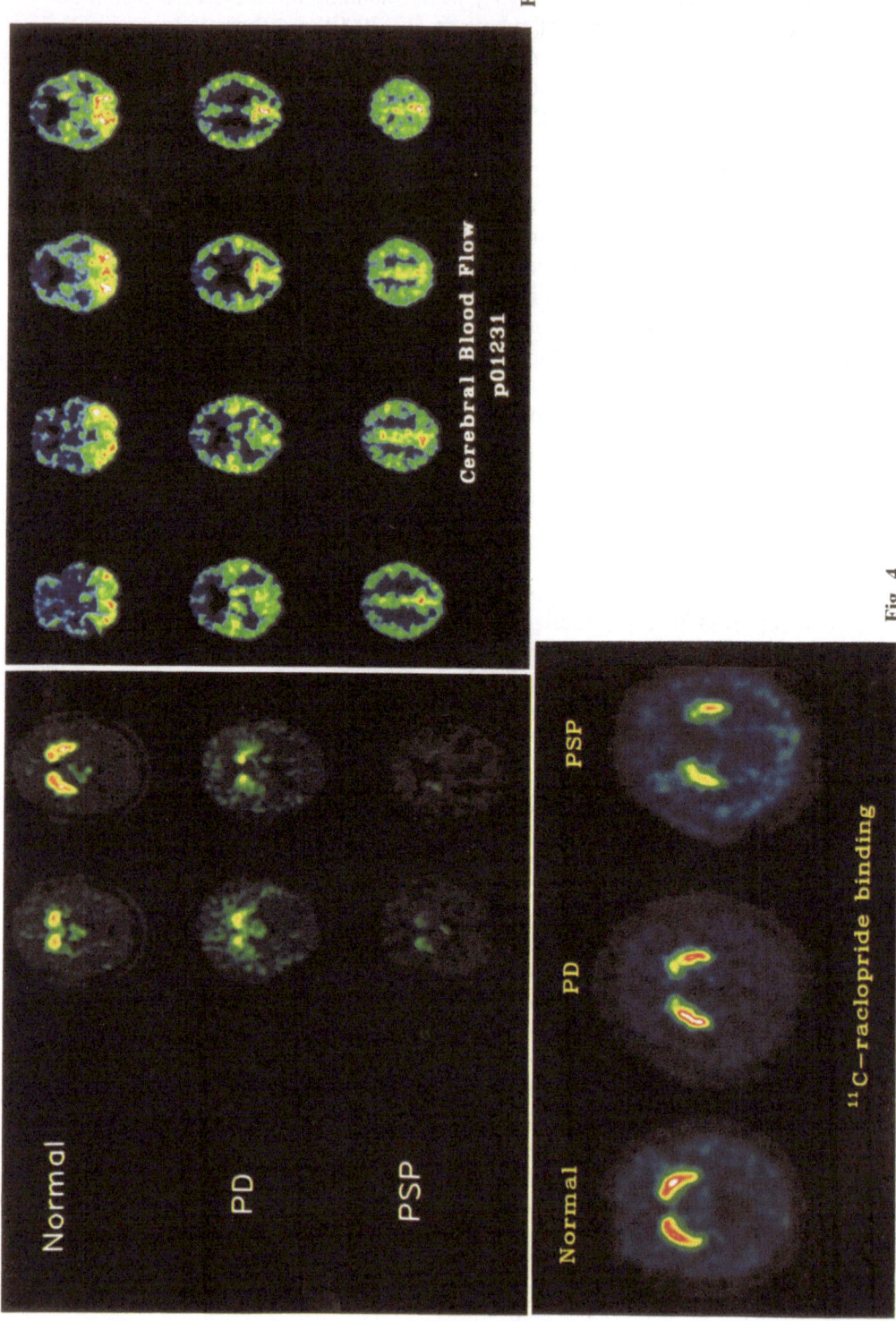

Fig. 5

Fig. 4

Fig. 1

Cerebral Blood Flow
p01231

¹¹C-raclopride binding

Normal

PD

PSP

PSP

PD

Normal

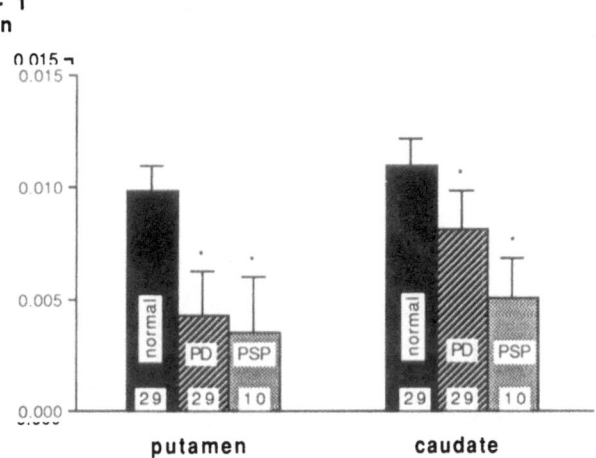

$Ki \ min^{-1}$

Fig. 2. Mean putamen and caudate [18]F-dopa influx constants for 29 normal, 29 PD, and 10 PSP cases. Putamen and caudate tracer uptake were reduced to 37% and 47% of normal, respectively, in PSP. In PD putamen tracer uptake was 44% of normal, while cuadate uptake is relatively preserved (74% of normal). * p < .05 Student's t test with Bonferroni's correction

Baron et al. (1986) were the first workers to study striatal $D_2$ binding potential in PSP with PET. They scanned seven PSP patients with the reversible $D_2$ antagonist [76]Br-bromospiperone (BSP). This tracer equilibrates throughout the brain after 4 hours and, as the cerebellum contains few $D_2$ receptors, the equilibrium cerebellar signal provides an estimate of non-specific BSP binding in striatum. These workers found a significant mean 24% fall in the equilibrium striatum:cerebellum tracer uptake ratio in PSP. Two of their seven PSP patients, however, had striatal:cerebellar BSP uptake ratios within the normal range and only three PSP patients showed a significant fall in striatal tracer binding. Six of the seven PSP patients were taking dopaminergic replacement medication at the time of PET. The effects of such medication on striatal $D_2$ binding remains unclear and it is

Fig. 1. PET images of striatal [18]F-dopa uptake, collected 30–90 minutes after tracer administration, for a normal subject and patients with PD and PSP. It can be seen that there is a uniform reduction of striatal signal in PSP, while head of caudate tracer uptake is relatively preserved in PD

Fig. 4. PET images of striatal [11]C-raclopride uptake, collected over 60 minutes after tracer administration, for a normal subject and patients with untreated PD and PSP. It can be seen that there is reduction of striatal signal in PSP, particularly from the head of caudate, while tracer uptake is normal in PD

Fig. 5. PET images of rCBF for a PSP patient. Striatal, thalamic, frontal, temporal, and parietal blood flow are significantly reduced, while cerebellar and occipital blood flow are relatively preserved

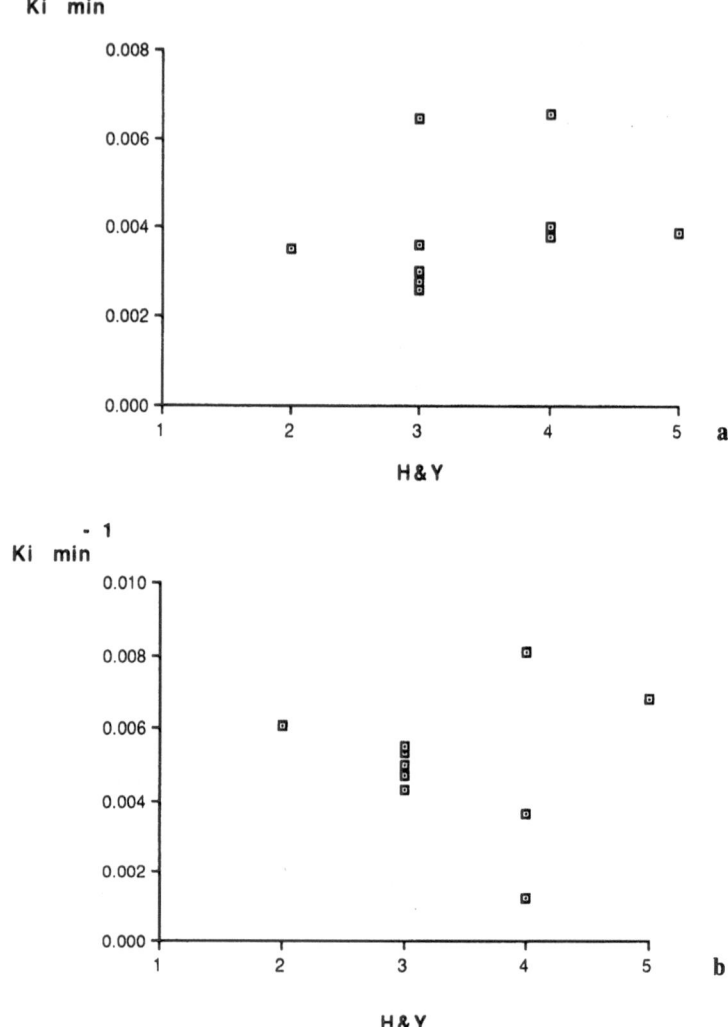

**Fig. 3.** Scatter diagrams showing a lack of correlation between putamen (**a**) and caudate (**b**) $^{18}$F-dopa influx constants and locomotor disability, as assessed on the Hoehn and Yahr scale, for individual PSP patients

possible that the medication could have in part contributed to the reduced striatal BSP uptake.

Wienhard et al. (1990) scanned two PSP patients using the irreversible D$_2$ antagonist $^{18}$F-fluoroethylspiperone. They found a mean 17% fall in caudate $^{18}$F-FESP binding potential (B$_{max}$/K$_d$) but did not report on putamen binding. Brooks et al. (1992) studied nine PSP patients with the more specific, reversible, D$_2$ antagonist $^{11}$C-raclopride. Four of their nine patients were taking medication at the time of PET. Figure 3 shows PET images of striatal $^{11}$C-raclopride activity, accumulated over 60 minutes following

**Table 3.** Mean striatal: cerebellar $^{11}$C-raclopride uptake ratios in PD and PSP

|  | | Caudate (mean ± SD) | Putamen (mean ± SD) |
|---|---|---|---|
| Controls | (8) | 3.78 ± 0.32 | 3.72 ± 0.36 |
| PD untreated | (6) | 3.79 ± 0.33 | 4.24 ± 0.59 |
| PD treated | (5) | **2.64 ± 0.48 | *3.04 ± 0.52 |
| PSP | (9) | **2.89 ± 0.64 | 3.40 ± 0.61 |

*p < 0.05 compared to normal } Student's t test
**p < 0.005 compared to normal

tracer administration, in a normal subject, and patients with untreated Parkinson's disease and PSP. It can be seen that there is a loss of $^{11}$C-raclopride binding in the PSP case, caudate being particularly affected. The PSP group showed a mean 24%, and 9%, fall in equilibrium caudate and putamen:cerebellum tracer uptake ratios, respectively, corresponding to a 32% and 12% loss of caudate and putamen $D_2$ binding sites — see Table 3. Five of the nine individual PSP patients had caudate, and three putamen, $^{11}$C-raclopride binding that was significantly reduced, but four of the nine had striatal tracer uptake that fell within the normal range. There was no correlation between striatal $^{11}$C-raclopride uptake and either disease duration or the presence or absence of treatment.

The above PET findings of moderately reduced striatal $D_2$ receptor binding in PSP are in broad agreement with post-mortem reports. Using $^3$H-spiperone, Bokobza et al. (1984) found that at autopsy five PSP patients showed mean 37%, and 30%, reductions in putamen and caudate dopamine $D_2$ receptor density, respectively. In a follow up series of seven PSP patients Ruberg et al. (1985) reported mean 42%, and 48%, losses of putamen and caudate $D_2$ sites, respectively, while Pierot et al. (1988) found mean 28% and 36% losses of putamen and caudate $D_2$ sites, respectively, in their eleven PSP patients.

To summarise, mean striatal $^{18}$F-dopa uptake is reduced to about 40% of normal in PSP and caudate and putamen are similarly affected, as would be predicted from pathological studies. Locomotor disability in PSP does not correlate with dopaminergic dysfunction suggesting that it arises as a consequence of loss of brainstem connections rather than dopamine terminals. Mean striatal $D_2$ receptor binding is moderately reduced in PSP, but individual patients may show normal levels of striatal $D_2$ receptor function. As these patients have L-dopa resistant akinetic-rigid syndromes, their lack of therapeutic response is likely to reflect degeneration of pallidal projections rather than loss of striatal dopamine receptors.

## Differential PET findings in parkinsonian syndromes

### *Metabolic studies*

While characteristic, the finding of reduced striatal and frontal glucose metabolism is not specific for PSP. De Volder et al. (1989) performed [18]FDG PET on seven patients with probable striatonigral degeneration (SND) and reported a similar pattern of functional involvement to that seen in PSP. Caudate and putamen metabolism were reduced to 64% and 54% of normal, respectively, while whole cortex and frontal metabolism were 80% and 72% of normal. As in PSP, premotor areas were marginally more affected than prefrontal cortex. Two of their seven SND patients had cerebellar ataxia and showed significantly reduced cerebellar metabolism. [18]FDG PET, therefore, does not provide a specific means of discriminating SND from PSP. By contrast, striatal and frontal blood flow and oxygen metabolism have been reported to be preserved in SND but significantly reduced in PSP (Brooks et al., 1992) — see Fig. 4. Further studies correlating regional cerebral metabolic and blood flow findings are required to establish whether SND and PSP can be reliably distinguished with functional imaging.

While PSP may be functionally difficult to distinguish from SND with PET, it can readily be distinguished from PD. In contrast to PSP patients, who have reduced striatal metabolism, patients with PD have normal or increased levels of striatal oxygen and glucose utilisation (Kuhl et al., 1984, 1985; Miletich et al., 1988; Wolfson et al., 1985). PSP patients show hypofrontality while cortical metabolism is normal in most non-demented PD patients (Kuhl et al., 1985). Demented PD patients show a pattern of dysfunction that is reminiscent of that seen in Alzheimer's disease: low metabolism in parietal and frontal association areas with preservation of primary sensorimotor and visual cortex function. The prevalence of dementia in PD is similar to the prevalence of concommitant Alzheimer pathology at post-mortem (about 20%). It is possible, therefore, that the pattern of metabolic dysfunction seen in PET scans of demented PD cases reflects the presence of coincident Alzheimer changes. Alternatively, the cortical hypofunction may be the result of cortical Lewy body disease.

Corticobasal degeneration is a condition characterised by asymmetric degeneration of motor cortical areas, the substantia nigra, and cerebellar dentate nuclei. The pathological hallmark is collections of swollen, achromatic, "Pick" cells in the absence of argyrophilic Pick bodies (Gibb et al., 1989). Posterior frontal, inferior parietal, and superior temporal, areas are targeted in contrast to Pick's disease where inferior frontal areas are most involved. The condition usually presents as a poorly L-dopa responsive akinetic-rigid syndrome involving one limb and subsequently generalises (Riley et al., 1990). Myoclonus is a frequent accompanying feature. Like PSP cases, CBD patients often have supranuclear gaze problems and bulbar dysfunction. As the condition progresses apraxia and cortical sensory loss are found in affected limbs, and the limb may manifest "alien" behaviour,

Ki min$^{-1}$

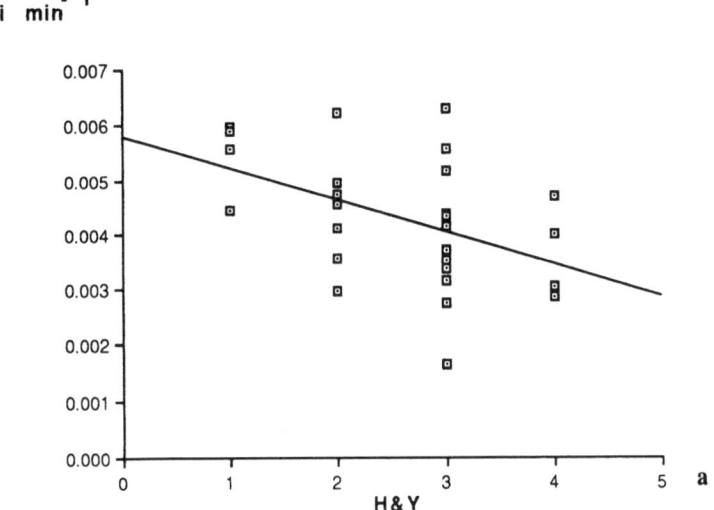

Ki min$^{-1}$

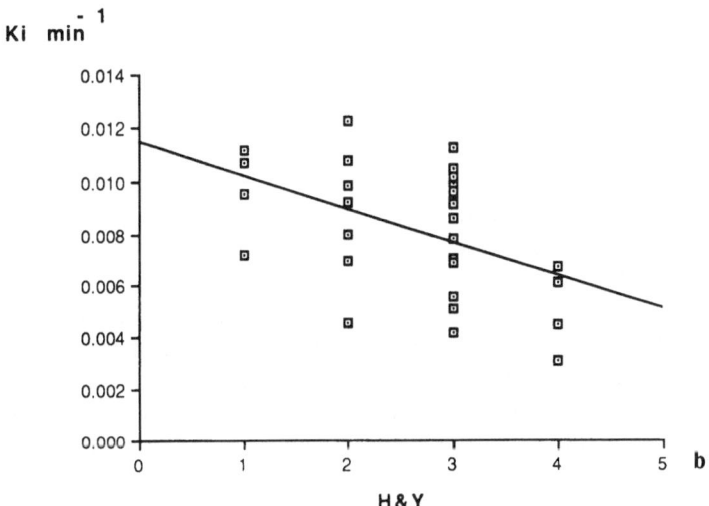

**Fig. 6.** Scatter diagrams showing an inverse correlation ($p < .01$ Kendall rank correlation statistics) between putamen (**a**) and caudate (**b**) [18]F-dopa influx constants and locomotor disability, as assessed on the Hoehn and Yahr scale, for individual PD patients

that is perform involuntary, but apparently purposeful, movements. PET findings are characteristic in CBD with asymmetric hypometabolism of inferior parietal, superior frontal, superior temporal, thalamic, and striatal areas being evident (Blin et al., 1990; Eidelberg et al., 1989; Sawle et al., 1991). This contrasts with PSP where striatal and frontal lobe function tends to be more symmetrically affected.

**Table 4.** PET findings in parkinsonian syndromes

|  | PD | PD (demented) | SND | PSP | CBD |
|---|---|---|---|---|---|
| FDG | normal | low parietal/ temporal | low striatal/ frontal | low striatal/ frontal | low striatal/ inferior parietal |
| F-dopa | putamen < caudate | putamen < caudate | putamen < caudate | putamen = caudate | putamen = caudate |
| striatal $D_2$ sites | normal or raised (untreated) normal or low (treated) | | normal or low | | ? |

*Dopaminergic studies*

While the nigra is uniformly involved in PSP, in PD the Lewy body degeneration targets the ventrolateral nigral dopaminergic projections to dorsal putamen (Fearnley and Lees, 1991; German et al., 1989) and putamen dopamine levels are half those found in the head of caudate (Kish et al., 1988). In SND the ventrolateral nigra is also targeted but involvement of dorsomedial nigral projections to caudate is more extensive than in PD (Goto et al., 1989).

Brooks et al. (1990) have compared [18]F-dopa PET findings in PSP and PD and found that while mean putamen [18]F-dopa influx was reduced to similar levels in these conditions (37% and 44% of normal, respectively) mean caudate tracer uptake was significantly more impaired in PSP than in PD (47% and 74% of normal, respectively) — see Fig. 2. 90% of individual PSP cases showed reduced caudate [18]F-dopa influx compared to only 25% of PD patients. Because of this differential involvement of caudate [18]F-dopa PET is able to distinguish PSP from PD patients 90% of the time (Burn et al., 1993). In PSP cases locomotor function appears to be unrelated to putamen and caudate [18]F-dopa uptake whereas in PD a clear correlation exists — see Fig. 5. Leenders et al. have also reported a correlation between striatal [18]F-dopa uptake and locomotor status in PD but not PSP (Leenders et al., 1986, 1988). It would appear that in PD locomotor disability is primarily determined by the extent of loss of nigro-striatal dopaminergic projections while in PSP it is determined by the extent of loss of pallidal and brainstem connections.

The possibility of distinguishing PSP from PD patients by PET measurements of striatal $D_2$ binding potential has also been investigated. Using [11]C-raclopride Brooks et al. (1992) found that the mean putamen:cerebellar tracer uptake ratio was raised by 11% in untreated PD patients while caudate tracer uptake remained normal. In contrast, PSP patients showed a decrease in mean putamen and caudate:cerebellar raclopride uptake ratios of 9% and 24%, respectively. Chronically treated PD patients who had developed fluctuating therapeutic responses, however, showed similar levels of striatal [11]C-raclopride uptake to that of the PSP patients as downregulation of striatal $D_2$ binding sites had occurred. While PET measurements of striatal $D_2$ receptor density with $^{11}$C-raclopride are potentially capable of distinguishing PSP from drug-naive PD patients, in practice there is an overlap between the untreated PD and PSP ranges of striatal tracer uptake. This PET finding is in line with autopsy studies. A 28%–48% loss of caudate and putamen $D_2$ binding sites has been reported in PSP (Bokobza et al., 1984; Pierot et al., 1988; Ruberg et al., 1985). Striatal $D_2$ receptor density has been found to be normal or raised in drug-naive PD patients and normal or reduced in treated PD cases (Bokobza et al., 1984; Cortes et al., 1989; Guttman and Seeman, 1986; Lee et al., 1978; Pierot et al., 1988; Quik et al., 1979).

## Conclusions

In this chapter the ways in which PET can be used to demonstrate the patterns of functional disruption of regional cerebral metabolism and the dopaminergic system in PSP has been presented. Table 4 summarises reported PET findings in PSP and compares these with other parkinsonian disorders.

## References

Alexander GE, Delong MR, Strick PL (1986) Parallel organization of functionally segregated circuits linking basal ganglia and cortex. Ann Rev Neurosci 9: 357–381

Baron JC, Maziere B, Loc'h C, et al (1986) Loss of striatal (76Br)bromospiperone binding sites demonstrated by positron tomography in progressive supranuclear palsy. J Cereb Blood Flow Metab 6: 131–136

Bhatt MH, Snow BJ, Martin WRW, Peppard R, Calne DB (1991) Positron emission tomography in progressive supranuclear palsy. Arch Neurol 48: 389–391

Blin J, Baron JC, Dubois P, et al (1990) Positron emission tomography study in progressive supranuclear palsy. Arch Neurol 47. 747–752

Blin J, Vidhailhet M, Pillon B, et al (1992) Corticobasal degeneration: decreased and asymmetrical glucose consumption as studied by PET. Mov Disord 7: 348–354

Bokobza B, Ruberg M, Scatton B, Javoy-Agid F, Agid Y (1984) 3H-spiperone binding, dopamine and HVA concentrations in Parkinson's disease and supranuclear palsy. Eur J Pharmacol 99: 167–175

Brooks DJ, Ibañez V, Sawle GV, et al (1990) Differing patterns of striatal $^{18}$F-dopa uptake in Parkinson's disease, multiple system atrophy and progressive supranuclear palsy. Ann Neurol 28: 547–555

Brooks DJ, Ibañez V, Sawle GV, et al (1992) Striatal $D_2$ receptor status in Parkinson's disease, striatonigral degeneration, and progressive supranuclear palsy, measured with $^{11}$C-raclopride and PET. Ann Neurol 31: 184–192

Burn DJ, Sawle GV, Brooks DJ (1993) The differential diagnosis of Parkinson's disease, multiple system atrophy, and Steele-Richardson-Olszewski syndrome: discriminant analysis of striatal 18F-dopa PET data. J Neurol Neurosurg Psychiatry (in press)

Cortes R, Camps M, Gueye B, Probst A, Palacios JM (1989) Dopamine receptors in the human brain: autoradiographic distribution of $D_1$ and $D_2$ sites in Parkinson syndrome of different etiology. Brain Res 483: 30–38

D'Antona R, Baron JC, Samson Y, et al (1985) Subcortical dementia: frontal cortex hypometabolism detected by positron tomography in patients with progressive supranuclear palsy. Brain 108: 785–800

De Volder AG, Francard J, Laterre C, et al (1989) Decreased glucose utilisation in the striatum and frontal lobe in probable striatonigral degeneration. Ann Neurol 26: 239–247

Dubas F, Gray F, Escourolle R (1983) Maladie de Steele-Richardson-Olszewski sans ophthalmologie. Rev Neurol 139: 407–416

Eidelberg D, Dhawan V, Moeller JR, et al (1991) The metabolic landscape of cortico-basal ganglionic degeneration: regional asymmetries studied with position emission tomography. J Neurol Neurosurg Psychiatry 54: 856–862

Fearnley JM, Lees AJ (1991) Ageing and Parkinson's disease: substantia nigra regional selectivity. Brain 114: 2283–2301

Fearnley JM, Revesz T, Brooks DJ, Frackowiak RSJ, Lees AJ (1991) Diffuse Lewy body disease presenting with a supranuclear downgaze palsy. J Neurol Neurosurg Psychiatry 54: 159–161

Foster NL, Gilman S, Berent S, Morin EM, Brown MB, Koeppe RA (1988) Cerebral hypometabolism in progressive supranuclear palsy studied with positron emission tomography. Ann Neurol 24: 399–406

Foster NL, Gilman S, Berent S, et al (1992) Progressive subcortical gliosis and progressive supranuclear palsy can have similar clinical and PET abnormalities. J Neurol Neurosurg Psychiatry 55: 707–713

German DC, Manaye K, Smith WK, Woodward DJ, Saper CB (1989) Midbrain dopaminergic cell loss in Parkinson's disease: computer visualization. Ann Neurol 26: 507–514

Gibb WRG, Luthert P, Marsden CD (1989) Corticobasal degeneration. Brain 112: 1171–1192

Goffinet AM, De Volder AG, Gillain C, et al (1989) Positron tomography demonstrates frontal lobe hypometabolism in progressive supranuclear palsy. Ann Neurol 25: 131–139

Goto S, Hirano A, Matsumoto S (1989) Subdivisional involvement of nigrostriatal loop in idiopathic Parkinson's disease and striatonigral degeneration. Ann Neurol 26: 766–770

Guttman M, Seeman P (1986) Dopamine D2 receptor density in Parkinsonian brain is constant for duration of disease, age, and duration of L-dopa therapy. Adv Neurol 46: 51–57

Hughes AJ, Daniel SE, Kilford L, Lees AJ (1992) The accuracy of the clinical diagnosis of Parkinson's disease: a clinicopathological study of 100 cases. J Neurol Neurosurg Psychiatry 55: 181–184

Jackson JA, Jankovic J, Ford J (1983) Progressive supranuclear palsy: clinical features and response to treatment in 16 patients. Ann Neurol 13: 273–278

Jellinger K, Riederer P, Tomananga M (1980) Progressive supranuclear palsy: clinico-pathological and biochemical studies. J Neural Transm [Suppl] 16: 111–128

Johnson KA, Sperling RA, Holman BL, Nagel JS, Growdon JH (1992) Cerebral perfusion in progressive supranuclear palsy. J Nucl Med 33: 704–709

Karbe H, Grond M, Huber M, Herholz K, Kessler J, Heiss WD (1992) Subcortical damage and cortical dysfunction in progressive supranuclear palsy demonstrated by positron emission tomography. J Neurol 239: 98–102

Kish SJ, Chang LJ, Mirchandani LJ, Shannak K, Hornykiewicz O (1985) Progressive supranuclear palsy: relationship between extrapyramidal disturbances, dementia, and brain neurotransmitter markers. Ann Neurol 18: 530–536

Kish SJ, Shannak K, Hornykiewicz O (1988) Uneven pattern of dopamine loss in the striatum of patients with idiopathic Parkinson's disease. N Engl J Med 318: 876–880

Kiyosawa M, Baron JC, Hamel E, et al (1989) Time course of effects of unilateral lesions of the Nucleus Basalis of Meynert on glucose utilisation by the cerebral cortex. Positron emission tomography in baboons. Brain 112: 435–455

Koeppen AH, Hans MB (1976) Supranuclear ophthalmoplegia in olivopontocerebellar degeneration. Neurology 26: 764–768

Kuhl DE, Metter EJ, Benson DF, et al (1985) Similarities of cerebral glucose metabolism in Alzheimer's and Parkinsonian dementia. J Cereb Blood Flow Metab 5 [Suppl] 1: S169–S170

Kuhl DE, Metter EJ, Riege WH (1984) Patterns of local cerebral glucose utilisation determined in Parkinson's disease by the 18F-fluorodeoxyglucose method. Ann Neurol 15: 419–424

Laplane D, Levasseur M, Pillon B, et al (1989) Obsessive-compulsive and other behavioural changes with bilateral basal ganglia lesions. Brain 112: 699–725

Leenders KL, Frackowiak RS, Lees AJ (1988) Steele-Richardson-Olszewski syndrome. Brain energy metabolism, blood flow and fluorodopa uptake measured by positron emission tomography. Brain 111: 615–630

Leenders KL, Palmer A, Turton D, et al (1986) DOPA uptake and dopamine receptor binding visualized in the human brain in vivo. In: Fahn S, Marsden CD, Jenner P, Teychenne P (eds) Recent developments in Parkinson's disease. Raven Press, New York, pp 103–113

Lee T, Seeman P, Rajput A, Farley IJ, Hornykiewicz O (1978) Receptor basis for dopaminergic supersensitivity in Parkinson disease. Nature 273: 59–60

Lees AJ (1986) The Steele-Richardson-Olszewski syndrome (progressive supranuclear palsy). In: Marsden CD, Fahn S (eds) Movement disorders, vol 2. Butterworths, London, pp 273–287

Maher ER, Lees AJ (1986) The clinical features and natural history of the Steele-Richardson-Olszewski syndrome (progressive supranuclear palsy). Neurology 36: 1005–1008

McCulloch J, MacKenzie ET, Cudennec A, Duverger D, Degueurce A, Scatton B (1984) Influences of the raphe nuclei on brain glucose utilisation. Soc Neurosci Abstr 10: 218

Miletich RS, Bankiewicz R, Plunkett R, et al (1988) L-[18F]6-Fluorodopa PET images of catecholaminergic tissue implants in hemi-parkinsonian monkeys. Neurology 38: S145–S140 (abstract)

Perkin GD, Lees AJ, Stern GM, Kocen RS (1978) Problems in the diagnosis of progressive supranuclear palsy (Steele-Richardson-Olszewski syndrome). Can J Neurol Sci 6: 167–173

Pappata S, Mazoyer B, Tran Dinh S, Cambon H, Levasseur M, Baron JC (1990) Effects of capsular or thalamic stroke on metabolism in the cortex and cerebellum: a positron tomography study. Stroke 21: 519–524

Pierot L, Desnos C, Blin J, et al (1988) D1 and D2-type dopamine receptors in patients with Parkinson's disease and progressive supranuclear palsy. J Neurol Sci 86: 291–306

Quik M, Spokes E, MacKay A, Bannister R (1979) Alterations in $^{3}$H-spiperone binding in human caudate nucleus, substantia nigra and frontal cortex in the Shy Drager syndrome and Parkinson's disease. J Neurol Sci 43: 429–437

Rajput AH, Rozdilsky B, Rajput A (1991) Accuracy of clinical diagnosis in Parkinsonism — a prospective study. Can J Neurol Sci 18: 275–278

Riley DE, Lang AE, Lewis A, et al (1990) Cortical-basal ganglionic degeneration. Neurology 40: 1203–1212

Ross-Russell R (1980) Supranuclear palsy of eyelid closure. Brain 103: 71–82

Ruberg M, Javoy-Agid F, Hirsch E, et al (1985) Dopaminergic and cholinergic lesions in progressive supranuclear palsy. Ann Neurol 18: 523–529

Savaki HE, Graham DI, Grome JJ, McCulloch J (1984) Functional consequences of unilateral lesion of the locus coeruleus: a quantitative [$^{14}$C]2-deoxyglucose investigation. Brain Res 292: 239–249

Sawle GV, Brooks DJ, Marsden CD, Frackowiak RSJ (1991) Corticobasal degeneration: a unique pattern of regional cortical oxygen metabolism and striatal fluorodopa uptake demonstrated by positron emission tomography. Brain 114: 541–556

Steele JC, Richardson JC, Olszewski J (1964) Progressive supranuclear palsy. A heterogeneous degeneration involving the brain stem, basal ganglia, and cerebellum, with vertical gaze and pseudobulbar palsy. Arch Neurol 10: 333–359

Taniwaki T, Hosokawa S, Goto I, et al (1992) Positron emission tomography (PET) in "pure akinesia". J Neurol Sci 107: 34–39

Wienhard K, Coenen HH, Pawlik G, et al (1990) PET studies of dopamine receptor distribution using [18F]fluoroethylspiperone: findings in disorders related to the dopaminergic system. J Neural Transm [Gen Sect] 81: 195–213

Will RG, Lees AJ, Gibb W, Barnard RO (1988) A case of progressive subcortical gliosis presenting clinically as Steele-Richardson-Olszewski syndrome. J Neurol Neurosurg Psychiatry 51: 1224–1227

Wolfson LI, Leenders KL, Brown LL, Jones T (1985) Alterations of regional cerebral blood flow and oxygen metabolism in Parkinson's disease. Neurology 35: 1399–1405

Author's address: Dr. D. J. Brooks, MRC Cyclotron Unit, Hammersmith Hospital, Du Cane Road, London W12 0HS, United Kingdom.

# Morphological aspects

# The neuropathology of progressive supranuclear palsy

**P. L. Lantos**

Department of Neuropathology, Institute of Psychiatry, London,
United Kingdom

**Summary.** The macroscopical, histological, ultrastructural and immuno-cytochemical features of progressive supranuclear palsy (PSP) are reviewed. Recent investigations have revealed important differences in the distribution, ultrastructure and immunocytochemical profile of neurofibrillary tangles in PSP and in Alzheimer's disease. Cortical involvement, as demonstrated by the presence of tangles and neuropil threads, has extended the neuropathological spectrum of PSP. Quantitative assessments of neuronal populations show neuronal loss, not only in various nuclei of the brainstem, diencephalon and cerebellum, but also in other areas, including the nucleus basalis of Meynert, substantia nigra and neostriatum.

A new classification, based on neuropathological criteria, is suggested in order to take into consideration the phenotypic heterogeneity of PSP. This new classification distinguishes three types: typical, atypical and combined cases. Typical (Type 1) cases conform to the original definition of PSP. Type 2, atypical cases are variants of the histological changes characteristic of PSP: either the severity or the distribution of abnormalities, or both of these deviate from the typical pattern. Cases with combined pathology belong to type 3 group: in these the typical pathology of PSP is accompanied by lesions characteristic of another neurodegenerative or vascular disease.

Progressive supranuclear palsy (PSP) has a short, but interesting history. PSP was defined as a clinical and neuropathological entity in 1964. Steele, Richardson and Olszewski whose names later served to be the eponymous alternative for the designation of the disease, reported nine cases of a progressive brain disease characterised by supranuclear ophthalmoplegia, pseudobulbar palsy, dysarthria and dystonic rigidity of the neck and upper trunk. Other, less constant cerebellar and pyramidal signs, and a relatively mild dementia also occurred. All their patients were male and the disease usually started in the sixth decade leading, through a progressive course, to death within five-to-seven years. The neuropathological picture was rather similar in the seven patients who had died. There was evidence of neuronal loss, gliosis, neurofibrillary tangles, granulovacuolar degeneration and demyelination in various regions of the basal ganglia, brainstem and cerebellum. The distribution of the histological abnormalities showed a consistent pattern: the globus pallidus, subthalamic nucleus, red nucleus, substantia nigra, superior colliculi, cuneiform and subcuneiform nuclei, periaqueductal grey matter, pontine tegmentum and the dentate nucleus were most severely affected. An inflammatory response was extremely rare. The authors drew attention to the striking histopathological similarities between PSP, postencephalitic parkinsonism and the parkinsonism-dementia complex of Guam, but emphasised the different distribution of the lesions. Although unable to establish the aetiology of the disease, they suggested a degenerative process or viral infection (Steele et al., 1964). Interestingly, the first clinical descriptions of the disease date back to the early years of this century (Posey, 1904; Spiller, 1905), six decades before PSP was defined as a clinicopathological entity.

Following the seminal publication, new cases have been reported from all over the world, and these have been reviewed (Barr, 1986; Kristensen, 1985; Steele, 1975). However, PSP remained a diagnostic curiosity of limited interest. The last decade witnessed an increased awareness of this disorder, as research on the cellular and molecular biology of neurodegenerative diseases has made spectacular advances. In particular, interest has focused on the neurofibrillary tangle of PSP, an histological hallmark which, although different in both structure and biochemistry from those of Alzheimer's disease, has been subjected to intense investigations. The occurrence of this cytoskeletal abnormality is a common feature of several neurodegenerative diseases and the study of PSP may further the insight into the pathogenesis of other, more common disorders, including Alzheimer's disease.

The aim of this review is to give an up-to-date account of pathology, including research on the neurofibrillary tangle, and to draw attention to the neuropathological heterogeneity of PSP.


## Epidemiology

The precise incidence of PSP is not known and only a few epidemiological data are available. Although it is a rare disease, a literature survey carried out in 1984, two decades after the disease had been defined, yielded a total of 325 cases (Kristensen, 1985). Apparently, four percent of patients with parkinsonian signs have PSP (Jackson et al., 1983) and the incidence in Western Australia is four cases per $10^6$ population per year (Mastaglia et al., 1973). A minimum estimate of prevalence in New Jersey was 1.39 per $10^6$ (Golbe et al., 1988). There is a slight male preponderance: 60 percent of patients are men and 40 per cent women (Kristensen, 1985), although in a British review of 52 cases, 22 were men and 30 women (Maher and Lees, 1986).

In a large series of 202 patients, the average age of onset was 59.6 years, ranging from 12 to 80 years. However, the disease most often starts during or after the sixth decade: the age of onset in 71% of patients was between 50 and 65 years (Kristensen, 1985), thus PSP is a disease of the presenium. The duration varies from one to 23 years, the average survival time of 73 cases was 5.7 years (Kristensen, 1985). There is no obvious connection between the age of onset and the duration of the disease (Hynd et al., 1982). The patients die, as they often do suffering from neurodegenerative disorders, of intercurrent infections, usually bronchopneumonia, aspiration or inanition.

## Neuropathology

### *Macroscopical appearances*

On gross examination the brain usually shows only minor abnormalities or may even appear normal. The brain weight may be slightly reduced, but quite often is within normal limit. On coronal slices the cortex and the white matter are unremarkable. The lateral ventricles, the third ventricle and the cerebral aqueduct may all be slightly or moderately enlarged, but without the overall impression of diffuse severe atrophy which often occurs in Alzheimer's disease. However, the midbrain is shrunken, particularly the superior colliculi, the mesencephalic tegmentum and periaqueductal grey matter. The substantia nigra is usually pale and the red nuclei may occasionally be discoloured. The pontine tegmentum is atrophied and the locus coeruleus is paler than usual. The subthalamic regions and the inner segment of the globus pallidus may be atrophied and discoloured. In the cerebellum, the superior cerebellar peduncles and the dentate nuclei are shrunken, while the hilus appears discoloured. Thus gross pathology, when present, is limited to the diencephalon, brainstem and cerebellum.

### *Histology, ultrastructure and immunocytochemistry*

The histological abnormalities are more striking than the otherwise rather bland gross morphology would suggest and include neurofibrillary tangles, neuronal loss and glial changes. Granulovacuolar degeneration, myelin loss and perivascular cuffing are inconstant findings.

The presence of neurofibrillary tangles (NFT) is the single most important lesion of PSP (Figs. 1–3). They are present in the substantia nigra, subthalamic nucleus, nucleus basalis of Meynert, globus pallidus, pretectal region, tegmentum of the midbrain and pons including the periaqueductal grey matter, locus coeruleus, raphe nuclei, and in the nuclei of various cranial nerves. NFTs occur less frequently in the red nucleus, corpus striatum, thalamus and inferior olives, and their distribution has been semiquantitatively assessed (Jellinger et al., 1980). They also develop in the cortex.

NFTs are easily discernible on routine haematoxylin and eosin-stained sections, but their presence is accentuated in silver-impregnated preparations, for example Marsland-Glees, modified Bielschowsky or Gallyas (Figs. 1 and 2). Morphologically two subtypes can be distinguished: the globose type is far more common than the flame-shaped form. This structural variation is influenced by the configuration of the affected neurons. Electron microscopy reveals the NFTs to be composed predominantly of bundles of

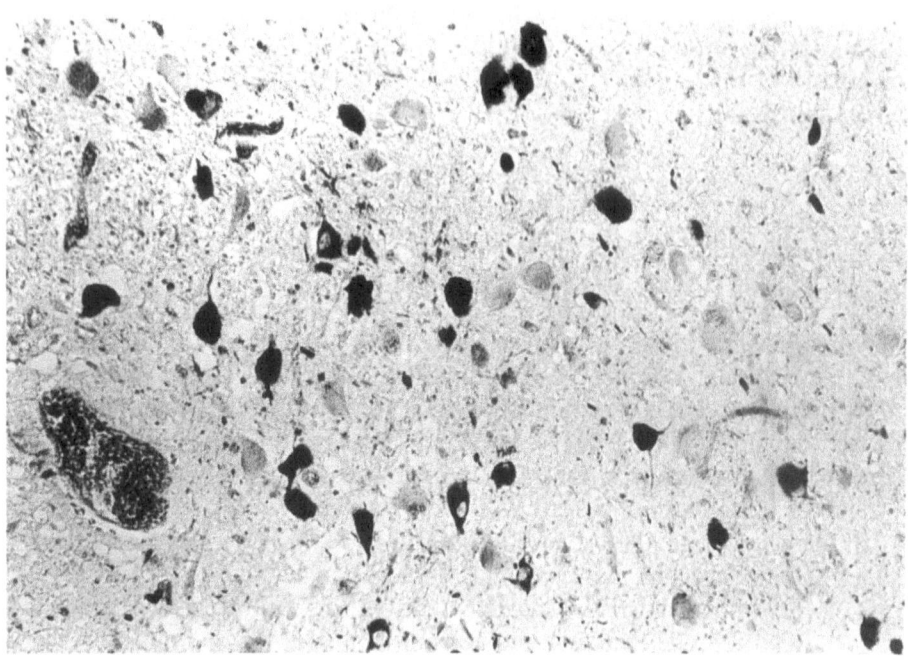

**Fig. 1.** Neurofibrillary tangles in the locus coeruleus demonstrated by the silver impregnation technique of Gallyas. ×160

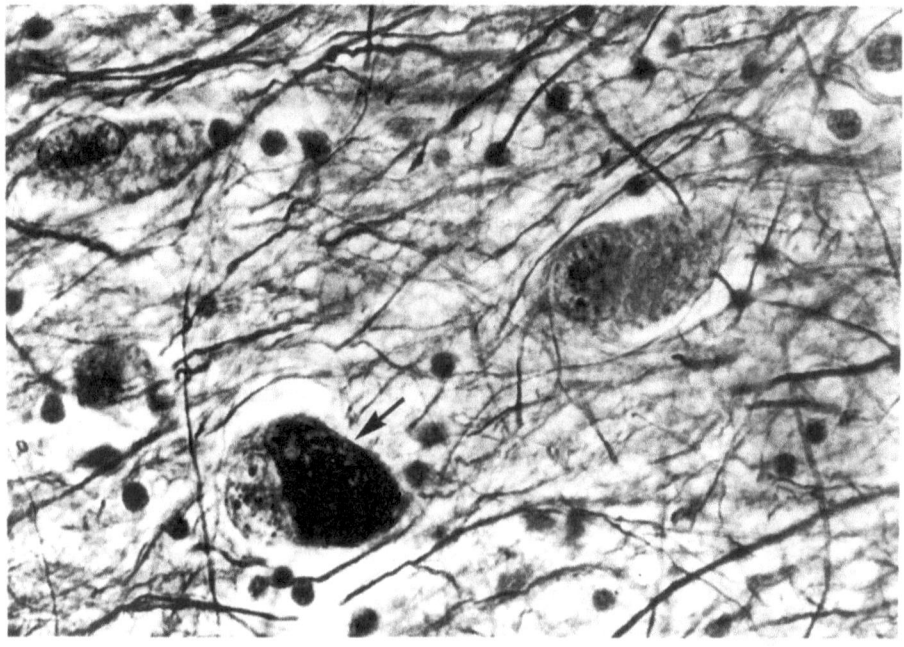

**Fig. 2.** Neurofibrillary tangle (arrow) in the nucleus basalis of Meynert. Silver impregnation according to Marsland and Glees. ×640

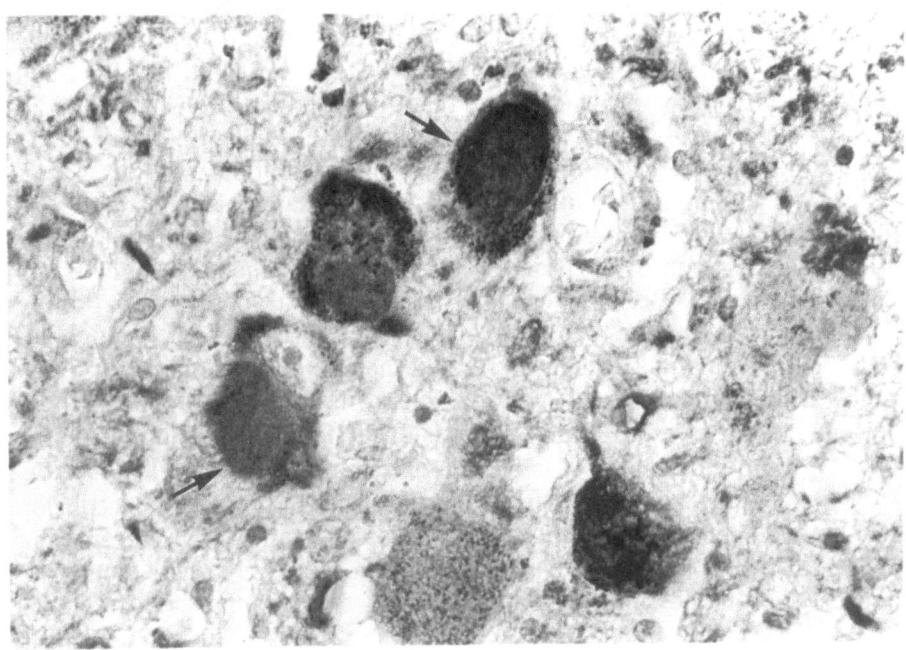

**Fig. 3.** Neurofibrillary tangles (arrows) in the locus coeruleus give positive reaction with an antibody to tau protein. Avidin-biotin complex method. ×640

straight filaments of indeterminate length and 15 nm in diameter (Tellez-Nagel and Wisniewski, 1973). These are, in turn, composed of six or more protofilaments of 2–5 nm (Montpetit et al., 1985). In addition to straight filaments, twisted or paired filaments have been also described, and it was suggested that the former represents only a stage in the formation of the latter (Tomonaga, 1977; Yagishita et al., 1979).

Immunocytochemically, the NFTs of PSP give positive reactions with antibodies to tau protein (Bancher et al., 1987; Pollock et al., 1986; Probst et al., 1988; Fig. 3) and to isolated paired helical filaments (Probst et al., 1988). The immunoreactivity of NFTs in PSP to components of neurofilaments and ubiquitin is more controversial. Mouse monoclonal antibodies to phosphorylated epitopes on high and medium molecular weight neurofilament subunits do not or only weakly stain the subcortical NFTs (Probst et al., 1988). Moreover, 12 anti-neurofilament monoclonal antibodies specific for multiphosphorylation repeat domains or other phosphate-dependent and independent epitopes did not bind NFTs in PSP (Schmidt et al., 1988). Immunostaining with antibodies to ubiquitin gives either negative or weakly positive reaction (Leigh et al., 1989; Lennox et al., 1988; Love et al., 1988), although an immuno-electron microscope study has demonstrated labelling of straight filaments (Manetto et al., 1988).

These investigations reveal differences as well as similarities in the antigenicity of NFTs in Alzheimer's disease and PSP, in addition to the previously established structural discrepancy. Alzheimer NFTs readily stain with ubiquitin and with monoclonal antibodies to the medium and heavy molecular weight neurofilament subunits. There are also differences in immunoreactivity to tau, indicating that the mechanisms which lead to NFT formation may be different in PSP from those in normal ageing and Alzheimer's disease (Shin et al., 1991). The distribution patterns of tau-positive neurons are different from those in ageing and Alzheimer's disease (Shin et al., 1991), confirming early observations which emphasised the different distribution of NFTs in the brainstem in Alzheimer's disease and PSP (Jellinger, 1971). Moreover, there is a discrepancy between the profiles of abnormal tau in the two neurodegenerative diseases. An immunoblot study of PSP brain homogenates revealed that of the abnormal tau protein triplets, tau 55, tau 64 and tau 69, which are constantly found in both Alzheimer's disease and Down's syndrome, the first was missing, while the other two were detected only in smaller amounts (Flament et al., 1991).

There is neuronal loss in various nuclei of the brainstem, diencephalon and cerebellum. The severity of nerve cells loss is related to NFT formation: areas with the most severe neuronal depletion show the largest number of tangles. Neuronal loss may also occur in other areas. In five cases of typical PSP there was an age dependent neuronal loss ranging from 12.6 percent to 54.1 percent in the nucleus basalis of Meynert when compared with controls (Tagliavini et al., 1984). The neuronal depletion of the substantia nigra affects chiefly the medial portion as a quantitative study of 14 cases of PSP has recently revealed (Fearnley and Lees, 1991). Neuronal loss (Hirsch et al., 1987; Zweig et al., 1987) and NFT formation (Zweig et al., 1987) were observed in the pedunculopontine nucleus. In PSP a significant neuronal loss of 60% in the pars compacta of this nucleus was associated with tangle formation in 40–64% of the surviving neurons (Jellinger, 1988). Since both this nucleus and the nucleus basalis are cholinergic, this pathology could be particularly relevant to the development of subcortical dementia associated with PSP. Moreover, there is a loss of cholinergic neurons in the pontine reticular formation as indicated by quantitative immunocytochemistry of choline acetyltransferase (Malessa et al., 1991). A selective decrease of large neurons occurs in the neostriatum: the number of these cells is reduced to 40 percent and 30 percent of control values in the head of the caudate nucleus and putamen, respectively. The population of small nerve cells, in contrast, is well preserved (Oyanagi et al., 1988). Nerve cell loss is also found in the locus coeruleus (Mann et al., 1983), but this report has not been confirmed (Tomonaga, 1983).

In the dentate nucleus, in addition to neuronal loss and NFTs (Fig. 4), the so-called grumose degeneration around dentate neurons may also develop (Fig. 5). These are eosinophilic granular structures which in silver impregnation appear as argentophilic rings or knobs. Antibodies to phosphorylated neurofilaments positively stained some of these structures. Electron microscopy revealed that they consisted of clusters of axon terminals and pre-terminal axons. Their origin could not be identified with certainty, but they are most likely to have derived from Purkinje cells. Many of these

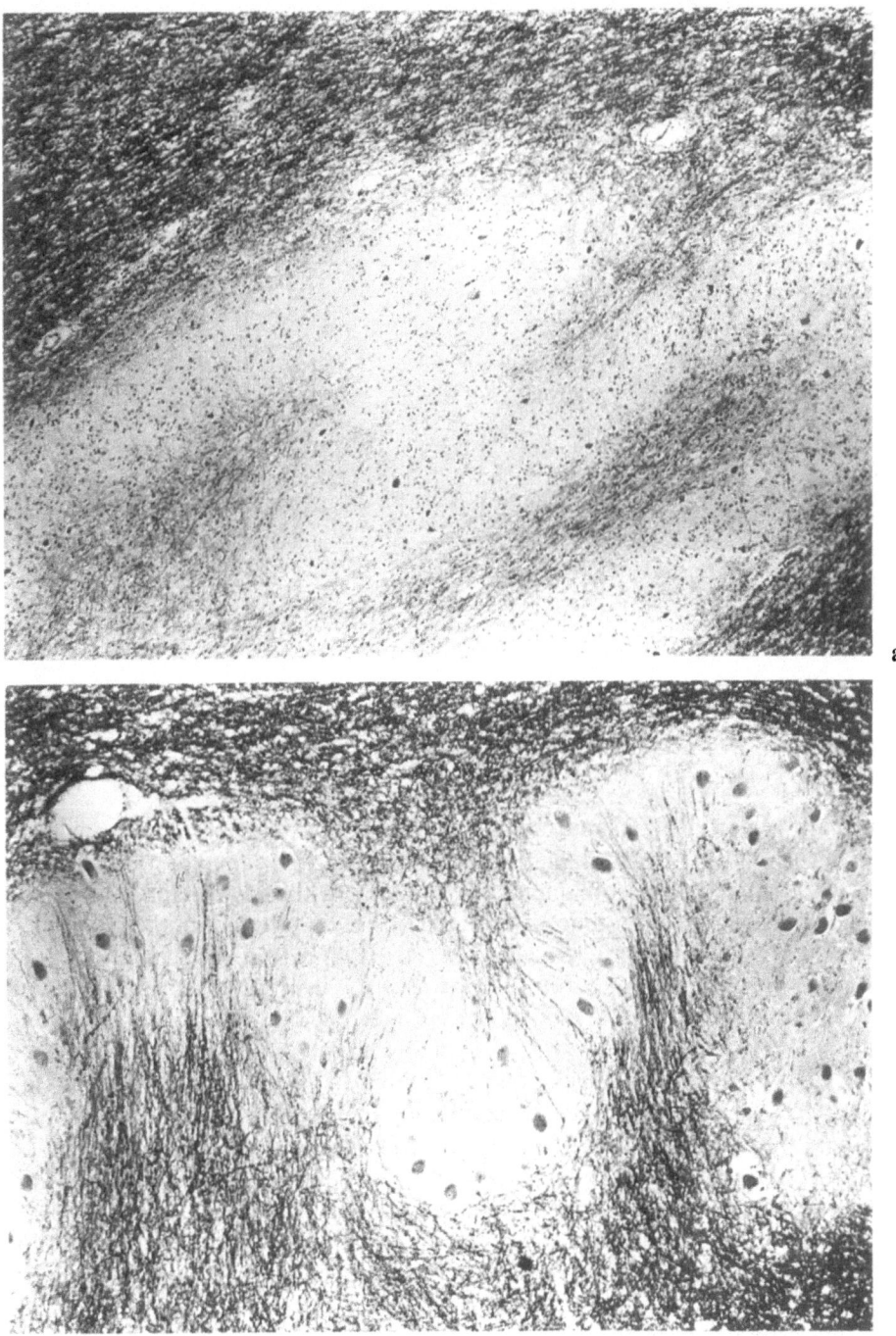

**Fig. 4.** Neuronal loss in the dentate nucleus of the cerebellum in PSP (**a**) compared with control (**b**). Cresyl violet and luxol fast blue. ×60

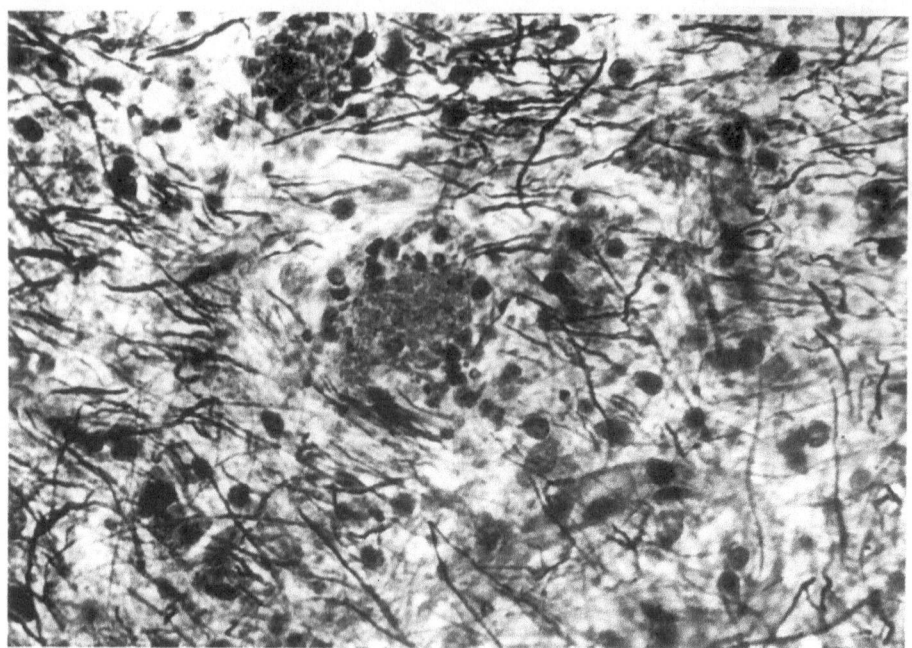

**Fig. 5.** "Grumose degeneration" of the dentate nucleus. Silver impregnation according to Marsland and Glees. ×700

structures were swollen and contained an admixture of organelles including mitochondria, neurofilaments, synaptic vesicles, lamellar bodies and vacuoles. In some, neurofilaments predominated; these were the structures which appeared argentophilic and stained positively with antibodies to phosphorylated neurofilaments (Mizusawa et al., 1989). In areas with extensive grumose degeneration, neuronal loss was also severe and surviving nerve cells were atrophic. Other neurons were swollen or showed central chromatolysis (Arai, 1989).

Hypertrophy of the olivary nucleus may also occur, caused most likely by the lesions in the dentate nucleus of the cerebellum and in the central tegmental tract of the pons. Ultrastructurally, the enlarged neurons of the olives contain masses of neurofilaments, concentric laminated bodies and paired helical filaments mixed with straight tubules (Yagashita et al., 1986). These neurons show positive staining with two monoclonal antibodies which recognise phosphorylated epitopes of the heavy molecular weight subunit of neurofilaments (Giaccone et al., 1988).

Interestingly, in two cases the spinal cord was also involved: there were NFTs in the anterior, posterior and lateral horns, in Clarke's column and in the intermediate grey matter. The posterior horns were most severely affected (Kato et al., 1986).

The most controversial aspect of the topography of PSP lesions is the involvement of the cerebral cortex. PSP was originally defined as a disease affecting subcortical grey matter, particularly in the diencephalon, brain-stem and cerebellum. That the cortex may be affected has been realised only more recently (Takahashi et al., 1989; Hauw et al., 1990). Since neuronal loss and NFTs also occur in Alzheimer's disease and, to a lesser degree in normal ageing, it is important to define the cortical lesions of PSP. Neocortical NFTs, demonstrated by Bodian's silver impregnation and tau immunolabelling, were found in all five cases studied (Hauw et al., 1990). However, these differed from the NFTs of Alzheimer's disease and ageing in their lobar, laminar and cellular distribution. In PSP they were most frequent in the precentral gyrus and not in the association areas, usually developed in large pyramidal neurons and small nerve cells, sparing cell populations preferentially involved in Alzheimer's disease and were most numerous in the deepest layers of the cortex, leaving layer III relatively normal. Ultrastructural examination confirmed these differences by revealing straight tubules of 15 nm in the cortical neurons of PSP (Takahashi et al., 1989).

Whilst NFTs remain the histological hallmark of PSP, more recently neuropil threads, demonstrated by the silver impregnation of Gallyas and antibodies to both tau protein and isolated paired helical filaments, have been extensively found in the subcortical grey matter (Probst et al., 1988), neocortex and hippocampus (Hauw, 1990). Moreover, bundles of positive axons were present in white matter tracts which interconnect subcortical nuclei (Probst et al., 1988). These findings clearly indicate that PSP affect the brain more extensively and more profoundly than hitherto thought: the cortex may also be involved in the pathological process disrupting the cytoskeletal system more severely, resulting not only in NFTs, but also in neuropil threads.

Deposition of amyloid βA4 protein may variably occur in PSP (Hauw et al., 1990; Mann and Jones, 1990; Tan et al., 1988), but plaque formation as seen in Alzheimer's disease, is clearly not part of the pathological spectrum. βA4 deposition tend to be diffuse and this phenomenon is likely to be associated with ageing.

Although granulovacuolar degeneration was described in the original report (Steele et al., 1964), this is an inconsistent finding. There may be more than one granulovacuole in the same cell and occasionally the same neuron may contain NFTs and granulovacuoles. Unlike in Alzheimer's disease in which granulovacuoles occur in the hippocampus, in PSP their distribution reflects the basic topography and granulovacuolar degeneration develops particularly in the subthalamic nucleus, substantia nigra, red nucleus, locus coeruleus, basis pontic and dentate nucleus (Hynd et al., 1982).

Gliosis is usually part of the histological picture and astrocytosis is particularly noticeable in areas whose neuronal population has become depleted. Immunocytochemistry using antibodies to glial fibrillary acidic protein strikingly demonstrates this increase in astrocytes (Fig. 6). Increase of microglial cells may also occur. Glial nodules, perivascular cuffing and neuronophagia are uncommon (for review see Kristensen, 1985).

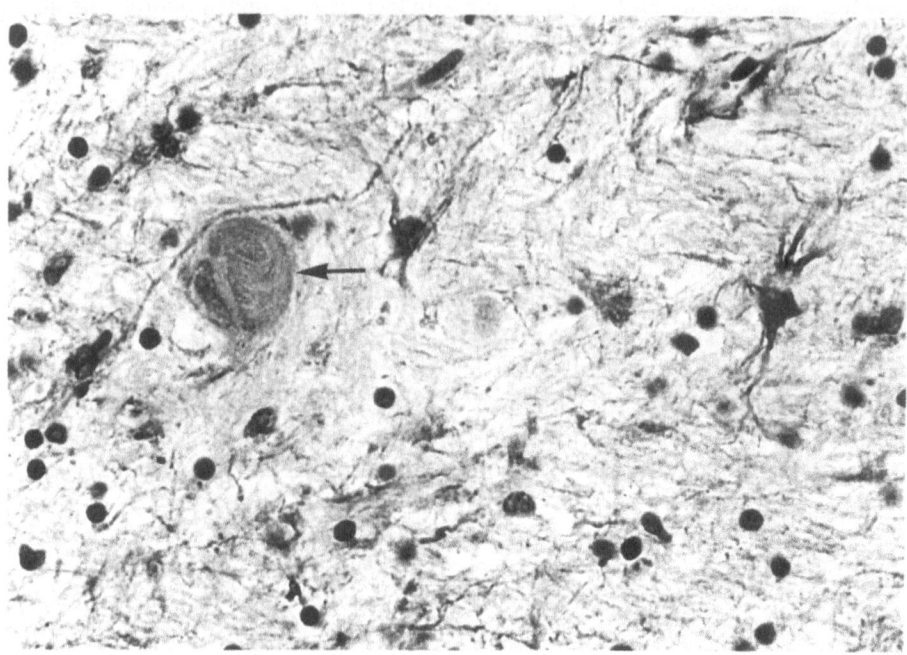

**Fig. 6.** Immunocytochemistry for glial fibrillary acidic protein demonstrates astrocytes in the globus pallidus. A neurofibrillary tangle (arrow) is also present. Avidin-biotin complex method. ×640

## Discussion

PSP is a relatively new, complex and somewhat enigmatic neurodegenerative disorder. Recent investigations, focused on the structural configuration and chemical composition of NFTs, the cellular hallmark of the disease, which are likely to hold the key to the understanding of the pathogenesis and perhaps the cause of PSP.

The distribution of NFTs indicates that this cytoskeletal abnormality affects neuronal populations more extensively than originally appreciated. The occurrence of NFTs in the cerebral cortex is of particular relevance, since it raises two important problems. The first is the neuropathological differential diagnosis, since PSP is most frequent in the fifth and sixth decades when tangle formation may occur both in normal ageing and in Alzheimer's disease. However, there are obvious differences in the distribution, structural configuration and chemical composition, between the tangles of PSP and Alzheimer's disease. Moreover, plaque formation, the other histological hallmark of Alzheimer disease is not a prominent feature of PSP, although βA4 deposition may occur. The second issue is a clinical one. Considering neocortical involvement, the concept of subcortical dementia in PSP should be reassessed. The possibility that cortical pathology may contribute to cognitive deterioration cannot be excluded. Another more recently described structural abnormality which may also be associated with the development of dementia is the widespread occurrence of neuropil threads. These were found in the cortex by one group (Hauw et al., 1990), but not by another (Probst et al., 1988). Further neuropathological investigations of the cerebral cortex may clarify this issue by establishing the full range of cortical involvement.

Neuropathological examinations have already revealed that PSP does not always follow the clinical pattern, as originally defined, but may also present, clinically and pathologically, in different, atypical forms. The recognition of these variations is, without doubt, of paramount importance in the complex differential diagnosis of neurodegenerative disorders. That the cortex may be involved in PSP is now little disputed, but severe cortical atrophy with marked neuronal loss accompanied by gliosis as reported in a woman of 60 years of age is extremely rare (Akashi et al., 1989).

The typical pathology of PSP may be associated with Alzheimer type changes and the co-existence of two neurodegenerative diseases, resulting in dementia, may present differential diagnostic problems (Milder et al., 1984). A recent study have found neuropathological features of both PSP and Alzheimer's disease in two patients who had classical clinical features of PSP, and showed evidence of severe, progressive dementia (Cruz-Sanchez et al., 1992). In another case of a woman of 65 years of age, the limbic system was particularly affected: there were many tangles with neuronal loss in the hippocampus, parahippocampal gyrus, amygdala, subcallosal area, anterior perforated substance, cingulate gyrus and anterior olfactory nucleus (Takahashi et al., 1987). However, the precise assessment and distribution of Alzheimer type pathology in these cases are of great importance, since ageing changes which particularly involve the medial temporal structures should be excluded, before the double pathology of Alzheimer's disease and PSP is accepted.

PSP may also be complicated by the occurrence of Lewy bodies both in the brainstem nuclei and cerebral cortex (Mori et al., 1986). PSP may share neuropathological features with post-encephalitic Parkinson's syndrome. In a review of six cases of PSP without ophthalmoplegia few neurofibrillary tangles were found, and lesions in the tectal, periaqueductal and reticular structures were less severe than in 10 clinically and pathologically typical cases of PSP. Moreover, the abnormalities were more severe in the substantia nigra, globus pallidus and the subthalamic nucleus, changes similar to those of pallido-luyso-nigral atrophy (Dubas et al., 1983). That the histological features of PSP may be intermingled with those of pallido-luyso-nigral atrophy has been demonstrated by a more recent case of a patient whose parkinsonism was dominated by akinesia (Yamamoto et al., 1991).

Clinically typical cases of PSP may also show unusual neuropathology, including extensive cortical and subcortical gliosis (Will et al., 1988) as demonstrated by a case subsequently classified as progressive subcortical gliosis. Two patients, who also suffered from carcinomas, clinically developed gait disturbances, dementia and/or dysarthria, but no abnormalities of eye movements. Neuropathological changes were more limited than

**Table 1.** Neuropathological classification of PSP

| | |
|---|---|
| Type 1 | Typical cases |
| Type 2 | Atypical cases |
| Type 3 | Combined cases |

usual and subsequently these cases were considered as early forms of PSP (Kleinschmidt-DeMasters, 1989). Recently, we noted the coincidence of a progressive extrapyramidal syndrome and dementia in three members of a family. One of the patients died and his brain showed features of PSP. This case is of particular importance, since it suggests that PSP may occur as a familial disease or alternatively, there is a novel form of familial dementia which shows pathological features of PSP (Brown et al., 1993).

These examples clearly indicate that PSP may present a wider neuro-pathological spectrum than previously considered. Whilst the typical examples cause no diagnostic problems, it is the rare atypical case which presents a real challenge to the neuropathologist. These wide morphological variations, including the type, distribution and severity of lesions, warrant a neuropathological classification of PSP cases. The following simple classification distinguishes three types: typical, atypical and combined cases (Table 1). Typical cases conform to the original definition of PSP (Steele et al., 1964). Type 2 cases are variants of the histological changes characteristic of PSP: either the severity or the distribution of abnormalities, or both of these deviate from the typical pattern. In this group often there is more severe involvement of the cortex. Cases with combined pathology belong to type 3 group: in these the typical pathology of PSP is accompanied by lesions characteristic of another neurodegenerative disease (e.g., the presence of Lewy bodies or Alzheimer type changes over and above the level expected in normal ageing) or cerebral vascular disease.

Neuropathologically, the differential diagnosis of PSP should include Alzheimer's disease, Parkinson's disease, post-encephalitic parkinsonism, the parkinsonism-dementia complex of Guam, corticobasal degeneration and multiple system atrophy. The differential diagnosis of PSP will be discussed in another chapter.

The aetiology of PSP remains an enigma. Steele, Richardson and Olszewski (1964) originally considered either a primary degeneration or a post-infection process, most likely caused by a virus. However, subsequent investigations did not support the infectious hypothesis: there is no inflammatory response associated with traditional viruses and no history of a previous encephalitis, the ultrastructure of neurofibrillary tangles is different from those to be seen in other post-infectious diseases and finally PSP could not be transmitted to primates. Similarly, evidence for environmental factors, including geographic, climatic, dietary or occupational factors is missing and no particular neurotoxin has been identified. Head injury and electroconvulsive therapy have been associated with PSP in a few cases, although no conclusive evidence has been reached. Genetic factors also do not appear to play an important role, but family history has been reported (Kristensen, 1985; Brown et al., 1993).

Current research has yielded interesting results on neurotransmitter and neuropeptide abnormalities in PSP (see Javoy-Agid, this volume, chapter 16). However, the basic underlying cellular and molecular mechanisms remain to be elucidated. The selective vulnerability of the neuronal groups affected in the disease and the chemical composition of the tangles are the likely keys to the enigma of PSP.

## Acknowledgements

This work was partly supported by a grant from the Medical Research Council. The author wishes to thank Dr. N. J. Cairns and Mr. A. Chadwick of the MRC Alzheimer's Disease Brain Bank, Department of Neuropathology, Institute of Psychiatry, and Mrs. E. Kemp for her secretarial assistance. The antibody to tau protein was kindly provided by Prof. B. H. Anderton, Department of Neuroscience, Institute of Psychiatry, London.

## References

Akashi T, Arima K, Maruyama N, Ando S, Inose T (1989) Severe cerebral atrophy in progressive supranuclear palsy: a case report. Clin Neuropathol 8: 195–199

Arai N (1989) "Grumose degeneration" of the dentate nucleus. A light and electron microscopic study in progressive supranuclear palsy and dentatorubropallidoluysial atrophy. J Neurol Sci 90: 131–145

Bancher C, Lassmann H, Budka H, et al (1987) Neurofibrillary tangles in Alzheimer's disease and progressive supranuclear palsy: antigenic similarities and differences. Acta Neuropathol 74: 39–46

Barr AN (1986) Progressive supranuclear palsy. In: Vinken PJ, Bruyn GW, Klawans HJ (eds) Handbook of clinical neurology, vol 5(49). Extrapyramidal disorders. Elsevier, Amsterdam, pp 239–254

Brown J, Lantos P, Stratton M, Roques P, Rossor M (1993) A familial dementia with progressive supranuclear palsy pathology. J Neurol Neurosurg Psychiatry 56: 473–476

Cruz-Sanchez FF, Rossi ML, Cardozo A, Deacon P, Tolosa E (1992) Clinical and pathological study of two patients with progressive supranuclear palsy and Alzheimer's changes. Antigenic determinants that distinguish cortical and subcortical neurofibrillary tangles. Neurosci Lett 136: 43–46

Dubas F, Gray F, Escourolle R (1983) Maladie de Steele-Richardson-Olszewski sans ophtalmoplégie. Six cas anatomo-cliniques. Rev Neurol 139: 407–416

Fearnley JM, Lees AJ (1991) Ageing and Parkinson's disease: substantia nigra regional selectivity. Brain 114: 2283–2301

Flament S, Delacourte A, Veray M, Hauw J-J, Javoy-Agid F (1991) Abnormal tau protein in progressive supranuclear palsy. Similarities and differences with the neurofibrillary degeneration of the Alzheimer type. Acta Neuropathol 81: 591–596

Giaccone G, Tagliavini F, Street JS, Ghetti B, Bugiani O (1988) Progressive supranuclear palsy with hypertrophy of the olives. An immunocytochemical study of the cytoskeleton of argyrophilic neurons. Acta Neuropathol 77: 14–20

Golbe LI, Davis PH, Schoenberg BS, Duvoisin RC (1988) Prevalence and natural history of progressive supranuclear palsy. Neurology 38: 1031–1034

Hauw J-J, Verny M, Delaère P, Cervera P, He Y, Duyckaerts C (1990) Constant neurofibrillary changes in the neocortex in progressive supranuclear palsy. Basic differences with Alzheimer's disease and aging. Neurosci Lett 119: 182–186

Hirsch EC, Graybiel AM, Duyckaerts C, Javoy-Agid F (1987) Neuronal loss in the pedunculopontine tegmental nucleus in Parkinson disease and in progressive supranuclear palsy. Proc Natl Acad Sci USA 84: 5976–5980

Hynd EW, Pirozzolo FJ, Maletta GJ (1982) Progressive supranuclear palsy. Int J Neurosci 16: 87–98

Jackson JA, Jankovic J, Ford J (1983) Progressive supranuclear palsy: clinical features and response to treatment in 16 patients. Ann Neurol 13: 273–278

Jellinger K, Riederer P, Tomonaga M (1980) Progressive supranuclear palsy: clinico-pathological and biochemical studies. J Neural Transm [Suppl] 16: 111–128

Jellinger K (1971) Progressive supranuclear palsy (subcortical argyrophilic dystrophy). Acta Neuropathol 19: 347–352

Jellinger K (1988) The pedunculopontine nucleus in Parkinson's disease, progressive supranuclear palsy and Alzheimer's disease. J Neurol Neurosurg Psychiatry 52: 540–543

Kato T, Hirano A, Weinberg MN, Jacobs AK (1986) Spinal cord lesions in progressive supranuclear palsy: some new observations. Acta Neuropathol 71: 11–14

Kleinschmidt-DeMasters BK (1989) Early progressive supranuclear palsy: pathology and clinical presentation. Clin Neuropathol 8: 79–84

Kristensen MO (1985) Progressive supranuclear palsy—20 years later. Acta Neurol Scand 71: 177–189

Leigh PN, Probst A, Dale GE, et al (1989) New aspects of pathology of neuro-degenerative disorders as revealed by ubiquitin antibodies. Acta Neuropathol 79: 61–72

Lennox G, Lowe J, Morrell K, Landon M, Mayer RJ (1988) Ubiquitin is a component of neurofibrillary tangles in a variety of neurodegenerative diseases. Neurosci Lett 94: 211–217

Love S, Saitoh T, Quijada S, Cole GM, Terry RD (1988) Alz-50, ubiquitin and tau immunoreactivity of neurofibrillary tangles, Pick bodies and Lewy bodies. J Neuropathol Exp Neurol 47: 393–405

Maher ER, Lees AJ (1986) The clinical features and natural history of the Steele-Richardson-Olszewski syndrome (progressive supranuclear palsy). Neurology 36: 1005–1008

Malessa S, Hirsch EC, Cervera P, et al (1991) Progressive supranuclear palsy: loss of choline acetyltransferase-like immunoreactive neurons in the pontine reticular formation. Neurology 41: 1593–1597

Manetto V, Perry G, Tabaton M, et al (1988) Ubiquitin is associated with abnormal cytoplasmic filaments characteristic of neurodegenerative disease. Proc Natl Acad Sci USA 85: 4501–4505

Mann DMA, Yates PO, Hawkes J (1983) The pathology of the human locus coeruleus. Clin Neuropathol 2: 1–7

Mann DMA, Jones D (1990) Deposition of amyloid (A4) protein within the brains of persons with dementing disorders other than Alzheimer's disease and Down's syndrome. Neurosci Lett 109: 68–75

Mastaglia FL, Grainger K, Kee F, Sadka M, Lefroy R (1973) Progressive supranuclear palsy (the Steele-Richardson-Olszewski syndrome): clinical and electrophysiological observations in eleven cases. Proc Aust Assoc Neurol 10: 35–44

Milder DG, Elliott CF, Evans WA (1984) Neuropathological findings in a case of coexistent progressive supranuclear palsy and Alzheimer's disease. Clin Exp Neurol 20: 181–187

Mizusawa H, Yen S-H, Hirano A, Llena JF (1989) Pathology of the dentate nucleus in progressive supranuclear palsy: a histological, immunohistochemical and ultra-structural study. Acta Neuropathol 78: 419–428

Montpetit V, Clapin DF, Guberman A (1985) Substructure of 20 nm filaments of progressive supranuclear palsy. Acta Neuropathol 68: 311–318

Mori H, Yoshimura M, Tomonaga M, Yamanouchi H (1986) Progressive supranuclear palsy with Lewy bodies. Acta Neuropathol 71: 344–346

Oyanagi K, Takahashi H, Wakabayashi K, Ikuta F (1988) Selective decrease of large neurons in the neostriatum in progressive supranuclear palsy. Brain Res 458: 218–223

Probst A, Langui D, Lautenschlager C, Ulrich J, Brion JP, Anderton BH (1988) Progressive supranuclear palsy: extensive neuropil threads in addition to neurofibrillary tangles. Very similar antigenicity of subcortical neuronal pathology in progressive supranuclear palsy and Alzheimer's disease. Acta Neuropathol 77: 61–68

Pollock NJ, Mirra SS, Binder LI, Hansen LA, Wood JG (1986) Filamentous aggregates in Pick's disease, progressive supranuclear palsy, and Alzheimer's disease share antigenic determinants with microtubule-associated protein, tau. Lancet ii: 1211

Posey WC (1904) Paralysis of the upward movement of the eyes. Ann Ophthal 13: 523–531

Schmidt MK, Lee VM-Y, Hurtig H, Trojanowski JQ (1988) Properties of antigenic determinants that distinguish neurofibrillary tangles in progressive supranuclear palsy and Alzheimer's disease. Lab Invest 59: 460–466

Shin R-W, Kitamoto T, Tateishi J (1991) Modified tau is present in younger non-demented persons: a study of subcortical nuclei in Alzheimer's disease and progressive supranuclear palsy. Acta Neuropathol 81: 517–523

Spiller WG (1905) The importance in clinical diagnosis of paralysis of associated movements of the eyeballs (Blick-Lahmung), especially of upward and downward associated movements. J Nerv Ment Dis 32: 417–448 and 497–530

Steele JC, Richardson JC, Olszewski T (1964) Progressive supranuclear palsy. Arch Neurol 10: 333–359

Steele JC (1975) Progressive supranuclear palsy. In: Vinken PJ, Bruyn GW, deJong JMB (eds) Handbook of clinical neurology, vol 22. System disorders and atrophies, part II North Holland, Amsterdam, pp 217–229

Takahashi H, Oyanagi K, Takedo S, Hinokuma K, Ikuta F (1989) Occurrence of 15-nm—wide straight tubules in neocortical neurons in progressive supranuclear palsy. Acta Neuropathol 79: 233–239

Takahashi H, Takeda S, Ikuta F, Homma Y (1987) Progressive supranuclear palsy with limbic system involvement: report of a case with ultrastructural investigation of neurofibrillary tangles in various locations. Clin Neuropathol 6: 271–276

Tagliavini F, Pilleri G, Bouras C, Constantinidis J (1984) The basal nucleus of Meynert in patients with progressive supranuclear palsy. Neurosci Lett 44: 37–42

Tan N, Mastaglia FL, Masters CL, Beyreuther K, Kakulas BA (1988) Amyloid (A4) protein deposition in brain in progressive supranuclear palsy (PSP). Alzheimer Dis Assoc Dis 2: 264

Tellez-Nagel I, Wisniewski HM (1973) Ultrastructure of neurofibrillary tangles in Steele-Richardson-Olszewski syndrome. Arch Neurol 29: 324–327

Tomonaga M (1977) Ultrastructure of neurofibrillary tangles in progressive supranuclear palsy. Acta Neuropathol 37: 177–181

Tomonaga M (1983) Neuropathology of the locus coeruleus: a semi-quantitative study. J Neurol 230: 231–240

Will RG, Lees AJ, Gibb W, Barnard RO (1988) A case of progressive subcortical gliosis presenting clinically as Steele-Richardson-Olszewski syndrome. J Neurol Neurosurg Psychiatry 51: 1224–1227

Yagishita S, Itoh Y, Amano N, Nakano T, Saitoh A (1979) Ultrastructure of neurofibrillary tangles in progressive supranuclear palsy. Acta Neuropathol 48: 27–30

Yagashita S, Itoh Y, Nakano T (1986) Hypertrophy of the olivary nucleus. An ultrasctructural study. Acta Neuropathol 69: 132–138

Yamamoto Y, Kawamura J, Hashimoto S, Nakamura H, Kobashi Y, Ichijima K (1991)
    Pallido-nigro-luysian atrophy, progressive supranuclear palsy and adult onset
    Hallervorden-Spatz disease: a case of akinesia as a predominant feature of par-
    kinsonism. J Neurol Sci 101: 98–106
Zweig RM, Whitehouse PJ, Casanova MF, Walker LC, Jankel WR, Price DL (1987)
    Loss of pedunculopontine neurons in progressive supranuclear palsy. Ann Neurol
    22: 18–25

Author's address: Dr. P. L. Lantos, Department of Neuropathology, Institute of
Psychiatry, London SE5 8AF, United Kingdom.

# Neurofibrillary pathology in progressive supranuclear palsy (PSP)

## J. Cervós-Navarro and K. Schumacher

Institute of Neuropathology, Free University of Berlin, Federal Republic of Germany

**Summary.** In progressive supranuclear palsy (PSP), globose neurofibrillary tangles (NFT) are found in the subcortical areas and occasionally in the central cortex and spinal cord. An inverse relationship was found between the degree of neuronal loss and the presence of NFT.

It has been postulated that NFT comes first and atrophy as a secondary event. Others authors have reported that the neurologic findings are associated with the presence of carcinomas and the CNS changes can be assessed as a paraneoplastic effect. In PSP the neuritic changes are mainly located in the basal ganglia and composed of straight filaments and tubules, different from the paired helical filaments found in the Alzheimer's disease, suggesting that they are formed of a new type of fibrous protein. In addition immunohistochemistry preparations using antibodies against tau and ubiquitin reveal an antigenic profile similar to early NFT in dementia of Alzheimer's type. These findings support the hypothesis that these changes may reflect different types of non-specific cytoskeletal disorganization.

Neuropathologically apart from mild to moderate neuronal loss and gliosis in various parts of the basal ganglia, brain stem and cerebellum the Steele-Richardson-Olszewski syndrome is characterized by the presence of neurofibrillary tangles (NFT) in various nerve cells of the affected areas. Most of the neurofibrillary tangles are subcortical i.e. of globose type. (Fig. 1A) and differed from the tangles in the cortex (Fig. 1B).

Generally, it is said that the number of NFT-bearing cells is proportional to cell loss (Steele et al., 1964; Kristensen, 1985), however, a certain inverse relationship was noted between the degree of neuronal loss and that of neuronal "Alzheimerization". Occasionally nerve cell wastage has been reported most severe where neurofibrillary degeneration was rare or absent: conversely, many tangles were found where neuronal dropout was slight or absent.

Seitelberger (1969), in a study of heterogeneous system degeneration, puts forward the idea that neurofibrillary degeneration comes first and atrophy is only a secondary phenomenon. Kleinschmidt-DeMasters (1989) reported two early cases of PSP, lacking ophthalmoplegia and unrecognized in life. The patients developed their neurologic findings in association with, or exacerbated by, carcinomas an observation made elsewhere and speculated to be a paraneoplastic effect (Jankovic, 1985). The neuropathologic findings were limited to the presence of numerous NFTs (up to 25 per slide). In early cases, however, the NFTs are prominent even before moderate or severe cell loss had occurred. Anzil (1969) postulated that the proportion of cell loss to NFT-bearing cells depends on the time period in which the disorder is analyzed.

### Localization

The main localization of tangles in PSP are the third nuclear complex, n. supratrochlearis (Fig. 1A), n. centralis superior, locus coeruleus and nucleus basalis of Meynert. A minor number of tangles are also present in thalamus, globus pallidus, subthalamic nucleus, pons, inferior olivary nucleus and Fascia dentata (Arima et al., 1992a). In the cerebellum the nucleus dentatus is often involved (Mizusawa et al., 1989).

In some cases, NFTs have also been described in the cerebral cortex (Takahashi et al., 1989). However only in few cases the number has been considerable (Dubas and De Recondo, 1983; Ishino and Otsuki, 1976; Ishino et al., 1987). A quantitative regional and laminar analysis of the distribution of the lesions in the cortex in PSP showed the neurofibrillary tangle distribution in the cerebral cortex largely confined to the hippocampal formation, especially in the parahippocampal gyri and other limbic areas (Amando et al., 1989). In particular, the granule cell layer of the dentate gyrus is mostly involved. In the prefrontal and inferior temporal cortex, neurofibrillary tangles were predominantly distributed in layers II

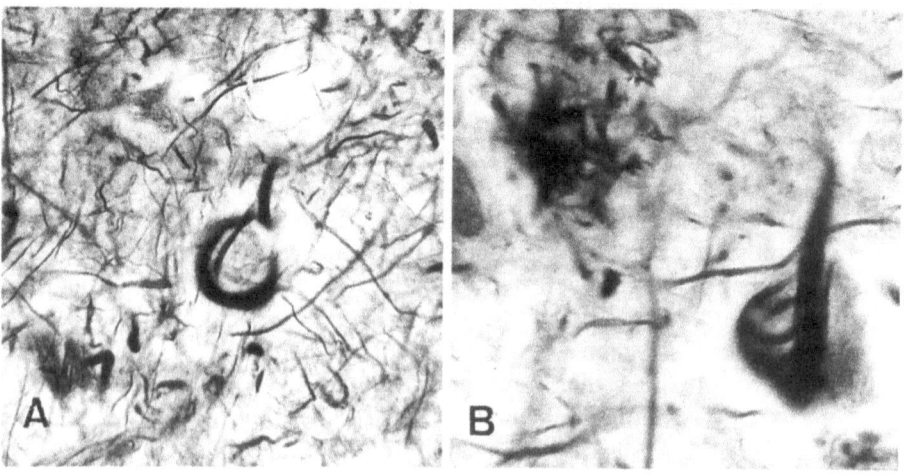

**Fig. 1. A** Flame-shaped argentophilic neurofibrillary tangle in a pyramidal neuron of the frontal cortex (Bielschowsky's stain, original magnification 400). **B** Neurofibrillary tangle of globose type, in a median raphe neuron (Bielschowsky's stain, original magnification 400)

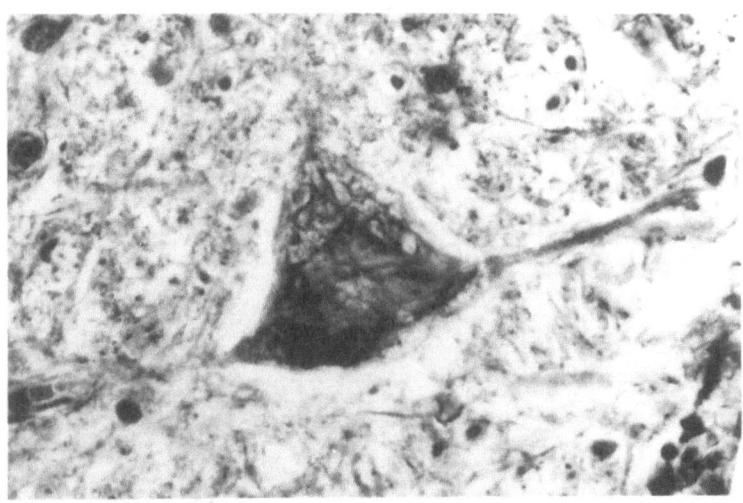

**Fig. 2.** Neurofibrillary tangle in a motorneuron of the spinal cord, moderate argentophilia of the neurofibrils forming the whorls of the tangle (Bielschowsky's stain, original magnification 400)

and III. In addition, there were moderate-to-high neurofibrillary tangle densities in the primary motor cortex (Hof et al., 1992).

In the spinal cord neurofibrillary tangles were seen most frequently in the posterior horn as well as in a few neurons, in the anterior horn (Fig. 2), lateral horn, Clark's column, and intermediate gray (Kato et al., 1986).

In addition to the perikaryal staining Probst et al. (1988) reported extensive networks of Gallyas-positive, tau- and PHF-immunoreactive neurites in subcortical gray areas containing NFT, and bundles of positive axons in white matter tracts interconnecting subcortical nuclei of PSP. The short, curvilinear neuritic profiles were described first by Braak et al. (1986) as neuropil threads in Alzheimer's Disease (AD). Nelson et al. (1989) found that treatment of frozen sections with phosphatase enhanced staining of NFT and neuropil threads by Tau-1. Both NFT and neuropil threads were immunoreactive with the ubiquitin antiserum. The results suggest that extensive neuritic degeneration in the neuropil is not unique to AD.

However, in contrast to AD where neuropil threads are located in the cortex and contain paired helical filaments, the neuritic change in PSP is in the basal ganglia and is composed of straight filaments.

### Ultrastructure

The ultrastructural appearance of the classical type of neurofibrillary tangles (Fig. 3) has been found to be similar in senile and presenile dementia (Terry and Wisniewski, 1970; Wisniewski et al., 1970), Parkinson-dementia complex of Guam (Hirano et al., 1961), Down's syndrome (Olson and

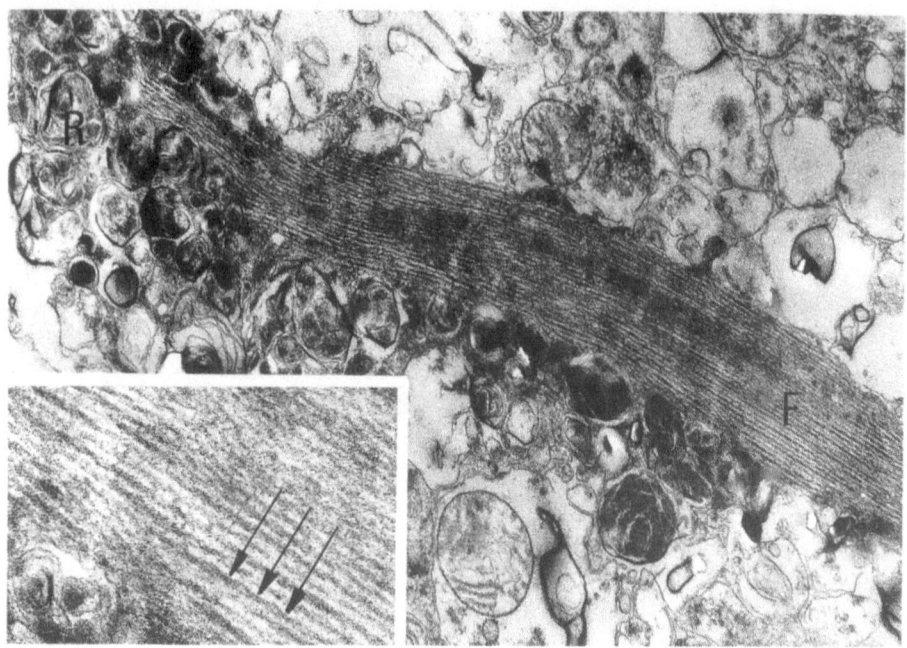

**Fig. 3.** Densely packed paired helical filaments in a neuronal dendrite in Alzheimer dementia. Inset shows the constriction of the fibrils (arrowheads), (original magnification 80,000)

Shaw, 1969; Ohara, 1972; Schochet et al., 1973; Burger and Vogel, 1973), and amyotrophic lateral sclerosis (Meyers et al., 1974).

However, electron microscopic observations of neurofibrillary tangles in the Steele-Richardson-Olszewski syndrome reported 15 nm straight tubules (Fig. 4), an ultrastructure different from the tangles found in any other disease. This observation seemed to indicate that a new type of fibrous protein, different from that seen in tangles in other disorders, accumulates in the nerve cells in the PSP. The occurence of neurofibrillary tangles (NFT) composed of straight tubules was therefore regarded as the hallmark of the progressive supranuclear palsy (Tellez-Nagel and Wisniewski, 1973; Powell et al., 1979; Roy et al., 1974; Bugiani et al., 1979; Jellinger et al., 1980).

However, Tomonaga (1977) and Yagishita et al. (1979) reported cases of PSP in which some of the NFT were composed of paired helical filaments of Alzheimer type. Ghatak et al. (1980) described in some cases twisted fibrillary structures mixed with 15 nm straight filaments. Takauchi et al. (1983) observed paired helical filaments in the globus pallidus, subthalamic nucleus, substantia nigra, and pontine tegmentum of a typical case of PSP. Each filament had a diameter of 10–12 nm and showed central low density and a smooth contour. The thickest portion of a pair was 22–24 nm in diameter. Its periodicity of twist differed from the periodicity of the twisted tubules in Alzheimer disease in which a periodicity of 80 nm is predominant.

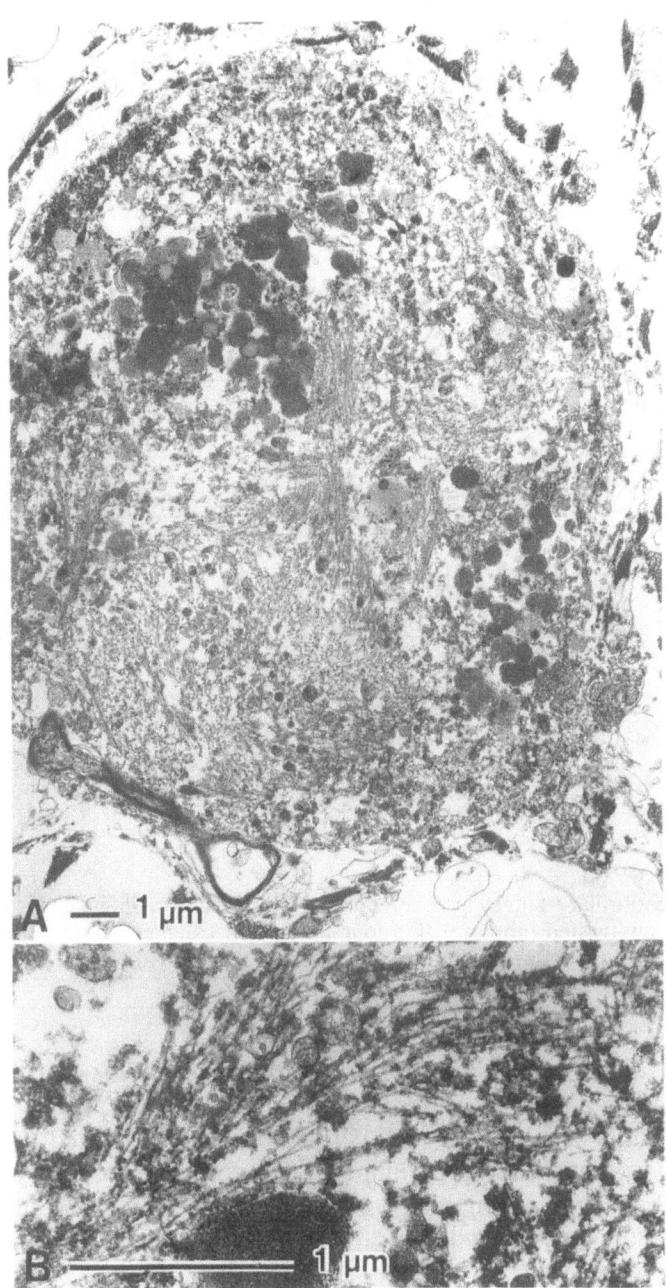

**Fig. 4.** Fine structure of a PSP-type NFT in the inferior olivary nucleus. **A** A PAS-type NFT penetrates through the lipofuscin and other cell organells. It consists of fine filamentous structures that run in a near-parallel fashion and form loose bundles. **B** High-power view of A. Filamentous structures are tubular, smooth in their outer surface, and 13–15 nm in diameter. Bars = 1 m. (After Arima et al., 1992b)

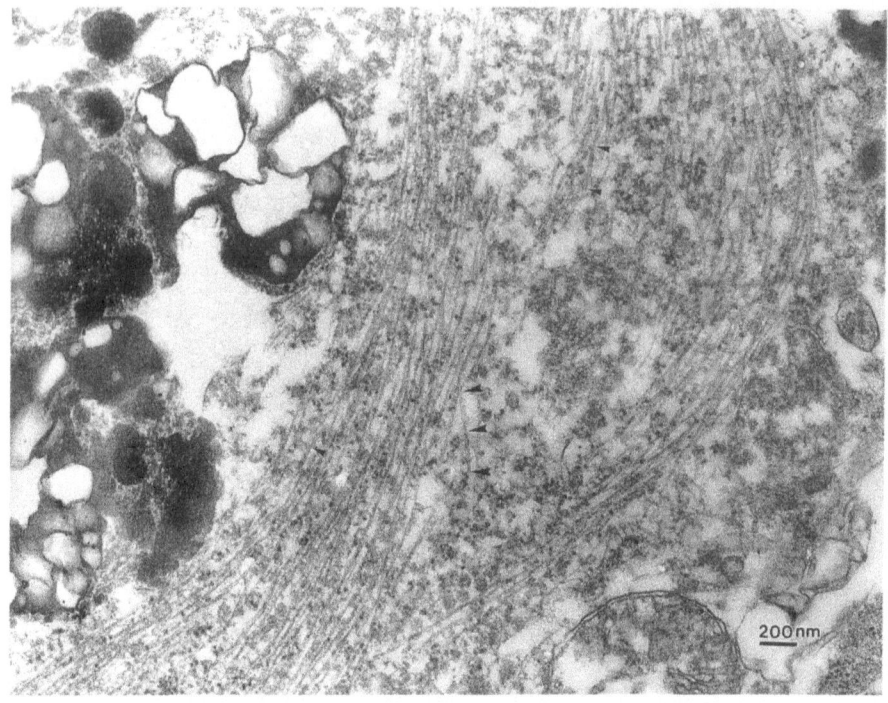

**Fig. 5.** Neurofibrillary tangle in a cerebellar dentate neuron showing 15-nm straight tubules and few twisted tubules with a long periodicity of about 200 nm (arrowheads). Bar = 200 nm. (After Yamamoto, 1990)

The ultrastructural finding is unusual in the neurofibrillary pathology of PSP, and has not been confirmed. In the cases of Amando et al. (1989) ultrastructurally, tangles of frontal cortex consisted of twisted tubules and those of hippocampal and parahippocampal cortices consisted of straight and twisted tubules, beeing observed separately in neurons. Yamamoto et al. (1992) found straight and twisted tubules in the nucleus dentatus of the cerebellum (Fig. 5).

## Immunohistochemistry

Despite their morphological differences, paired helical filaments and the straight fibers of PSP are made of substances that are closely related to some normal brain elements and to each other.

Yen et al. (1983) tested antiserum raised against human brain microtubule fraction that binds specifically to tangles of Alzheimer's dementia labelled also the tangles composed of 15 nm straight fibers in PSP. Similar results have been found in antibodies against a large variety of neurofilament associated antigens (Tabaton et al., 1988; Cammarata et al., 1990; Gheuens et al., 1991). Particularly, similar as in Alzheimer Dementia NFTs in PSP strongly react to antibodies against tau. However, Galloway (1988) as well as Flament et al. (1991) showed that abnormal tau types produced in PSP are different from those in AD. Although the presence of some ubiquitin epitopes has been described in PSP tangles (Manetto et al., 1988), the ubiquitin epitope, an antigenetic determinant which appears in late stages of tangle maturation in AD, is absent in PSP tangles. PSP tangles, thus, reveal an antigenetic profile, similar to early tangles in Alzheimer Dementia (Bancher et al., 1987).

Amyloid P (AP) component is present in all types of systemic amyloid deposits. The globose tangles in the brain stem from the PSP are also stained by the AP antiserum. Electron microscopy confirms that AP immunoreactivity is associated with a variety of abnormal filaments (Kalaria et al., 1991).

Perry et al. (1992) found that the NFTs in progressive supranuclear palsy contained heparinase sensitive basic fibroblast growth factor (bFGF) binding sites indicating that heparin sulfate proteoglycans (HSPG) interactions and possible role in the formation of intraneuronal inclusions are not limited to Alzheimer's disease.

### Diagnostic value and clinical relevance

The presence of neurofibrillary tangles is not a specific abnormality and can be found in many neurologic disorders (Hirano, 1970; Wisniewski and Terry, 1970). Alzheimer's tangles composed of masses of paired helical filaments occur in a number of not obviously related disorders of the adult and the aged, such as encephalitic parkinsonism. They are also observed in children and adolescence with chronic SSPE. In the infantile leukodystrophy of sudanophilic type the Alzheimer's tangles were present in the cerebral cortex, as well as in the brain stem, basal ganglia, and hypothalamus including the n. basalis of Meynert (Harada et al., 1986).

The characteristic globose neurofibrillary tangles can be found in other conditions than PSP and take their shape from the type of neuron that bears them (Steele, 1972). Hence, other disorders of neurofibrillary degeneration that involve the brain stem have similar neuropathological findings on light microscopy. In cases of presenile AD numerous globose neurofibrillary tangles were described in the rostral mesencephalon (Hunter, 1985), substantia nigra and nucleus centralis superior (Tabaton et al., 1985), locus coeruleus and nucleus supratrochlearis (Shortridge et al., 1985). Hain et al. (1990) stressed the fact that neocortical neurofibrillary tangles in progressive supranuclear palsy differed from AD or age-related changes in their localization being most frequent in the precentral gyrus (Brodmann's area 4) whereas associated areas are predominantly lesioned in AD. Furthermore, they affected mainly large pyramidal neurons and small cells, relatively sparing the cell population selectively involved in AD and they predominated in layers V and VI of area 4, whereas in AD are more dense in layers III and V.

Masliah et al. (1991) reported 6 patients with progressive dementia that neuropathologically showed argyrophilic grains and subcortical tangles (not composed of PHF) in a distribution similar to PSP.

Since the first description of Rebeiz et al. (1967) many patients with clinical and pathological characteristic for PSP presented also lesions of corticonigral degeneration with neuronal achromasia and Pick cells have been reported (Paulus and Selium, 1990). Cases with the pathological diagnosis of progressive supranuclear palsy and coincident Pick bodies, balloon cells, and Alzheimer's changes were reported by different authors. Arima et al. (1992b) demonstrated in a single case neurofibrillary tangles of PSP type in the Edinger-Westphal nucleus, locus coeruleus, cerebellar dentate nucleus, inferior olivary nucleus (Fig. 5) as well as the posterior horn of the spinal cord and Pick bodies in the atrophied cerebral cortex and red nucleus. This cases as well as others reinforce the likelihood that certain neuropathologic changes once assumed entity-specific may be relatively nonspecific degenerative alterations. Despite their pathognomonic significance in certain disorders the different structures may reflect a form of cytoskeletal disorganization, which is not entirely restricted to a single disease entity.

Neurofibrillary tangles were considered to be responsible for memory disturbance and dementia-like symptoms, which can be correlated clinically with disorders of attentiveness, wakefulness and consciousness, indicating reticular activating system lesions. However, Ishino et al. (1987) are of the opinion that it is difficult to consider that it influences mental symptoms, because most of the tangles showed argyrophilic filaments which occupied only a part of the neuronal cytoplasm or thickened fibers which coiled around the well-preserved nucleus.

## Astrocytic tangles

Abe et al. (1992) described a large number of argyrophilic structures best detected with Galyas method, in addition to neurofibrillary tangles. Electronmicroscopically they consisted to straight tubular structures (15–20 nm in diameter) and were immunoreactive for anti-tau antibody as well as glial fibrillary acidic protein (GFAP). The authors thought that they were located in astrocytes and called them astrocytic tangles (ACT). Their distribution corresponded to the lesions associated with PSP and the interconnecting fascicles between them. They postulated that the appearance of ACT is one of the characteristic features in PSP. Yamada et al. (1992) identified many Tau-positive glia with paired nuclei and astrocyte type morphology in three brains from patients with PSP. They were positive by Bielschowsky's and Bodians's silver staining as well as by immunostaining with Tau-2, Alz-50, anti GFAP and anti-paired helical filament antibodies, but not with anti-ubiquitin antibody. They were predominantly localized in the striatum, thalamus and frontal cortex but were not seen in white matter and were not plentiful in areas of heavy neuronal degeneration. Electron microscopy clearly showed the nuclear pairing and localized the Tau protein to bundles suggestive of microtubules in the cytoplasm and proximal processes. Such glial cells were rarely seen in cases of other neurodegenerative diseases or neurologically normal controls. These data suggest that there is an unusual gliotic reaction in PSP in brain areas which show relatively little neuronal loss.

## References

Abe H, Yagishita S, Amano N, Bise K (1992) Ultrastructural and immunohistochemical study of "astrocytic tangles" (ACT) in patients with progressive supranuclear palsy. Clin Neuropathol 11: 278

Amano N, Iwabuchi K, Yokoi S, Yagishita S, Itoh J, Saitoh A, Nagatomo H, Matsushita M (1989) The reappraisal study of the ultrastructure of Alzheimer's neurofibrillary tangles in three cases of progressive supranuclear palsy. No To Shinkei Brain and Nerve 41: 35–44

Anzil AP (1969) Progressive supranuclear palsy. Case report with pathological findings. Acta Neuropathol (Berl) 14: 72–76

Arima K, Murayama S, Oyanagi S, Akashi T, Inose T (1992a) Presenile dementia with progressive supranuclear palsy tangles and Pick bodies: an unusual degenerative disorder involving the cerebral cortex, cerebral nuclei, and brain stem nuclei. Acta Neuropathol (Berl) 84: 128–134

Arima K, Oyanagi S, Akashi T, Sakata C, Sunohara N, Inose T (1992b) Neurofibrillary tangles of progressive supranuclear palsy in the dentate fascia: an ultrastructural study of a case. Neuropathology 12: 51–57

Bancher C, Lassmann H, Budka H, Grundke-Iqbal I, Iqbal K, Wiche G, Seitelberger F, Wisniewski HM (1987) Neurofibrillary tangles in Alzheimer's disease and progressive supranuclear palsy: antigenetic similarities and differences. Acta Neuropathol (Berl) 74: 39–46

Braak H, Braak E, Iqbal IG, Iqbal K (1986) Occurrence of neuropil threads in the senile human brain and in Alzheimer's disease: a third location of paired helical filaments outside of neurofibrillary tangles and neuritic plaques. Neurosci Lett 65: 351–355

Bugiani O, Macardi GL, Brusa A, Ederli A (1979) The fine structure of subcortical neurofibrillary tangles in progressive supranuclear palsy. Acta Neuropathol (Berl) 45: 147–152

Burger PC, Vogel FS (1973) The development of the pathologic changes of Alzheimer's dementia and senile dementia in patients with Down's syndrome. Am J Pathol 73: 457–476

Cammarata S, Mancardi G, Tabaton M (1990) Formic acid treatment exposes hidden neurofilament and tau epitopes in abnormal cytoskeletal filaments from patients with progressive supranuclear palsy and Alzheimer's disease. Neurosci Lett 115: 351–355

Dubas F, De Recondo J (1983) Paralysie supra-nucléaire progressive. Maladie de Steele, Richardson, Olszewski. Encyc Méd Chir (Paris) Neurologie, 17062 B-10

Flament S, Delacourte A, Verny M, Hauw JJ, Javoy-Agid F (1991) Abnormal Tau proteins in progressive supranuclear palsy. Similarities and differences with the neurofibrillary degeneration of the Alzheimer type. Acta Neuropathol 81: 591–596

Galloway PG (1988) Antigenic characteristics of neurofibrillary tangles in progressive supranuclear palsy. Neurosci Lett 91: 148–153

Ghatak NR, Nochlin D, Hadfield MG (1980) Neurofibrillary pathology in progressive supranuclear palsy. Acta Neuropathol (Berl) 52: 73–76

Gheuens J, Cras P, Perry G, Boons J, Ceuterickdegroote C, Lubke U, Mercken M, Tabaton M, Gambetti PL, Vandermeeren M, Mulvihill P, Siedlak S, Vanheuverswijn H, Martin JJ (1991) Demonstration of a novel neurofilament associated antigen with the neurofibrillary pathology of Alzheimer and related diseases. Brain Res 558/N1: 43–52

Harada AK, Krucke GL, Mancardi TI (1986) Alzheimer's tangles in sudanophilic leucodystrophy. J Neuropathol Exp Neurol 45: 349

Hauw JJ, Verny M, Delaere P, Cervera P, He Y, Duyckaerts C (1990) Constant neurofibrillary changes in the neocortex in progressive supranuclear palsy. Basic differences with Alzheimer's disease and aging. Neurosci Lett 119: 182–186

Hirano A (1970) Neurofibrillary changes in conditions related to Alzheimer's disease. In: Wolstenholme GEW, O'Connor (eds) Alzheimer's disease and related conditions. Ciba Foundation Symposium. Churchill, London, pp 185–207

Hirano A, Malamud N, Kurland LT (1961) Parkinsonism-dementia complex, an endemic disease in the island of Guam. II. Pathological features. Brain 84: 662–679

Hof PR, Delacourte A, Bouras C (1992) Distribution of cortical neurofibrillary quantitative analysis of six cases. Acta Neuropathol (Berl) 84: 45–51

Hunter S (1985) The rostral mesencephalon in Parkinson's disease and Alzheimer's disease. Acta Neuropathol (Berl) 68: 53–58

Ishino H, Otsuki S (1976) Frequency of Alzheimer's neurofibrillary tangles in the cerebral cortex in progressive supranuclear palsy (subcortical argyrophilic dystrophy). J Neurol Sci 28: 306–316

Ishino H, Sasaki T, Yamashita K, Seno H, Kodaka H, Yoshinaga J, Ideshita H, Yamanaka T, Hikiji A (1987) A case of progressive supranuclear palsy with fibrillary gliosis of the midbrain and pontine reticular formation. Clin Neuropathol 6: 61–66

Jankovic J (1985) Progressive supranuclear palsy: paraneoplastic effect of bronchial carcinoma. Neurology 35: 446–447

Jellinger K, Riederer R, Tomonaga M (1980) Progressive supranuclear palsy. Clinical, pathological and biochemical studies. J Neural Transm [Suppl] 16: 111–128

Kalaria RN, Galloway PG, Perry G (1991) Widespread serum amyloid P immunoreactivity in cortical amyloid deposits and the neurofibrillary pathology of Alzheimer's disease and degenerative disorders. Neuropathol Appl Neurobiol 17: 189–201

Kato T, Hirano A, Jacobs AK, Weinberg AK, Weinberg MN (1986) Spinal cord lesions in progressive supranuclear palsy: some new observations. J Neuropathol Exp Neurol 45: 377

Kleinschmidt-DeMasters BK (1989) Early progressive supranuclear palsy: pathology and clinical presentation. Clin Neuropathol 8: 79–84

Kristensen MO (1985) Progressive supranuclear palsy — 20 years later. Acta Neurol Scand 71: 177–189

Masliah E, Hansen LA, Quijada S, et al (1991) Late-onset dementia with argyrophilic grains and subcortical tangles or atypical progressive supranuclear palsy. Ann Neurol 29: 389–396

Meyers K, Dorenkamp DG, Suzuki K (1974) Amyotrophic lateral sclerosis with diffuse neurofibrillary changes. Arch Neurol (Chic) 30: 84–89

Nelson SJ, Yen SH, Davies P, Dickson DW (1989) Basal ganglia neuropil threads in progressive supranuclear palsy. J Neuropathol Exp Neurol 48: 324

Ohara PT (1972) Electron microscopical study of the brain in Down's syndrome. Brain 95: 681–684

Olson MI, Shaw C-M (1969) Presenile dementia and Alzheimer's disease in mongolism. Brain 92: 147–156

Perry G, Richey P, Siedlak SL, Galloway P, Kawai M, Cras P (1992) Basic fibroblast growth factor binds to filamentous inclusions of neurodegenerative diseases. Brain Res 579: 350–352

Probst A, Langui D, Lautenschlager C, Ulrich J, Brion JP, Anderton BH (1988) Progressive supranuclear palsy: extensive neuropil threads in addition to neurofibrillary tangles. Acta Neuropathol (Berl) 77: 61–68

Roy S, Datta CK, Hirano A, Ghatak NR, Zimmermann HM (1974) Electronmicroscopic study of neurofibrillary tangles in Steele-Richardson Olszewski syndrome. Acta Neuropathol (Berl) 29: 175–179

Seitelberger F (1969) Heterogenous system degeneration. Subcortical argyrophilic dystrophy. Acta Neurol 24: 276–284

Shortridge BA, Vogel FS, Burger PC (1985) Topographic relationship between neurofibrillary change and acetylcholinesterase rich neurons in the upper brainstem of patients with senile dementia of the Alzheimer's type and Down's syndrome. Clin Neuropathol 4: 227–237

Steele JC (1972) Progressive supranuclear palsy. Brain 95: 693–704

Steele JC, Richardson JC, Olszewski J (1964) Progressive supranuclear palsy. Arch Neurol 10: 333–359

Tabaton M, Schenone A, Romagnoli P, Mancardi GL (1985) A quantitative and ultrastructural study of substantia nigra and nucleus centralis superior in Alzheimer's disease. Acta Neuropathol (Berl) 68: 218–223

Tabaton M, Whitehouse PJ, Perry G, Davies P, Autilio-Gambetti L, Gambetti P (1988) Alz 50 recognizes abnormal filaments in Alzheimer's disease and progressive supranuclear palsy. Ann Neurol 24: 407–413

Takahashi H, Oyanagi K, Takeda S, Hinokuma K, Ikuta F (1989) Occurence of 15-nm-wide straight tubules in neocortical neurons in progressive supranuclear palsy. Acta Neuropathol (Berl) 79: 233–239

Takauchi S, Mizuhara T, Miyoshi K (1983) Unusual paired helical filaments in progressive supranuclear palsy. Acta Neuropathol (Berl) 59: 225–228

Téllez-Nagel I, Wisniewski HM (1973) Ultrastructure of neurofibrillary tangles in Steele-Richardson-Olszewski syndrome. Arch Neurol 29: 324–327

Terry RD, Wisniewski HM (1970) The ultrastructure in the neurofibrillary tangles and the senile plaque. In: Wolstenholme GEW, O'Connor M (eds) Alzheimer's disease and related conditions. Ciba Foundation Symposium. Churchill, London, pp 145–165

Tomonaga M (1977) Ultrastructure of neurofibrillary tangles in progressive supranuclear palsy. Acta Neuropathol (Berl) 37: 177–181

Wisniewski HM, Terry RD (1970) An experimental approach to the morphogenesis of neurofibrillary degeneration and the argyrophilic plaque. In: Wolstenholme GEW, O'Connor M (eds) Alzheimer's disease and related conditions. Ciba Foundation Symposium. Churchill, London, pp 223–248

Wisniewski HM, Terry RD, Hirano A (1976) Neurofibrillary pathology. J Neuropathol Exp Neurol 29: 163–176

Yamada T, McGeer PL, McGeer EG (1992) Appearance of paired nucleated, Tau-positive glia in patients with progressive supranuclear palsy in brain tissue. Neurosci Lett 135: 99–102

Yamamoto T, Kawamura J, Hashimoto S, Nakamura M, Iwamoto H, Kobashi Y, Ichijima K (1992) Pallido-nigro-luysian atrophy, progressive supranuclear palsy and adult onset Hallervorden-Spatz disease: a case of akinesia as a predominant feature of parkinsonism. J Neurol Sci 101: 98–106

Yen S-H, Horoupian DS, Terry RD (1983) Immunocytochemical comparison of neur-
    ofibrillary tangles in senile dementia of Alzheimer type, progressive supranuclear
    palsy, and postencephalitic Parkinsonism. Ann Neurol 13: 172–175

Authors' address: Dr. J. Cervós-Navarro, Institute of Neuropathology, Free University of Berlin, Hindenburgdamm 30, D-12200 Berlin, Federal Republic of Germany.

# Antigenic determinant properties of neurofibrillary tangles
## Relevance to progressive supranuclear palsy

### F. F. Cruz-Sánchez

Neurological Tissue Bank, Hospital Clinic, University of Barcelona, Spain

**Summary.** Neuronal cytoskeleton is composed of microfilaments, neurofilaments and microtubules which show distinctive ultrastructural characteristics. Different groups of antibodies against neurofilaments and microtubule associated proteins which were grouped according to their specificity for proteins of perykarium, axons and/or dendrites have been produced. A 8.6 KD polypeptide called ubiquitin has been recognized as one of the heat shock proteins. Ubiquitin is implicated in the non-lysosomal degradation of abnormal proteins and other proteolytic intracellular mechanisms.

Several immunohistological studies on Alzheimer's disease (AD)-neurofibrillary tangles (NFTs) demonstrated that antibodies for different normal cytoskeletal components bind to NFTs-bearing neurons. AD-NFTs could be also demonstrated using antibodies for the beta-amyloid protein. The production and accumulation of abnormal proteins such as those observed in AD-NFTs induce a ubiquitin-mediated degradative pathway to remove them. It has been demonstrated that ubiquitin is covalently associated with insoluble neurofibrillary material of AD-NFTs.

Topographical differences in the distribution of NFTs underscore that different neuronal populations including neocortical neurones are affected in progressive supranuclear palsy (PSP) and AD. Differences in the molecular composition of PSP-NFTs highlighted by immunochemical studies induce us to speculate that different physio- and aetiopathogenetic mechanisms are operative in the production of PSP-NFTs.

## Introduction

Neurofibrillary tangles (NFTs) are found in various neurodegenerative processes (Wisniewski et al., 1979). In the brainstem they appear to be the main morphological change in progressive supranuclear palsy (PSP) (Steele et al., 1964). Cortical NFTs have also been described in PSP (Hauw et al., 1990; Cruz-Sánchez et al., 1992a).

As it has been reported in another chapter of this book, PSP-NFTs consist ultrastructurally of straight filaments with an average diameter of 150 Angstroms found in neuronal perikarya. In Alzheimer's disease (AD)

NFTs are composed of 100 Ansgtroms thick twisted tubules (Wiesniewski et al., 1970). It has been demonstrated that AD-NFTs are composed of normal neuronal cytoskeletal components (Perry et al., 1985).

PSP differs from AD both clinically and morphologically. NFTs could however be the substrate linking these two conditions from the pathophysiolocal point of view. Much effort has gone into defining the molecular composition of NFTs in PSP and AD thus permitting to differentiate possible pathophysiological mechanisms in their production.

### Normal components of the neuronal cytoskeleton

On the main, neuronal cytoskeleton is composed by microfilaments, neurofilaments and microtubules which show distinctive ultrastructural characteristics. Microfilaments, (composed by polymers of actin with associated actin-regulatory proteins) and neurofilaments are predominantly neuronal structural elements. Microtubules are also likely to be important in mediating bidirectional transport of organelles and play a major role in the transport of new components down axons and dendrites (Burgoyne and Cambray-Deakin, 1988).

Studies directed to the recognition of the molecular composition of neuronal cytoskeleton have demonstrated that neurofilaments are oligomeric protein filaments composed of three major polypeptides with molecular weights of 200, 150 and 70 KD. Structurally, each polypeptide has a central core region with an alpha-helical configuration able to form alpha-helical coiled-coils. COOH-terminal regions of these filaments are heterogenous which might explain biochemical and immunological properties (Carden et al., 1985). Most of these terminals can be separated in fragments by chymotryptic digestion. Neurofilament proteins are modified during their lifetime by a succession of protein kinase and phosphates (Nixon, 1993). The study of most of these fragments demonstrated that fractions with 200 and 150 KD show the highest numbers of phosphorylation sites (Carden et al., 1985). Local regulation of phosphorylation events could account for variations in the size, morphology and dynamics of the neurofilament network in different regions of the neurons (Nixon, 1993).

The apparent greater plasticity of the neurofilament network in regions like the perikaryon, initial segment and nodes along the axon may provide some insight into the vulnerability of these regions in neurofibrillary diseases (Nixon, 1993).

Microtubules are composed by many proteins beside tubulin (McKeithen and Rosenbaum, 1984). These microtubule-associated proteins (MAPs) are classified on the basis of their molecular weight and their stability to heat treatment (Yen et al., 1987). MAPs of 200–250 KD are heat stable and are designated MAPs-1. MAPs in the low molecular weight range (52–68 KD) are also heat-stable and are referred to as "Tau" proteins (Herzog and Weber, 1978). High molecular weight MAPs show a preferential localization to microtubules of cell bodies and dendrites. In contrast, low molecular weight MAPs (Tau) are highly enriched in parallel fibres and appear to be absent from dendrites (Burgoyne and Cambray-Deakin, 1988). Other authors have demonstrated that Tau protein is enriched in neurons and is segregated to the axonal compartement (Kosik, 1989) and also in dendrites after dephosphorylation (Riederer and Innocenti, 1991).

MAPs also show phosphorylation sites and Tau is a phosphoprotein. Phosphorylation is a major mechanism in the cellular function of many proteins including in the exchange of proteins. Phosphorylation of Tau has been shown to affect the assembly of tubulin and the interaction of microtubules with actin filaments (Grundke-Iqbal et al., 1986; Selkoe, 1987). Phosphorylation is the only known post-translational modification of Tau. Only likely function of Tau phosphorylation is the regulation of its binding to microtubules (Kosik, 1993). Thus, one function of Tau is to alter cell morphology by traducing microtubule elongation into a specific cellular shape change (Kosik, 1993).

## Immunocytochemical properties of normal cytoskeletal proteins

Immunohistochemical studies demonstrated that a large number of antibodies, most of them monoclonal, are specific for neural elements in the central nervous system. Sternberger et al. (1985) identified different groups of antibodies against neurofilaments which were grouped according to their specificity for proteins of perykarium, axons and/or dendrites. Subsequently, Sternberger and Sternberger (1983) demonstrated that most of the differences in the specifity of these antibodies depended on phosphorylation. According to these results, neurofilaments in dendrites, perykaria and proximal axons are non-phosphorylated and phosphorylation occurs during transport along the axon, thus phosphorylated neurofilaments are more compact than non-phosphorylated ones (Sternberger et al., 1985).

The microtubule-associated family of proteins consists of several electrophoretically heterogeneous phosphoproteins that co-assemble with tubulin and promote its polymerization when microtubules are purified by repetitive cycles of temperature-dependent assembly and deassembly (Selkoe, 1987). Microtubule preparations prepared by various methods have been shown to contain many proteins beside tubulin and numerous antibodies have been produced against these proteins which could be distinguished on the basis of different molecular weights of the antigens recognized. Some monoclonal antibodies raised against three or more supernatant proteins in the 50–70 KD range from Tris buffer extracts of both normal and Alzheimer brains (Yen et al., 1987) have been called anti-Tau antibodies. Wolozin and co-workers (1986) produced an antibody designated Alz-50 which identifies proteins of 68 KD. This antibody is directed against proteic fractions of Alzheimer brains and proteins of 59 KD found in normal cerebral cortex which are similar to Tau in normal brains.

## Ubiquitin-protein conjugates

Mammalian cells respond to being exposed to temperatures different from their normal physiological one by activating a set of genes producing proteins which are often called "stress proteins". A 8.6 KD polypeptide called ubiquitin has been recognized as one of these heat shock proteins in chicken embryo fibroblasts (Bond and Schlesinger, 1985). This polypeptide has also been related to different intracellular stress responses. The covalent bond of ubiquitin to various target proteins within the cell represents a regulatory processe (Haas and Bright, 1985). The formation of ubiquitin-protein conjugates is a step in the selective ATP-dependent degradation of abnormal and short lived proteins (Hersko et al., 1980; Haas and Bright, 1985). Thus, ubiquitin is implicated in the non-lysosomal degradation of abnormal proteins and other proteolytic intracellular mechanisms (Rechsteiner, 1991). Following the marking of a protein with ubiquitin, the protein moiety of the conjugate is selectively degraded with the release of free and reutilizable ubiquitin (Chiechanover, 1993).

There is a constant association between filamentous inclusions and the presence of ubiquitin (Lowe et al., 1993). The identification of ubiquitin in cellular inclusions (Lowe et al., 1988) suggests that different pathophysiological phenomena involving cytoskeletal proteins may activate the production of ubiquitin through the participation of a second messeger. Several monoclonal and polyclonal antibodies raised against polypeptides with biochemical characteristics of ubiquitin have been produced and used to recognize the presence of ubiquitin in cellular abnormalities found in various pathological conditions including neurological disorders (Lowe et al., 1988; Manetto et al., 1988; Leigh et al., 1989).

## Immunohistochemical properties of AD-NFTs

The neuronal accumulation of 10 nm paired-helical filaments (PHFs) is the most characteristic ultrastructural feature of AD-NFTs. PHFs are made up of 8 protofilaments which differ ultrastucturally from normal neurofilaments (Wen and Wiesniewski, 1985). A large variety of heterogeneous filaments (including 10 to 15 nm straight filaments) has also been detected ultrastructurally in AD-NFTs. These filaments may be the result of molecular modifications of component proteins and/or stages in the evolution of filamentous pathology (Price et al., 1986). The insolubility of PHFs and the difficulty in purifying and isolating their polypeptides are the main characteristics of NFTs which hinder the recognition of their molecular components. Immunochemical studies including immunocytochemical ones using monoclonal and polyclonal antibodies showed that normal cytoskeletal components are important antigenic constituents of AD-NFTs.

Most of the polypeptides that compose PHFs share some antigenic determinants in common with neurofilaments. Different immunohistological studies on AD-NFTs demonstrated that phosphorylated antibodies bind to NFTs-bearing neurons (Anderton et al., 1982). Some of these studies compared results with adjacent silver impregnated sections (Connolly et al., 1987). Phosphorylated neurofilaments are not normally present within perykaria (Sternberger et al., 1985). Phosphorylated neurofilament-immunoreactive PHFs may represent an abnormal phosphorylation of neurofilaments present in perykaria and an early stage in the abnormal accumulations of cytoskeletal components. Immunoelectron microscopic studies of NFTs demonstrated that straight filaments share all their identified epitopes and solubility properties with phosphorylated antibodies identified in PHFs. These features have been considered as indirect evidences that PHFs and straight filaments are related (Perry et al., 1987).

NFTs could differ morphologically and these differences could be related to their molecular composition. Schmidt and co-workers (1988) demonstrated that there are at least two populations of morphologically and immunohistologically distinct NFTs.

Intracellular-NFTs (I-NFTs) are present in degenerated neurons and extracellular-NFTs (E-NFTs) are bands of PHFs enveloped by glial fibrillary acidic protein (GFAP)-positive astrocytic processes. Different immunohistological determinants among NFTs may reflect important differences in the pathogenesis of NFTs (Schmidt et al., 1988). In the study of Schmidt and co-workers (1988), a monoclonal antibody (Mab) against rat neurofilament proteins which recognize epitopes of high and medium molecular weight was found in I-NFTs. A Mab raised to a bovine GFAP (which also recognize human GFAP) stained only E-NFTs which were consistently negative with the first Mab. Thioflavin S or Congo red techniques were used to counterstain immunostained sections in order to test for the presence of E- and I-NFTs (Schmidt et al., 1988).

Immunohistological studies with monoclonal and polyclonal antibodies have demonstrated that PHFs are also composed of polypeptides from 45 to 250 KD different from neurofilaments. Yen and co-workers (1987) raised a panel of antibodies against different types of MAPs raised against AD-NFTs. These antibodies also bound to a group of MAPs referred to as "Tau" which recognize a microtubule-associated phosphoprotein of low molecular weight.

Antibodies to Tau also bind to MAPs from normal human brains assembled in vitro but identically treated Alzheimer brain preparations have to be dephosphorylated to be properly recognized by the same antibody. This suggests that Tau in Alzheimer brains is an abnormally phosphorylated protein of PHF (Grundke-Iqbal et al., 1986). However, some tangles recognized by anti-Tau without prior dephosphorylation, may represent earlier stages in the process than those recognized only after dephosphorylation (Grundke-Iqbal, 1986).

AD-NFTs could be demonstrated using Congo Red and Thioflavin S stains and antibodies to "congophilic-associated antigens" (MacDonald and Esiri, 1986). This suggest that AD-NFTs also contain the beta-amyloid protein. Masters and co-workers (1985) demonstrated that AD-NFTs contain the same protein as amyloid plaque cores and blood vessels. It has been suggested that amino terminal heterogeneity of beta protein interacting with cytoskeletal proteins, in a unique environment as is the cytoplasm may account for the peculiar assembly of this amyloid subunit to form PHFs in AD (Castaño and Frangione, 1988).

The production and accumulation of abnormal proteins such as those observed in AD-NFTs induce a ubiquitin-mediated degradative pathway to remove them. However, the removal is ineffective (Perry et al., 1987). Several antibodies have been used to demonstrate that ubiquitin is covalently associated with insoluble neurofibrillary material of AD-NFTs (Perry et al., 1987; Mori et al., 1987; Manetto et al., 1988; Leigh et al., 1989). Mori and co-workers (1987) identified two molecular fragments in PHFS as derived from ubiquitin by protein sequencing. According to these results, these authors suggested that PHFs are not inert elements within degenerating neurons; tangle-bearing neurons are actively responding by ubiquitinating PHFs neurons.

### Antigenic properties that distinguish AD- and PSP-NFTs

PSP-NFTs are composed of 15 nm straight filaments (Tellez-Nagel and Wisniewski, 1973) differing from AD-NFTs. Several studies have demonstrated that PSP-NFTs also contain modified normal cytoskeletal components. However, controversy on the topic is evident in the literature.

A comparative study between NFTs in AD, PSP, postencephalitic parkinsonism and cases with nigrostriatal degeneration and olivopontocerebellar atrophy and normal elderly brains was carried out by Yen and co-workers (1983) using a polyclonal antiserum which recognizes a two-cycle-purified human brain microtubule fraction. NFTs in PSP, postencephalitic parkinsonism and AD share their immunological properties with components present in microtubule fractions of normal brain Tabaton and co-workers (1983) quantitatively assessed the antigenic properties of NFTs located in neurons of the tegmental nuclei of the pontine raphe in PSP and AD. These properties were then compared to those of NFTs of AD located in hippocampal neurons. For this study, these authors used a panel of antibodies including anti-PHFs (especially anti-Tau); anti-tau-1; anti-200 KD-neurofilaments and anti-ubiquitin antibodies. Results demonstrated that AD- and PSP-NFTs share antigenic properties when they are located in the same neuronal population.

In a case of PSP, Dickson and co-workers (1985) demonstrated a common antigenic determinant between AD- and PSP-NFTs using Mabs raised against a brain homogenate from a case of AD. Antibodies used stained only the excluded proteins at the top of the stacking gel on electrophoresis of proteins from AD brain and did not stain any proteins on immunoblots of normal brain or of neurofilament preparation from rat or bovine brain. These results demonstrated that AD- and PSP-NFTs show a common component which might be Tau type protein.

In another study, Probst and co-workers (1988) found a very similar antigenicity of subcortical neuronal pathology in PSP and AD using a panel of antibodies including antisera to Tau and to isolated PHFs and Mabs to phosphorylated epitopes on 210 and 155 KD neurofilament subunits.

Schmidt and co-workers (1988) by means of a large panel of Mabs described some properties of antigenic determinants that distinguish AD- and PSP-NFTs. In this study PSP-NFTs were consistently detected by Mabs specific for PHFs and Tau proteins. The antibodies also detected cortical and subcortical AD-NFTs. In contrast, none of Mabs to neurofilament proteins reacted with PSP-NFTs. These results led the authors to conclude that differences observed in both types of NFTs may reflect the effect of different pathological events specific to PSP and AD, or the selective formation of NFTs in different groups of neurons.

The Mab Alz 50 which recognizes a 68 KD protein present in cerebral tissue of patients with AD was found in neurons with and without NFTs from 8 confirmed cases of PSP (Tabaton et al., 1988) using inmunohisto-chemical, immunochemical and immunoelectron microscopic studies. These authors found that in neurons lacking NFTs Alz 50 was associated with straight filaments of various size but of smaller diameter than those of PSP. The excess of Alz 50 antigen in AD tissue correlates with a more severe fibrillary pathology in AD than in PSP. According to these results, Tabaton and co-workers (1988) concluded that the antigen recognized by Alz 50 is characteristic of neurons forming abnormal filaments.

In 1991, Flament and co-workers demonstrated that AD- and PSP-NFTs show different types of abnormal Tau accumulation. For these authors, differences in the pathological profile of Tau-variants may reflect different aetiopathogenetic pathways or, at least, the formation of different types of aggregates of Tau filaments.

Topographical studies have demonstrated that PSP-NFTs are predomin-antly seen in layers V and VI within large and small neurons whereas AD-NFTs are most numerous in layers III and V and affect middle size neurons (Hauw et al., 1990; Cruz-Sánchez et al., 1992a). Cruz-Sánchez and co-workers (1992a) described two cases of PSP associated with clinical and pathological diagnosis of AD. The immunohistological study of these cases demonstrated that PSP-NFTs showed antigenic determinants different from AD-NFTs based on the expression of ubiquitin. Cortical AD-NFTs were strongly positive with an antisera to ubiquitin whereas brainstem PSP-NFTs were consistently negative. These results confirm previous results from Tabaton and co-workers (1983) in cases with combined changes and defini-tive diagnosis of AD and PSP.

We have had the opportunity to study (Cruz-Sánchez et al., 1993) histologically and immunohistologically the distribution of NFTs in brain tissue from 13 cases clinically diagnosed as suffering from PSP. A panel of antibodies including one to the 155 kD (BF10) and another to the 210 kD (RT97) neurofilament subunits, beta amyloid-A4, ubiquitin and antibodies to 55 and 69 kD abnormal Tau proteins were used for the study. Brain

**Table 1.** Histological and immunohistological characteristics of NFTs in 13 cases of PSP, 5 cases of AD and controls

| NFTs | Biel. | CR | BF10 | RT97 | UBQ | 55TAU | 69TAU | A4 |
|---|---|---|---|---|---|---|---|---|
| Cort-PSP-NFTs | + | − | + | + | + | + | + | − |
| Brst-PSP-NFTs | + | − | + | + | − | − | + | − |
| Cort-AD-NFTs | + | + | + | + | + | + | + | + |
| Brst-AD-NFTs | | | | NF | | | | |
| Cort-Crl-NFTs | + | + | + | + | − | + | + | + |
| Brst-Crl-NFTs | | | | NF | | | | |

*Biel* Bielschowsky; *CR* alkaline congo red; *BF10* 155 kD neurofilament subunit; *RT97* 210 kD neurofilament subunit; *UBQ* ubiquitine; *55TAU* 55 kD abnormal TAU; *69TAU* 69 kD abnormal TAU protein; *A4* beta-amyloid
*Cort-PSP-NFTs* cortical-PSP-NFTs (layer V and VI);
*Brst-PSP-NFTs* brainstem-PSP-NFTs;
*Cort-AD-NFTs* cortical-AD-NFTs (layer III and V);
*Brst-AD-NFTs* brainstem-AD-NFTs; NF: absent.
*Cort-Crl-NFTs* cortical NFTs in control cases (layer III and V)
*Brst-Crl-NFTs* brainstem NFTs in control cases

tissue from patients suffering from AD and 5 controls of similar age were used. Sections were also stained with alkaline congo red and Bielschowsky.

Topographic analysis of NFTs was made on immunostained sections of the following areas: neocortex (including pre-central gyrus: Brodman area 4), hippocampus, putamen, thalamus (dorsal and ventral nuclei), pallidum, substantia nigra, locus coeruleus, dentate nucleus and nuclei pontis, tectum mesencephali, peri-aqueductal area, oculo-motor nuclei, vestibular nuclei, hypoglossal nuclei and inferior olives.

Immunohistological findings are summarized in Table 1.

#### Comments

PSP is a neurodegenerative condition which since its description in 1964 (Steele et al., 1964) has been studied from different viewpoints. Clinically it differs from AD in spite that in some cases severe dementia has been reported (Cruz-Sanchez et al., 1992a). The presence of NFTs could be the nexus between the pathophysiological mechanisms of the two conditions. Topographical differences in the distribution of NFTs underscore that different neuronal populations (Tabaton et al., 1983) including neocortical neurones (Fig. 1) are affected in both conditions (Hauw et al., 1990; Cruz-Sanchez et al., 1992a).

Biological studies of NFTs in various conditions reveal differences and similarities between the their composition (Yen et al., 1983; Wolozin, 1986; Shankar et al., 1989; Flamment et al., 1991). A particular problem relating to the study of NFTs is how to dissolve them to prepare them for biochemical

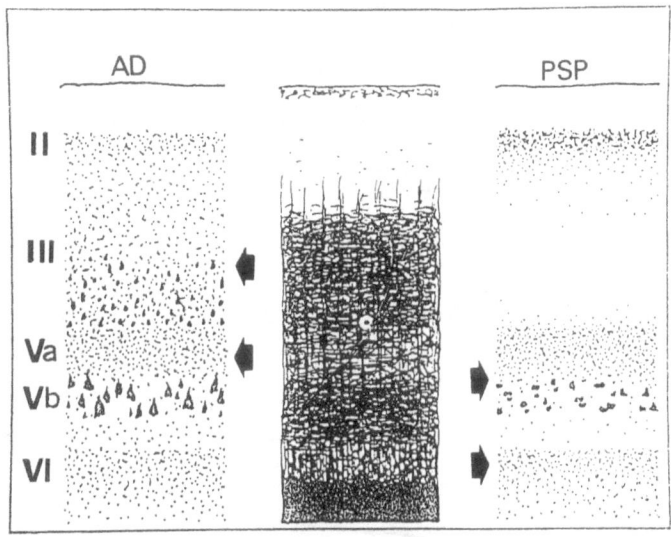

**Fig. 1.** Topographical distribution of cortical neurofibrillary tangles (NFTs) in Alzheimer's diseases (*AD*) and progressive supranuclear palsy (*PSP*). Note that PSP-NFTs are predominantly seen in layers V and VI within large and small neurons whereas AD-NFTs are most numerous in layers III and V and affect middle size neurons

or molecular studies (Price et al., 1986; Castaño and Frangione, 1988). Therefore immunohistological studies directed to the recognition of antigenic epitopes appear to be essential to establish differences between NFTs in several conditions particularly in PSP and AD (Figs. 2A, 2B, and 3). From this point of view PSP-NFTs have not, thus far, been demonstrated to include beta amyloid and ubiquitin conjugation may not have taken place in NFTs of brainstem neurones. This may be in favour of the existence of important differences in the pathophysiological mechanism causing NFTs in PSP and AD. On the other hand, alteration in the phosphorylation of neurofilaments and Tau proteins are present in both conditions but some of the underlying proteic components are different. For example the 55 Kd Tau protein which is only present in AD.

In relation to ubiquitin expression, experimental data suggested that ubiquitin is not simply targeting proteins for lysosomal degradations (Lowe et al., 1993). Ubiquitin may have a role in the deposition of amyloid in the brain causing accelerated deposition (Alizadeh et al., 1992). Co-secretion of peptide fragments and ubiquitin must be considered as a potential pathogenic factor in causing proteins to form amyloid (Lowe et al., 1993). A possible role of ubiquitin and amyloid deposition in cerebral angiopathy has been proposed (Cruz-Sanchez et al., 1992b). In "pure" cerebral amyloid angiopathy there may be a primary pathological process inducing ubiquitin production which may lead to enhance amyloid deposition. Similar features may occur in the production of AD-NFTs. In PSP-NFTs ubiquitin and

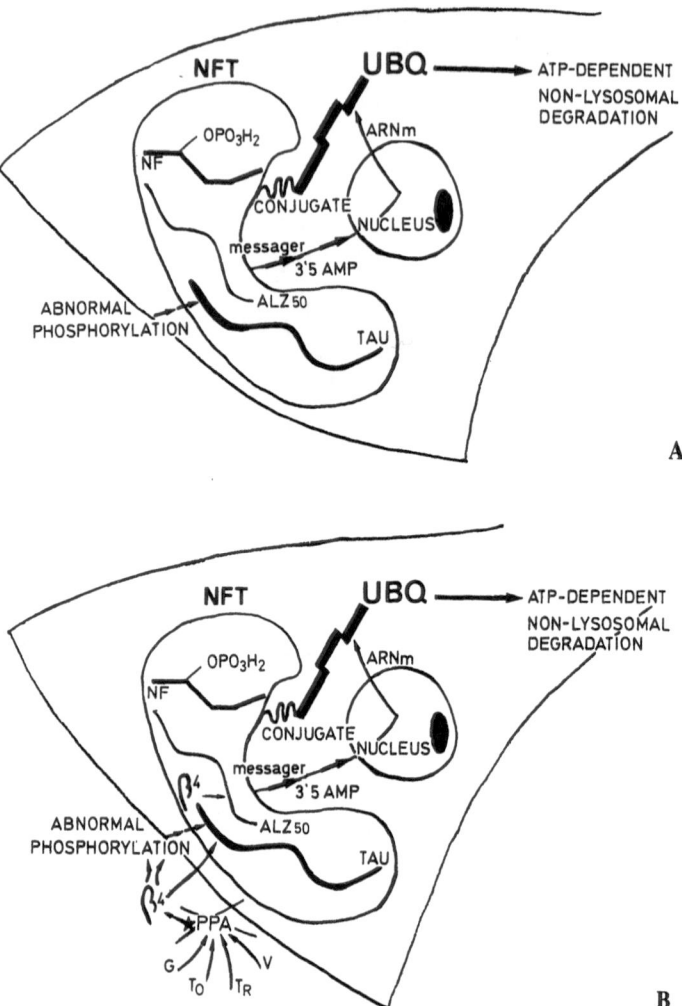

**Fig. 2.** Schemes representing the possible pathophysiological mechanism of production of neurofibrillary tangles (*NFTs*) in Alzheimer's disease (AD). **A** An abnormal phosphorylation produces the accumulation of neurofilament fragments and other phosphorylated proteins. Neurons respond by activating a set of genes producing ubiquitin. The formation of ubiquitin-protein conjugates is a stept in the selective ATP-dependent degradation of abnormal and short lived proteins. **B** Different aetiopathological factors like genetic (*G*), toxic (*To*), traumatic (*Tr*) and viral (*V*) could alterate the amyloid protein precursor (PPA) producing the accumulation of beta-amyloid protein (β4) and/or the abnormal phosphorylation. Ubiquitin could be a potential pathogenic factor in causing or accelerating the accumulation of beta-amyloid in AD

amyloid have not been demonstrated suggesting different mechanisms of production (Fig. 3).

Differences in the molecular composition highlighted by immunochemical studies induces us to speculate that different physio- and aetiopathogenetic mechanisms are operative in the production of PSP-NFTs.

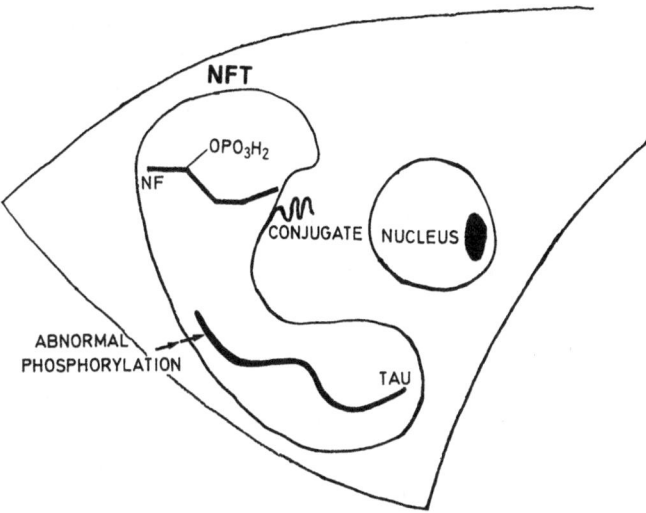

**Fig. 3.** Scheme representing the possible pathophysiological mecanism of production of neurofibrillary tangles (*NFT*s) in progressive supranuclear palsy (PSP). The accumulation of neurofilaments and tau is limitated to a shorter spectrum of proteins and the ubiquitin proteolytic intracellular mechanisms is not activated

Selective neuronal vulnerability, time span of evolution and possible different aetiological environmental factors may be peculiar to each of the two conditions.

According to the knowledge accumulated thus far, further studies will be necessary to recongize more components of NFTs related to the normal and or abnormal cytoskeleton and other NFTs components which could relate their production to toxic, infectious (virus), genetic and other causes of the diseases such as is suggested for AD. The presence of amyloid in AD is the specific element around which many studies revolve but despite this, no specific cause for the deposition of amyloid has been conclusively demonstrated.

Abnormal phosphorylation of proteins is a mechanism involved in the production of NFTs. The role of ubiquitination of proteins for degradation or for other reasons should be also further explored.

### References

Alizadeh-Khiavi K, Normand J, Chronopoulos S, Ali-kan Z (1991) Alzheimer's disease brain-derived ubiquitin has amyloid- enhancing factor activity: behavior of ubiquitin during accelerated amyloidogenesis. Acta Neuropathol (Berl) 81: 280–286

Anderton BH, Breinburg D, Downes MJ, Green PJ, Tomlison BE, Ulrich J, Wood JN, Kahn J (1982) Monoclonal antibodies show that neurofibrillary tangles and neurofilaments share antigenic determinants. Nature 298: 84–86

Bond U, Schlesinger MJ (1985) Ubiquitin is a heat shock protein in chicken embryo fibroblasts. Mol Cell Biol: 949–956

Burgoyne RD, Cambray-Deakin MA (1988) The cellular neurobiology of neuronal development: the cerebellar granule cell. Brain Res Rev 13: 77–101

Carden MJ, Schlaepfer WW, Y-Lee VM (1985) The structure, biochemical properties, and inmunogenicity of neurofilament peripheral regions are determined by phosphorylation state. J Biol Chem 260: 9805–9817

Castaño EM, Frangione B (1988) Biology of disease. Human amyloidosis, Alzheimer disease and related disorders. Lab Invest 58: 122–132

Ciechanover A (1993) The ubiquitin-mediated proteolytic pathway. Brain Pathol 3: 67–75

Connolly AAP, Anderton BA, Esiri MM (1987) A comparative study of a silver stain and monoclonal antibody reaction on Alzheimer's neurofibrillary tangles. J Neurol Neurosurg Psychiatry 50: 1221–1224

Cruz-Sánchez FF, Rossi ML, Cardozo A, Deacon P, Tolosa E (1992a) Clinical and pathological study of two patients with progressive supranuclear palsy and Alzheimer's changes. Antigenic determinants that distinguish cortical and subcortical neurofibrillary tangles. Neurosci Lett 136: 43–46

Cruz-Sánchez FF, Marin C, Rossi ML, Cardozo A, Ferrer I, Tolosa E (1992b) Ubiquitin in cerebral amyloid angiopathy. J Neurol Sci 112: 46–50

Cruz-Sánchez FF, Rossi ML, Tolosa E, Cardozo A (1994) Further pathological insight into progressive supranuclear palsy (submitted)

Dickson DW, Kress Y, Crowe A, Yen S-H (1985) Monoclonal antibodies to Alzheimer neurofibrillary tangles. 2. Demonstration of a common antigenic determinant between ANT and neurofibrillary degeneration in PSP. Am J Pathol 120: 292–303

Flament S, Delacourte A, Verny M, Hauw JJ, Javoy-Agid F (1991) Abnormal tau proteins in progressive supranuclear palsy. Similarities and differences with the neurofibrillary degeneration of the Alzheimer type. Acta Neuropathol (Berl) 81: 591–596

Grundke-Iqbal I, Iqbal K, Tung YC, Quinlan H, Wisniewski HM (1986) Abnormal phosphorylation of the microtubule associated protein (tau) in Alzheimer cytoeskeletal pathology. Proc Natl Acad Sci USA 83: 4913–4917

Haas AL, Bright PM (1985) The immochemical detection and quantification of intracellular ubiquitin-protein conjugates. J Biol Biochem 260: 12464–12473

Hauw JJ, Verny M, Delaére P, Cervera P, He Y, Duyckaerts C (1990) Constant neurofibrillary changes in the neocortex in progressive supranuclear palsy. Basic differences with Alzheimer's disease and aging. Neurosci Lett 119: 182–186

Hersko A, Ciechanover A, Heller H, Haas AL, Rose IA (1980) Proposed role of ATP in protein breakdown: conjugation of proteins with the multiple chains of the polypeptide of ATP-dependent proteolysis. Proc Natl Acad Sci USA 77: 1783–1786

Herzog W, Weber K (1978) Fractionation of brain microtubule associated proteins: isolation of two different proteins which stimulate tubulin polymerization in vitro. Eur J Biochem 92: 1–8

Kosik KS, Orecchio LD, Bakalis S, Neve RL (1989) Developmentally regulated expression of specific tau sequences. Neuron 2: 1389–1397

Kosik KS, Crandall JE, Mufson E, Neve RL (1989) Tau in situ hibridization in normal and Alzheimer brain. Localization in the somato dendritic compartment. Ann Neurol 26: 352–361

Kosik KS (1993) The molecular and cellular biology of Tau. Brain Pathol 3: 39–43

Leigh PN, Probst A, Dale GE, Power DP, Brion J-P, Dodson A, Anderton BH (1989) New aspects of the pathology of neurodegenrative disorders as revealed by ubiquitin antibodies. Acta Neuropathol (Berl) 79: 61–72

Lowe J, Blanchard A, Morrel K, Lennox G, Reynolds L, Billett M, Landon M, Mayer RJ (1988) Ubiquitin is a common factor in intermediate filament inclusion bodies of diverse type in man, including those of Parkinson's disease, Pick's disease and Alzheimer's disease, as well as Rosenthal fibres in cerebellar astrocytomas, cytoplasmic bodies in muscle and Mallory bodies in alcoholic liver disease. J Pathol 155: 9–15

Lowe J, Mayer RJ, Landon M (1993) Ubiquitin in neurodegenerative diseases. Brain Pathol 3: 55–65

Mannetto V, Perry G, Tabaton M, Mulvihill P, Fried VA, Smith HT, Gambetti P, Autulio-Gambetti L (1988) Ubiquitin is associated with abnormal cytoplasmic filaments characteristics of neurodegenerative diseases. Proc Natl Acad Sci USA 85: 4501–4505

MacDonald SM, Esiri MM (1986) Monoclonal antibody binding to congophilic elements in human Alzheimer brain. J Clin Pathol 39: 1199–1203

McKeithen T, Rosenbaum JL (1984) The biochemistry of microtubules: a review. In: Shay JW (ed) Cell and muscle motility, vol 5. The cytoskeleton. Plenum Press, New York, pp 225–288

Masters CL, Multhaup G, Simms G, Pottgieser J, Martins RN, Beyreuther K (1985) Neuronal origin of a cerebral amyloid: neurofibrillary tangles of Alzheimer's disease contain the same protein as the amyloid plaque core and blood vessels. EMBO J 4: 2757

Mori H, Kondo J, Ihara Y (1987) Ubiquitin is a component of paired helical filaments in Alzheimer's disease. Science 235: 1641–1644

Nixon RA (1993) The regulation of neurofilament protein dynamics by phosphorylation: clue to neurofibrillary pathobiology. Brain Pathol 3: 29–38

Perry G, Friedman R, Shaw G, Chau V (1985) Ubiquitin is detected in neurofibrillary tangles and senile plaque neurites of Alzheimer disease brains. Proc Natl Acad Sci USA 84: 3033–3036

Perry G, Mulvihill P, Manetto V, Autilio-Gambetti L, Gambetti P (1987) Inmunocytochemical properties of Alzheimer straight filaments. J Neurosci 7/11: 3736–3738

Price DL, Whitehouse PJ, Struble RG (1986) Cellular pathology in Alzheimer's and Parkinson's diseases. TINS: 29–33

Probst A, Langui D, Lautenschalager C, Ulrich J, Brion JP, Anderton BH (1988) Progressive supranuclear palsy: extensive neuropil threads in addition to neurofibrillary tangles. Very similar antigenicity of subcortical neuronal pathology in progressive supranuclear palsy and Alzheimer's disease. Acta Neuropathol (Berl) 77: 61–68

Rechsteiner M (1991) Natural substrate of the ubiquitin proteolytic pathway. Cell 66: 615–618

Riedeer BM, Innocenti GM (1991) Differential distribution of tau proteins in developing cat cerebral cortex and corpus callosum. J Neurosci 3: 1134–1145

Schmidt ML, Gur RE, Gur RC, Trojanowski JQ (1988) Intraneuronal and extracellular neurofibrillary tangles exhibit mutually exclusive cytoskeletal antigens. Ann Neurol 23: 184–189

Schmidt ML, Lee ML VM, Hurtig H, Trojanowsky JQ (1988) Properties of antigenic determinants that distinguish neurofibrillary tangles in PSP and Alzheimer's disease. Lab Invest 59: 460–465

Selkoe DJ (1987) Deciphering Alzheimer's disease: the pace quickens. TINS 10: 181–184

Shankar SK, Yanagihara R, Garruto RM, Grundke-Iqbal I, Kosik KS, Gajdusek DC (1989) Immunocytochemical characterization of neurofibrillary tangles in amyotrophic lateral esclerosis and parkinsonism-dementia of Guam. Ann Neurol 25: 146–151

Steele JC, Richardson JC, Olszewski J (1964) Progressive supranuclear palsy. Arch Neurol 10: 333–359

Sternberger LA, Sternberger NH (1983) Monoclonal antibodies distinguish phosporylated and non-phosphorylated forms of neurofilaments in situ. Proc Natl Acad Sci USA 80: 6126–6130

Sternberger NH, Sternberger LA, Ulrich J (1985) Aberrant neurofilament phosphorylation in Alzheimer disease. Proc Natl Acad Sci USA 82: 4274–4276

Tabaton M, Perry G, Autilio-Gambetti L, Manetto V, Gambetti P (1983) Influence of neuronal location on antigenic properties of neurofibrillary tangles. Ann Neurol 23: 604–610

Tabaton M, Withehouse PJ, Davies P, Autulio-Gambetti L, Gambetti P (1988) Alz50 recognize abnormal filaments in Alzheimer's and progressive supranuclear palsy. Ann Neurol 24: 407–413

Tellez-Nagel I, Wisniewski HM, Bronx NY (1973) Ultrastructure of neurofibrillary tangles in steele-Richardson-Olszewski syndrome. Arch Neurol 29: 324–327

Wen GY, Wisniewski HM (1984) Substructures of neurofilaments. Acta Neuropathol (Berl) 64: 339–343

Wiesniewski HM, Terry RD, Hirano A (1970) Neurofibrillary pathology. J Neuropathol Exp Neurol 29: 163–176

Wisniewski K, Jervis GA, Moretz RC, Wisniewski HM (1979) Alzheimer neurofibrillary tangles in diseases other than senile and presenile dementia. Ann Neurol 5: 288–294

Wolozin BL, Pruchnicki A, Dickson DW, Davies P (1986) A novel antigen in the Alzheimer brain. Science 232: 648–650

Yen SH, Dickson DW, Crowe A, Butler M, Shelanski ML (1987) Alzheimer's neurofibrillary tangles contain unique epitopes and epitopes in common with the heat-stable microtubule associated proteins tau and MAP2. Am J Pathol 126: 81–91

Yen S-H, Horoupian DS, Terry RD (1983) Immunocytochemical comparison of neurofibrillary tangles in senile dementia of Alzheimer type. Progresive supranuclear palsy, and post encephalitic parkinsonism. Ann Neurol 13: 172–175

Author's address: Dr. F. F. Cruz-Sánchez, Banco de Tejidos Neurológicos, Servicio de Neurologia, Hospital Clínico, Villarroel 170, Barcelona 08036, Spain.

# Cortical tangles in progressive supranuclear palsy

**M. Verny, C. Duyckaerts, P. Delaère, Y. He,** and **J.-J. Hauw**

Laboratoire de Neuropathologie R. Escourolle, Hôpital de La Salpêtrière, Paris, France

**Summary.** Ten cases of PSP were examined for the presence of neocortical and hippocampal lesions. Samples from 10 cortical areas were stained by Bodian's method and by tau, ubiquitin and βA4 immunocytochemistry. For the sake of comparison, 5 Alzheimer's cases were studied with the same techniques.

Neocortical tangles, star-like tufts of fibers, and neuropil threads were seen in all the cases of PSP. They were stained by Bodian's technique and labelled by an anti-tau, but not by a polyclonal anti-ubiquitin antibody. Senile plaques (Bodian's technique), diffuse or focal amyloid deposits (β-A4 immunohistochemistry) were rare or absent.

The density of tangles was the highest in area 4 and the lowest in area 17. In area 4, the tangles were mainly located in layers V–VI. By contrast, the Alzheimer's tangles had a bimodal distribution (layers III and V–VI). These results favor the specificity of cortical alterations in PSP.

## Introduction

It has been considered for a long time that cerebral cortex was devoid of significant lesions in progressive supranuclear palsy (PSP), (Behrman et al., 1969; Blumenthal et al., 1969; Probst et al., 1975, 1988). However, even in the initial report (Steele et al., 1964), the presence of tangles in the neocortex and hippocampus was mentioned. In subsequent papers, these alterations were viewed as coincidental, due to aging or to concomitant Alzheimer's disease (Jellinger et al., 1980; Kish et al., 1985; Probst et al., 1975).

The specificity of cortical tangles in PSP has been more recently emphasized, on the basis of their microscopical aspect (Hauw et al., 1990; Hof et al., 1992; Ishino et al., 1976) and ultrastructure (Takahashi et al., 1989), topography, immunocytochemical (Hauw et al., 1990; Hof et al., 1992) and biochemical (Flament et al., 1991) characteristics.

We examined the cortex in 10 cases of PSP; tangles were found in all the cases, in association with neuropil threads and peculiar alterations of the astrocytic cytoskeleton. Our data support the specificity of these changes.

The frequency of these lesions could indicate that the clinical signs do not only depend on the subcortical changes.

## Cases and methods

### Cases

We studied 7 men and 3 women, aged 68.8 years as a mean at death (range: 58–82 years). The analysis of the clinical files was retrospective. The diagnosis of PSP had been clinically correct in 7/10 cases. In all the cases, some signs of parkinsonism were noted: akinesia (9/10), axial rigidity (6/10), cogwheel phenomenon (9/10), tremor (2/10). Dysarthria was noticed in 8/10 cases. Nine cases had a typical supranuclear palsy of the gaze. In one case, there was no paralysis of gaze. Six cases had frontal symptoms, 9 were intellectually impaired (4 diagnosed as demented). In one case, the intellectual status was not specifically mentioned. Sensitivity to DOPA-therapy was absent in 5/10 cases, transitory and slight in 2/10, and not mentioned in 3/10.

The neuropathological diagnosis of PSP relied on the presence of numerous tangles, neuronal loss and gliosis in several subcortical nuclei (brainstem reticular nuclei, pallidum, subthalamic nucleus, substantia nigra), as usually admitted (in the absence of formal diagnostic criteria).

Five prospectively assessed cases with Alzheimer's disease (cases 2722, 2782, 2812, 2825, 2942 of the Charles Foix Longitudinal study (Delaère et al., 1989; Duyckaerts et al., 1987)) were used to highlight the differences between the tangles seen in PSP and in Alzheimer's disease. These 5 cases were selected because they exhibited (relatively) high densities of tangles in area 4, allowing a better comparison with the PSP cases. They were 5 females (mean age at death: 88 years; range: 82 to 93) and were deeply demented (Blessed test score between 2 and 8).

### Post-mortem study

The brains were fixed in 4% buffered formaldehyde for at least 3 months.

### PSP cases

Ten cortical samples were taken in specific areas (Fig. 1). Samples of thalamus, subthalamic and lenticular nucleus, substantia innominata, midbrain, pons, medulla oblongata and cerebellum were systematically taken. All the samples were embedded in paraffin and 7 μm thick sections were obtained.

Hematoxylin-eosin and Bodian silver method coupled with luxol fast blue were performed on all the samples. Immunohistochemistry used the following antibodies: polyclonal anti-Tau (generous gift of J.-P. Brion, Brussels; Delaère et al., 1989), βA4 (generous gift of K. Beyreuther and C. Masters; Delaère et al., 1989) and ubiquitin (Dako®). Anti-tau immunohistochemistry was applied to the samples of areas 4, 8, 9, 23, 24, 39, and hippocampus; anti-βA4 and anti-ubiquitin, to the samples of areas 4, 22, and hippocampus.

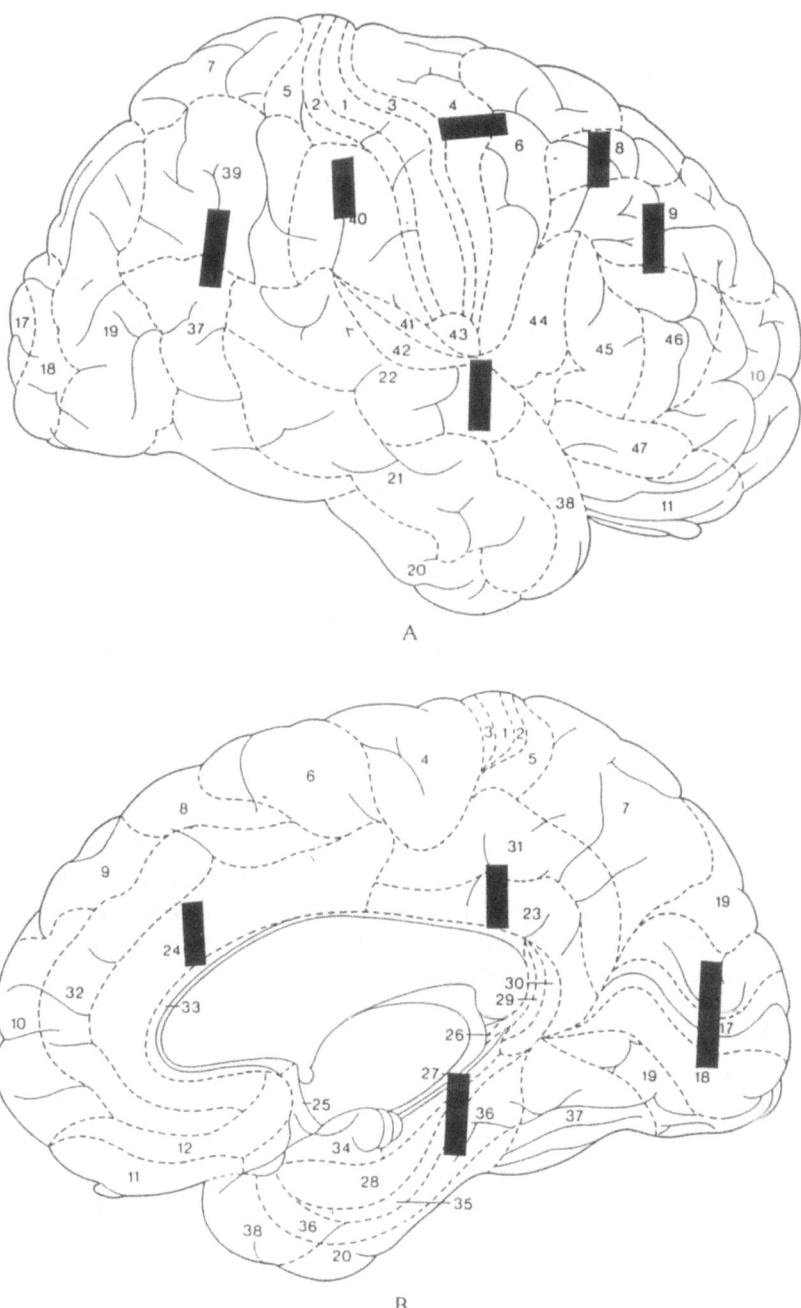

**Fig. 1.** Diagram showing the samples which were systematically obtained in the PSP cases

### Alzheimer's cases

Brodmann's area 4 was sampled as previously described. Section and staining protocols were the same as those applied to PSP cases. The results concerning other areas have been reported elsewhere (Delaère et al., 1989; Duyckaerts et al., 1987).

### Morphometry

The density of neocortical tangles and their location in the thickness of the neocortex (for area 4) was quantitatively assessed according to a technique detailed elsewhere (Duyckaerts et al., 1986; Hauw et al., 1990). The surface area and the form factor (4 $\pi$ area/perimeter$^2$) of the affected perikarya was calculated for 307 Alzheimer's disease and 537 PSP tangles.

## Results

Neocortical tangles were present in all the PSP cases of this series. They were seen after silver impregnation or tau-immunocytochemistry. The 2 techniques gave similar results both on qualitative and quantitative grounds. Tangles were most numerous in Brodmann's area 4 and rare in areas 17–18.

### Qualitative results

Silver-impregnated or tau positive tangles (Fig. 2a and b) were seen in large as well as small neurons; some Betz cells were affected. Tangles in PSP and in Alzheimer's disease had a different shape: PSP tangles were thinner and extended in more slender and numerous processes. Star-like tufts of fibers (Fig. 2c) sometimes centered by a nucleus, were seen mainly in severely affected regions. Numerous neuropil threads were observed in the layers where the tangles were abundant. Contrarily to Alzheimer's tangles, the PSP tangles were not labelled by polyclonal ubiquitin antibody. $\beta$A4 immunocytochemistry revealed only a few diffuse deposits, and extremely rare focal deposits.

### Quantitative results

#### (1) Density according to areas

The mean density of tangles in area 4 was 18.35/mm$^2$ ± 5.18, followed in decreasing order by Brodmann's area 39, area 8, subiculum, area 40, area 9, area 23, area 24, area 22 and finally area 17–18 where the density was down to 0.90 ± 0.34 tangles/mm$^2$. In the Alzheimer's cases, the density was the highest in area 40 (13.6 ± 20.5 tangles/mm$^2$) and the lowest in area 17 (1.1 ± 0.9 tangles/mm$^2$). Density of tangles in area 4 was 1.3 ± 1.4 tangles/mm$^2$.

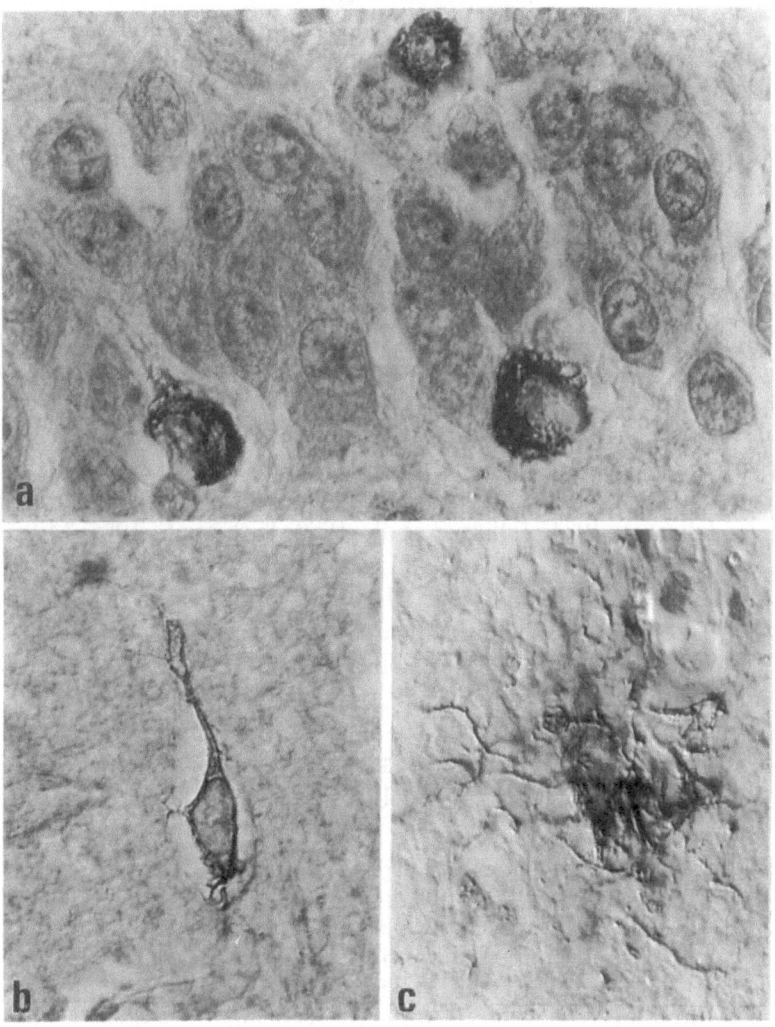

**Fig. 2.** Cortical neurofibrillary and glial fibrillary tangles in progressive supranuclear palsy. Staining by anti-tau immunohistochemistry counterstained by Harris haematoxylin. ×100 immersion objective (final magnification ×1,700). **a** tau positive inclusions in the dentate gyrus, **b** neurofibrillary tangle in the supramarginal gyrus, **c** glial fibrillary tangle in the supramarginal gyrus

## (2) Laminar topography

In area 4, the tangles mainly involved layers V and VI. The distribution was unimodal. By comparison, the Alzheimer's tangles, in the same area, were mainly located in layers III and V and their distribution was bimodal.

## (3) Area and form-factor

Mean surface area of the NFT containing perikarya was $122\,\mu m^2 \pm 6$ in PSP and $162\,\mu m^2 \pm 4.5$ in Alzheimer's disease ($p < 0.001$). The form factor of the NFT containing neurons was $0.54 \pm 0.007$ in PSP and $0.68 \pm 0.009$ in Alzheimer disease ($p < 0.001$), meaning that the shape of the perikaryon was more elongated in PSP.

### Discussion

These results emphasize the frequency and the specificity of cortical tangles in PSP. They suggest that the distribution (regional and laminar) of the tangles is not random.

The presence of cortical tangles was mentioned in the initial description of PSP (Steele et al., 1964) and in most subsequent reports (Behrman et al., 1969; Bugiani et al., 1979; Davis et al., 1985; Ghatak et al., 1980; Ishino et al., 1976; Jellinger, 1971; Jellinger et al., 1980; Kish et al., 1985; Mitsuyama et al., 1981; Probst et al., 1975; Rafal et al., 1981; Roy et al., 1974; Ruberg et al., 1985; Steele, 1972; Steele et al., 1964; Takahashi et al., 1987, 1989; Tellez-Nagel et al., 1973; Tomonaga, 1977; Yagishita et al., 1979; Zweig et al., 1987). At that time, the possibility that these alterations were of the same nature than those seen in the brainstem was underestimated. The opinion that these tangles were due to aging or to a coincidental Alzheimer's disease was expressed by several authors (Cruz-Sanchez et al., 1992; Jellinger et al., 1980; Kish et al., 1985; Probst et al., 1975). This was consistent with the opinion that the intellectual alterations of the disease was related to a subcortical mechanism (Albert et al., 1974).

Ishino et al. (1976) noticed the high density and the frequency of the PSP cortical tangles and suggested that they could be related to the disease process: one of the 2 patients was indeed 59 years old at death; senile plaques were absent in the 2 cases. The distribution of PSP tangles seemed distinctly different from the location of Alzheimer's tangles. The ultra-structural study of cortical PSP tangles by Takahashi et al. (1989) showed that they were made of 15 nm straight tubules, identical to those seen in the subcortical nuclei and considered specific of PSP (Powell et al., 1974; Roy et al., 1974; Tellez-Nagel et al., 1973). Biochemical studies have moreover suggested that the cortical PSP tangles were related to abnormally phos-phorylated tau proteins, whose chemical characteritics were different from those observed in Alzheimer's tangles (Flament et al., 1991). Finally, a recent neuropathological study insisted upon the presence of tangles in the dentate gyrus, which is usually spared in Alzheimer's disease (Hof et al., 1992).

The reactivity of PSP tangles to anti-ubiquitin antibodies has been dis-cussed. The presence of tufts of tau-positive filaments (Hauw et al., 1990) was also mentioned by Yamada et al. (1992) who showed that these were actually present in astrocytes (sometimes binucleated). The "tangles" seen in astrocytes ("glial fibrillary tangles") were shown to be abnormal straight tubules as in the neurons (Nishimura et al., 1992). The involvement of astrocytes suggests that the disease process is widespread and could in fact involve the cytoskeleton of a large variety of cell types.

Neuropil threads have been seen in subcortical nuclei (Probst et al., 1988) but were only rarely mentioned in the cortex (Hof et al., 1992). According to our observations, their presence is correlated with the presence of tangles.

Our studies (Hauw et al., 1990) have ascertained the specific location of the tangles in the neocortex: they mainly affect regions (such as area 4 and 39) which are known to be connected with the motor or oculomotor (Pierrot-Deseilligny, 1989; Pierrot-Deseilligny et al., 1989) system respectively and relatively spare the associative and sensory cortices. The negativity of the polyclonal ubiquitin antibody is in striking contrast with the results obtained in Alzheimer's disease (He et al., 1993; Lowe et al., 1988). It confirms the results obtained in the subcortical nuclei, where the negativity of polyclonal antiubiquitin has also been emphasized (Lennox et al., 1988). Three of the 4 monoclonal antibodies, tested by Tabaton et al. (1988) did not label the subcortical PSP tangles. Our quantitative data in area 4, as well as those obtained by Hof et al. (1992), suggest that the affected neurons belong mainly to the cortical efferent system (pyramids of layers V). This peculiar type of distribution might not be found in other cortical areas (Hof et al., 1992).

The presence of cortical lesions might suggest that the frontal symptoms, so frequently observed in PSP, could be related, at least in part, to direct injury of the cortex rather than to a subcortical mechanism. However, it should be stressed that the tangles are rare in the prefrontal cortex itself and mainly affect regions which are not considered as directly implicated in cognitive functions. These cortical lesions could then play a role in the "supranuclear" impairment of motor and oculomotor functions. The cortical alterations are, indeed, not randomly distributed but seem to affect a network of interrelated systems subserving these functions (Hauw et al., 1990). The clinico-pathological correlations in PSP might thus prove useful to unravel the physiological role of these anatomical systems, which are "labelled" by the cytoskeletal abnormalities.

## Acknowledgements

We thank Y. Agid, H. Beck, P. Brunet, F. Chain, D. Laplane and G. Rancurel who made the patients files available, J.-P. Brion, K. Beyreuther and C. Masters for the generous gift of their antibodies, and Mrs. N. Fenoy, C. Raiton and Mr. P. Miele for technical help.

## References

Albert ML, Feldman RG, Willis AL (1974) The "subcortical dementia" of progressive supranuclear palsy. J Neurol Neurosurg Psychiatry 37: 121–130

Behrman S, Caroll JD, Janota I, Matthews WB (1969) Progressive supranuclear palsy; clinico-pathological study of four cases. Brain 92: 663–678

Blumenthal H, Miller C (1969) Motor nuclear involvement in progressive supranuclear palsy. Arch Neurol 20: 362–367

Bugiani O, Mancardi GL, Brusa A, Ederli A (1979) The fine structure of subcortical neurofibrillary tangles in progressive supranuclear palsy. Acta Neuropathol (Berl) 45: 147–152

Cruz-Sanchez FF, Rossi ML, Cardozo A, Deacon P, Tolosa E (1992) Clinical and pathological study of two patients with progressive supranuclear palsy and Alzheimer's changes. Antigenic determinants that distinguish cortical and subcortical neurofibrillary tangles. Neurosci Lett 136: 43–46

Davis PH, Bergeron C, McLachlan DR (1985) Atypical presentation of progressive supranuclear palsy. Ann Neurol 17: 337–343

Delaère P, Duyckaerts C, Hauw J-J, Brion J-P, Poulain V (1989) Tau, paired helical filament and amyloid in the neocortex: a morphometric study of 15 cases with graded intellectual status in aging and senile dementia of Alzheimer type. Acta Neuropathol (Berl) 77: 645–653

Duyckaerts C, Brion J-P, Hauw J-J, Flament-Durand J (1987) Quantitative assessment of the density of neurofibrillary tangles and senile plaques in senile dementia of Alzheimer type. Comparison of immunochemistry with a specific antibody and Bodian's protargol method. Acta Neuropathol (Berl) 73: 167–170

Duyckaerts C, Hauw J-J, Bastenaire F, Piette F, Poulain C, Rainsard V, Javoy-Agid F, Berthaux P (1986) Laminar distribution of neocortical senile plaques in senile dementia of the Alzheimer type. Acta Neuropathol (Berl) 70: 249–256

Flament S, Delacourte A, Verny M, Hauw J-J, Javoy-Agid F (1991) Abnormal tau proteins in progressive supranuclear palsy. Similarities and differences with the neurofibrillary degeneration of the Alzheimer type. Acta Neuropathol (Berl) 81: 591–596

Ghatak NR, Nochlin D, Hadfield MG (1980) Neurofibrillary pathology in progressive supranuclear palsy. Acta Neuropathol (Berl) 52: 73–76

Hauw J-J, Verny M, Delaère P, Cervera P, He Y, Duyckaerts C (1990) Constant neurofibrillary changes in the neocortex in progressive supranuclear palsy. Basic differences with Alzheimer's disease and aging. Neurosci Lett 119: 182–186

He Y, Duyckaerts C, Delaère P, Piette F, Hauw J-J (1993) Alzheimer's lesions labelled by antiubiquitin antibodies: comparison with other staining techniques. A study of 15 cases with graded intellectual status in ageing and Alzheimer's disease. Neuropathol Appl Neurobiol 19: 364–371

Hof PR, Delacourte A, Bouras C (1992) Distribution of cortical neurofibrillary tangles in progressive supranuclear palsy: a quantitative analysis of six cases. Acta Neuropathol (Berl) 84: 45–51

Ishino H, Otsuki S (1976) Frequency of Alzheimer's neurofibrillary tangles in the cerebral cortex in progressive supranuclear palsy (subcortical argyrophyric dystrophy). J Neurol Sci 28: 309–316

Jellinger K (1971) Progressive supranuclear palsy (subcortical argyrophilic dystrophy). Acta Neuropathol (Berl) 19: 347–352

Jellinger K, Riederer P, Tomonaga M (1980) Progressive supranuclear palsy: clinico-pathological and biochemical studies. J Neural Transm [Suppl] 16: 111–128

Kish SJ, Chang LJ, Mirchandani L, Shannak K, Hornykiewicz O (1985) Progressive supranuclear palsy: relationship between extrapyramidal disturbances, dementia, and brain neurotransmitter markers. Ann Neurol 18: 530–536

Lennox G, Lowe J, Morrell K, Landon M, Mayer RJ (1988) Ubiquitin is a component of neurofibrillary tangles in a variety of neurodegenerative diseases. Neurosci Lett 94: 211–217

Lowe J, Blanchard A, Morrell K, Lennox G, Reynolds L, Billett M, Landon M, Doherty FJ, Mayer RJ (1988) Ubiquitin is a common factor in intermediate filament inclusion bodies of diverse type in man including those of Parkinson's, Pick's disease, and Alzheimer's disease as well as Rosenthal fibers in astrocytomas, cytoplasmic bodies in muscles, and Mallory bodies in alcoholic liver disease. J Pathol 155: 9–15

Mitsuyama Y, Seyama S (1981) Frequency of Alzheimer's neurofibrillary tangles in the brain of progressive supranuclear palsy, postencephalitic parkinsonism, Alzheimer's disease, senile dementia and non-demented elderly person. Folia Psychiatr Neurol Japonica 35: 189–204

Nishimura M, Namba Y, Ikeda K, Oda M (1992) Glial fibrillary tangles with straight tubules in the brains of patients with progressive supranuclear palsy. Neurosci Lett 143: 35–38

Pierrot-Deseilligny C (1989) Contrôle cortical des saccades. Rev Neurol (Paris) 145: 596–604

Pierrot-Deseilligny C, Rivaud S, Pillon B, Fournier E, Agid Y (1989) Lateral visually-guided saccades in progressive supranuclear palsy. Brain 112: 471–487

Powell HC, London GW, Lampert PW (1974) Neurofibrillary tangles in progressive supranuclear palsy; electron microscopic observations. J Neuropathol Exp Neurol 33: 98–106

Probst A, Dufresne JJ (1975) Paralysie supranucléaire progressive (ou dystonie oculofacio-cervicale). Schweiz Arch Neurol Neurochir Psychiatr 116: 107–134

Probst A, Langui D, Lautenschlager C, Ulrich J, Brion J-P, Anderton BH (1988) Progressive supranuclear palsy: extensive neuropil threads in addition to neurofibrillary tangles. Very similar antigenicity of subcortical neuronal pathology in progressive supranuclear palsy and Alzheimer's disease. Acta Neuropathol (Berl) 77: 61–68

Rafal RD, Grimm RJ (1981) Progressive supranuclear palsy: functionnal analysis of the response to methysergide and anti parkinsonian agents. Neurology 31: 1507–1518

Roy S, Datta CK, Hirano A, Ghatak NR, Zimmerman HM (1974) Electron microscopic study of neurofibrillary tangles in Steele-Richardson-Olszewski syndrome. Acta Neuropathol (Berl) 29: 175–179

Ruberg M, Javoy-Agid F, Lebalc'h N, Hirsch E, Scatton B, Lheureux R, Hauw J-J, Duyckaerts C, Gray F, Morel-Maroger A, Rascol A, Serdaru M, Agid Y (1985) Dopaminergic and cholinergic lesions in progressive supranuclear palsy. Ann Neurol 18: 523–529

Steele JC (1972) Progressive supranuclear palsy. Brain 95: 693–704

Steele JC, Richardson JC, Olszewski J (1964) Progressive supranuclear palsy; a heterogencous degeneration involving the brain stem, basal ganglia and cerebellum with vertical gaze and pseudobulbar palsy, nuchal dystonia and dementia. Arch Neurol 10: 333–359

Tabaton M, Perry G, Autillo-Gambetti L, Manetto V, Gambetti P (1988) Influence of neuronal location on antigenic properties of neurofibrillary tangles. Ann Neurol 23: 604–610

Takahashi H, Oyanagi K, Takeda S, Hinokuma K, Ikuda F (1989) Occurrence of 15-nm-wide straight tubules in neocortical neurons in progressive supranuclear palsy. Acta Neuropathol (Berl) 79: 233–239

Takahashi H, Takeda S, Ikuta F, Homma Y (1987) Progressive supranuclear palsy with limbic system involvement: report of a case with ultrastructural investigation of neurofibrillary tangles in various locations. Clin Neuropathol 6: 271–276

Tellez-Nagel I, Wisniewski HM (1973) Ultrastructure of neurofibrillary tangles in Steele-Richardson-Olszewski syndrome. Arch Neurol 29: 324–327

Tomonaga M (1977) Ultrastructure of neurofibrillary tangles in progressive supranuclear palsy. Acta Neuropathol (Berl) 37: 177–181

Yagishita S, Itoh Y, Amano N, Nakano T, Saitoh A (1979) Ultrastructure of neurofibrillary tangles in progressive supranuclear palsy. Acta Neuropathol (Berl) 48: 27–30

Yamada T, McGeer PL, McGeer EG (1992) Appearance of paired nucleated tau-
    positive glia in patients with PSP brain tissue. Neurosci Lett 135: 99–102
Zweig RM, Whitehouse PJ, Casanova MF, Walker LC, Jankel WR, Price DL (1987)
    Loss of pedunculopontine neurons in progressive supranuclear palsy. Ann Neurol
    22: 18–25

Authors' address: Dr. C. Duyckaerts, Laboratoire de Neuropathologie R. Escourolle, Hôpital de La Salpêtrière, 47, Blvd de l'Hôpital, F-75651 Paris, Cedex 13,
France.

# Vascular progressive supranuclear palsy

**J. Winikates** and **J. Jankovic**

Department of Neurology, Baylor College of Medicine, Houston, Texas, U.S.A.

**Summary.** *Background*: Progressive supranuclear palsy (PSP) is a neuro-logic syndrome of unknown cause. This idiopathic type of PSP is usually associated with characteristic clinical and pathological features. *Objective*: To assess evidence of cerebrovascular disease in a population of patients with clinically defined PSP, and to compare clinical and neuroimaging features in vascular versus idiopathic PSP. *Design and methods*: Using predetermined criteria, the records of 128 patients diagnosed with PSP were reviewed for evidence of vascular disease. *Results*: Thirty patients (23.3%) satisfied criteria for vascular PSP. The vascular group differed from the idiopathic group by asymmetric and predominantly lower body involvement ($p < 0.05$). Corticospinal signs, pseudobulbar signs, gait difficulties, de-mentia, and incontinence of bowels and bladder were also more common in the vascular group, but these differences did not reach statistical significance. *Conclusion*: PSP is a syndrome which can be caused by cerebro-vascular disease. In addition to an increased frequency of stroke risk factors and neuroimaging evidence of vascular disease, vascular PSP can be differen-tiated from idiopathic PSP by a higher degree of asymmetry, lower body involvement, and evidence of corticospinal and pseudobulbar signs.

## Introduction

In 1964, Steele, Richardson and Olszweski described a syndrome of pro-gressive supranuclear palsy (PSP) which has become recognized as distinct clinical-pathological entity, categorized as a parkinsonism plus syndrome (Steele et al., 1964; Jankovic, 1989; Jankovic et al., 1990). The cause of PSP is unknown, and it is generally viewed as an idiopathic neurodegenerative disorder. The relatively high frequency of a multi-infarct state in patients with clinically diagnosed PSP (Dubinsky, 1987; Moses, 1987; Tanner, 1987) suggests that a subtype of PSP may be due to vascular causes.

The existence of a vascular form of parkinsonism was proposed by Critchley as early as in 1929 (Critchley, 1929, 1981). In his discussion of arteriosclerotic parkinsonism, he emphasised the clinical characteristics of generalized rigidity, dementia, incontinence, pseudobulbar signs, and pyramidal tract signs; tremor was not prominent. Subsequent reports have refined the clinical features characteristically present in patients with vascular parkinsonism (Eadie and Sutherland, 1964; Friedman et al., 1986; Murrow et al., 1990; Parkes et al., 1974; Tolosa and Santamaria, 1984). In our earlier study (Dubinsky and Jankovic, 1987), we found that up to a third of patients with clinically diagnosed PSP could have a vascular etiology. The current study was designed to provide additional information about vascular PSP and to further characterize the syndrome.

## Methods

The records of 128 consecutive patients with the clinical diagnosis of PSP were reviewed. All patients were evaluated in the Movement Disorders Clinic at Baylor College of Medicine and were personally examined by one of the authors (JJ). The diagnosis of PSP was based on the following clinical criteria: 1. onset after age 40 years; 2. bradykinesia, axial rigidity, gait disorder, postural instability, and little or no tremor; 3. pseudobulbar signs including dysarthria, dysphagia, and emotional lability; and 4. marked vertical supranuclear gaze palsy with inability to move the eyes more than 15 degrees below the horizontal (Jankovic et al., 1990).

The following variables were analyzed: age, gender, duration of symptoms at time of initial visit, clinical characteristics, and evidence of vascular disease, based on the clinical description of the patient in the initial history and neurological examination, and follow up visits. Clinical characteristics noted were asymmetry of involvement, upper or lower body involvement. Tremor, gait disorder, rigidity, falling, and postural instability and other motor signs were rated on the Unified Parkinson Rating Scale (UPRS) (Fahn et al., 1987). Associated clinical findings of dementia, corticospinal signs, pseudobulbar signs (particularly emotional incontinence), bowel or bladder incontinence, orthostatic hypotension, history of seizures, history of neuropathy, and response to levodopa therapy.

Evidence of vascular disease was analyzed using a vascular rating scale. A point value was assigned for pathological, historical, and neuroimaging evidence of vascular disease. Two points were given for pathologically or angiographically proven diffuse vascular disease, 1 point for pathological evidence of both vascular and neurodegenerative PSP, and 1 point each for onset of symptoms within one month of a clinical stroke, a history of two or more strokes, a history of two or more risk factors for stroke, and neuroimaging evidence of diffuse vascular disease or vascular disease in two or more vascular territories. A point value of zero was assigned for the absence of each criteria. Thus, history of onset of symptoms of PSP beyond one month from a stroke, history of a single stroke, or history of a single stroke risk factor were scored as zero. This yielded a vascular score ranging from 0 to a maximum score of 6. A vascular score of 2 or more placed the patient in the vascular group; a patient with a vascular score of 0 or 1 was categorized as idiopathic.

The stroke risk factors analysed were based on the report of the Stroke Council of the American Heart Association (Dyken et al., 1984; Dyken, 1991). These included hypertension, smoking, diabetes mellitus, hyperlipidemia, the presence of any heart disease associated with stroke (coronary artery disease, atrial fibrillation, congestive heart failure, valvular heart disease, mitral prolapse, other arrhythmias), and other stroke risk factors, which included family history of stroke, peripheral vascular disease, polycythemia, and gout. The presence of carotid artery disease determined by ultrasound or angiography, in the absence of diffuse intracranial cerebrovascular disease was counted in this category, along with a history of carotid endarterectomy or the finding of a carotid bruit. Age and gender were not included separately as stroke risk factors, since they were recorded in the patient demographics.

Data were entered in a computerized data base (Paradox 3.5) for analysis. The mean and standard deviation of age and years duration of symptoms were compared by Student's t test. The various clinical characteristics were compared using the Chi squared function in a $2 \times 2$ matrix with the Yates correction. Statistical significance was assumed at a p value of $<0.05$.

## Results

Of the 128 patients, 30 (23.3%) satisfied criteria for vascular PSP, and the remaining 98 (76.7%) were classified as idiopathic. There was no difference between the two groups in terms of age (total = 69.2; vascular = 72.6; idiopathic = 68.2 years) or duration of symptoms (total = 5.2; vascular = 4.9; idiopathic = 5.3 years). There was a 61.7% male predominance in the total group; males comprised 73.3% of the vascular group and only 58.2% of the idiopathic group.

Only two clinical features distinguished the vascular group from the idiopathic group: asymmetry of clinical signs and predominance of lower body involvement (p < 0.05) (Table 1). Two additional features, presence of corticospinal tract signs and presence of pseudobulbar signs, were nearly twice as common in the vascular group compared to the idiopathic group, but this difference did not reach statistical significance. Similarly, tremor, gait disorder, falling, dementia, and urinary and fecal incontinence were higher in the vascular group, but none of these differences were statistically significant.

Two patients had pathologically proven vascular disease and 1 additional patient had diffuse intracranial cerebrovascular disease documented angiographically (Table 2).

Strokes occurred in both patient groups, but were significantly more common in the vascular group. Multiple strokes occurred only in the vascular group (Table 3). The vascular group had a higher frequency of hypertension and smoking. This difference was statistically significant. No statistically significant difference was noted between the two groups in reported frequency of diabetes mellitus, hyperlipidemia, or other stroke risk factors.

There was no difference between the two groups with respect to any particular type of heart disease. Coronary artery disease was present in 5 patients in the vascular group and in 3 of the idiopathic group, atrial fibrillation was present in 2 of the vascular group and in 3 of the idiopathic group, and other arrhythmias were present in 2 of the vascular group, and 4 of the idiopathic group. One patient in the idiopathic group had a history of complete heart block. Aortic valve thickening was noted in 1 patient in the vascular group, and two of the idiopathic group. Mitral valve prolapse was present in 1 patient in each group. No patient had a history of rheumatic valvular disease. A history of congestive heart failure was noted in two patients in each group.

Review of other stroke risk factors revealed no significant difference between the two groups. A family history of stroke was the most common other risk factor, and was reported in 3 patients in the vascular group and

**Table 1.** Clinical features: vascular versus idiopathic PSP

|  | Vascular | Idiopathic | Significance |
|---|---|---|---|
| Asymmetric | 44.8% | 19.8% | p < 0.05 |
| Upper body | 6.9 | 10.4 | NS |
| Lower body | 27.6 | 9.4 | p < 0.05 |
| Tremor | 17.2 | 9.4 | NS |
| Rigidity | 86.2 | 82.3 | NS |
| Gait disorder | 93.1 | 77.1 | NS |
| Postural instability | 93.1 | 87.5 | NS |
| Falling | 79.3 | 71.6 | NS |
| Dementia | 55.2 | 44.3 | NS |
| Seizures | 3.5 | 2.1 | NS |
| Corticospinal signs | 20.7 | 11.5 | NS |
| Neuropathy | 10.3 | 3.1 | NS |
| Pseudobulbar signs | 41.4 | 22.5 | NS |
| Incontinence | 20.7 | 13.4 | NS |
| Orthostatic hypotension | 0 | 8.2 | NS |
| Response to levodopa | 17.2 | 19.6 | NS |

**Table 2.** Categorization of patients

|  | Vascular (N = 29) | Idiopathic (N = 97) | Difference |
|---|---|---|---|
| Pathologically proven | 4 | 0 | p < 0.05 |
| Onset with stroke | 4 | 1 | p < 0.05 |
| History of 2 or more strokes | 3 | 0 | p < 0.05 |
| Neuroimaging with vascular disease | 26 | 19 | p < 0.001 |
| Stroke risk factors ≥2 | 26 | 19 | p < 0.001 |

**Table 3.** Stroke risk factors: vascular versus idiopathic PSP

|  | Vascular (N = 29) | Idiopathic (N = 97) | Difference |
|---|---|---|---|
| History single stroke | 5 | 4 | p < 0.05 |
| History multiple strokes | 3 | 0 | p < 0.05 |
| Hypertension | 18 | 31 | p < 0.01 |
| Diabetes | 1 | 6 | NS |
| Heart disease | 9 | 17 | NS |
| Smoking | 21 | 27 | p < 0.001 |
| Hyperlipidemia | 3 | 7 | NS |

12 in the idiopathic group. Carotid artery disease was found in 1 patient in each group, and one patient in each group had a history of carotid endarterectomy. Gout was reported in 4 patients of the idiopathic group, but in none in the vascular group. Peripheral vascular disease was noted in 4

Table 4. Neuroimaging findings: vascular versus idiopathic PSP

|  | Vascular | Idiopathic | Difference |
|---|---|---|---|
| Computed tomography performed | 15 | 46 | NS |
| MRI performed | 23 | 55 | NS (p = 0.08) |
| Evidence of vascular disease | 25 | 27 | p < 0.001 |
| Single vessel distribution | 2 | 8 | NS |
| Multiple vascular beds | 24 | 18 | p < 0.001 |
| Periventricular white matter changes | 24 | 21 | p < 0.001 |
| Internal capsule or centrum semiovale | 17 | 10 | NS |
| Basal ganglia | 6 | 14 | NS |
| Brain stem | 6 | 14 | NS |
| Hydrocephalus | 2 | 3 | NS |
| Atrophy | 23 | 56 | NS |

patients of the vascular group, and none of the idiopathic patients. One patient in the vascular group had documented diffuse systemic vasculitis, though no clear CNS vasculitis was noted. One vascular patient had a history of thrombophlebitis.

Although more patients in the vascular group were imaged by MRI (23 of 30 or 77%) as compared to the idiopathic groups (55 of 97 or 57%), this difference was not statistically significant. Approximately half of the patients in each group were imaged by CT, and a number of patients were imaged by both modalities. Findings from both modalities were recorded in a similar fashion. Evidence of vascular disease and involvement of multiple vascular distributions were more common in the vascular group. Periventricular white matter changes were particularly common in the vascular group. These differences were statistically significant. There was no difference in hydrocephalus or atrophy between the two groups. One patient in the idiopathic group had a normal angiogram, while 3 angiograms from patients in the vascular group showed intracranial disease: 1 diffuse, 1 single vessel occlusion, and 1 showed marked tortuosity of the vertebrobasilar system (Table 4).

Data on CSF analysis and EEG findings were collected in both patient groups. A total of 13 patients in the total series had CSF analysis. Two had low CSF homovanillic acid and 2 had low 5-hydroxy indole acetic acid. No other abnormal findings were noted. EEGs were performed in 47 patients in the total series. Twenty four were normal, and 18 showed only mild diffuse, slowing. The few focal abnormalities were confined to asymmetric temporal slowing. No epileptiform discharges were seen. Because these data were available in only a minority of the total patient group, they were not included in the statistical analysis.

**Illustrative cases**

*Case 1*

A 69 years old right handed man presented for evaluation of diplopia and forgetfulness. He had a 5 year history of progressive memory loss, unsteady gait with occasional falling, and episodes of tearfulness. Eight months prior to initial evaluation, he began to complain of intermittent diplopia. He had a history of hypertension for 5 years, a positive family history of stroke, and no history of diabetes, hyperlipidemia, smoking, or heart disease.

Initial examination revealed dementia, hypomimia, bradykinesia, rigidity, moderately severe in the neck, mild in the extremities, and postural instability. Gait was broad based and shuffling. No tremor was noted. Eye movements were markedly impaired in both vertical and horizontal planes, but the defect could be overcome by the oculocephalic maneuver. He was also noted to have an extensor plantar response on the left.

Neuroimaging by CT revealed diffuse atrophy of the cerebral hemispheres, but no midbrain atrophy. Prominent periventricular hyperlucent areas were noted bilaterally. An EEG showed diffuse slowing with no focal abnormalities. The CSF protein was 50, glucose 70, with no cells. HVA and 5-HIAA were not measured.

The patient experienced progressive dementia and gait difficulty which did not respond to levodopa. He expired in a nursing home 15 months after initial evaluation. Pathological examination of his brain revealed multiple lacunar infarcts varying from 0.1 to 1.0 cm in the centrum semiovale, diencephalon, mesenchephalon, and pons, worse on the right. No neurofibrillary tangles, neuronal cell loss, or gliosis of typical PSP were seen (Dubinsky et al., 1987).

*Case 2*

A 73 years old woman presented for evaluation of a four years history of "Parkinson's disease", poorly responsive to levodopa. Four years prior to evaluation, she developed bradykinesia, stiffness, gait difficulty, and frequent falling. A slight tremor was noted in the right hand, and stiffness was noted to be worse on the right side. About 2½ years before evaluation, she had worsening gait problems and had developed slurred speech. She then developed severe depression, which worsened further by 3 months before evaluation, by which time she was no longer able to walk. Levodopa had only minimal initial benefit, bromocriptine was discontinued due to hallucinations, and the patient complained of diffuse weakness on selegiline.

The patient had a 6 years history of hypertension, treated with calcium channel blockers. She had later developed orthostatic lightheadedness and urgency to urinate. Second surgical bladder suspension was complicated by a cardiac arrest. She had no history of dementia, stroke or TIA, heart disease, diabetes, hyperlipidemia, or smoking, and no family history of stroke, though she did have a half sister with "Alzheimer's disease".

On initial examination, she had orthostatic hypotension, blood pressure falling from 160/84 supine to 115/80 standing. She appeared depressed, but with normal cognitive function. Vertical ophthalmoparesis was evident on both down and up gaze, and horizontal eye movements were normal; abnormal eye movements could be overcome by the oculocephalic maneuver. Marked neck rigidity was noted (4+ on the UPRS scale), with only mild rigidity in the extremities (2+). Postural instability and gait difficulty were noted (3+), and a mild rest tremor was present in the right hand (1+).

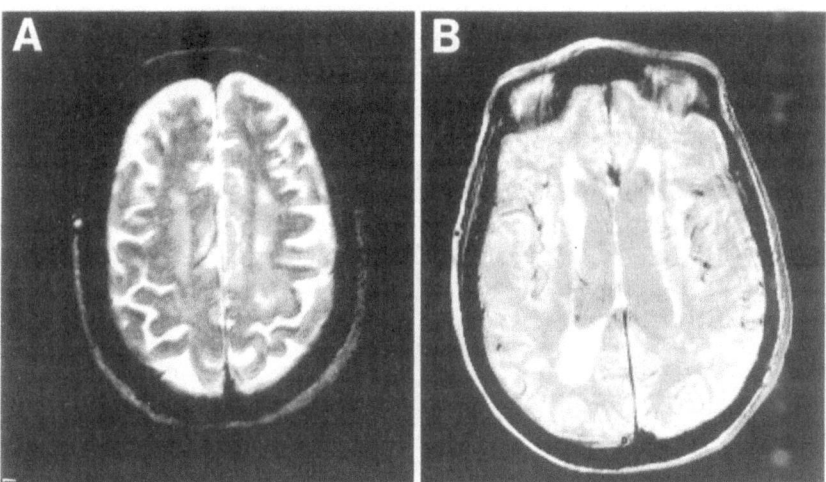

**Fig. 1.** Case 2: MRI of the brain demonstrating periventricular white matter changes

Follow up examinations over the next six months revealed progressive dementia, worsening rigidity, and worsening depression, with development of pseudobulbar affect and emotional incontinence.

MRI of the brain at the time of initial evaluation demonstrated periventricular white matter changes which were confluent, and particularly noted around the posterior horns of the lateral ventricles (Fig. 1).

## Case 3

A 72 years old right handed man with a history of cerebrovascular disease and hypertension presented for evaluation of "Parkinson's disease". He had two episodes of left sided weakness and numbness 8 and 6 years previously, both of which had resolved completely. Seventeen months prior to our evaluation he had another episode of left sided weakness which was mild, but persisted. This persistent difficulty prompted neurologic evaluation at another institution with findings of bilaterally increased tone, worse on the right, truncal rigidity, abnormal gait, and right hyperreflexia with an extensor plantar response on the right side. He was also noted to have anxiety and depression. He responded poorly to levodopa and sought further evaluation.

Initial examination dementia and an "astonished" facial expression with hypomimia. Vertical ophthalmoparesis was present, which could be overcome by the oculocephalic maneuver. He also had a low voice volume with monotonous speech, moderately severe neck rigidity (3+) with milder rigidity of the extremities, worse on the right, and in the lower extremities. His gait was broad based and unsteady, with some dragging of the right foot, and marked postural instability. No tremor was present. A right extensor plantar was noted. Follow up over the next three years revealed worsening gait difficulty and dementia, with poor response to levodopa.

Neuropsychological testing revealed mild depression and a variable pattern of cognitive defects felt to be consistent with multi-infarct dementia. An echocardiogram revealed concentric left ventricular hypertrophy consistent with chronic hypertension. MRI of the brain revealed diffuse atrophy of the cerebral hemispheres and diffuse periventricular and subcortical white matter changes on spin echo and T2 weighted images (Fig. 2).

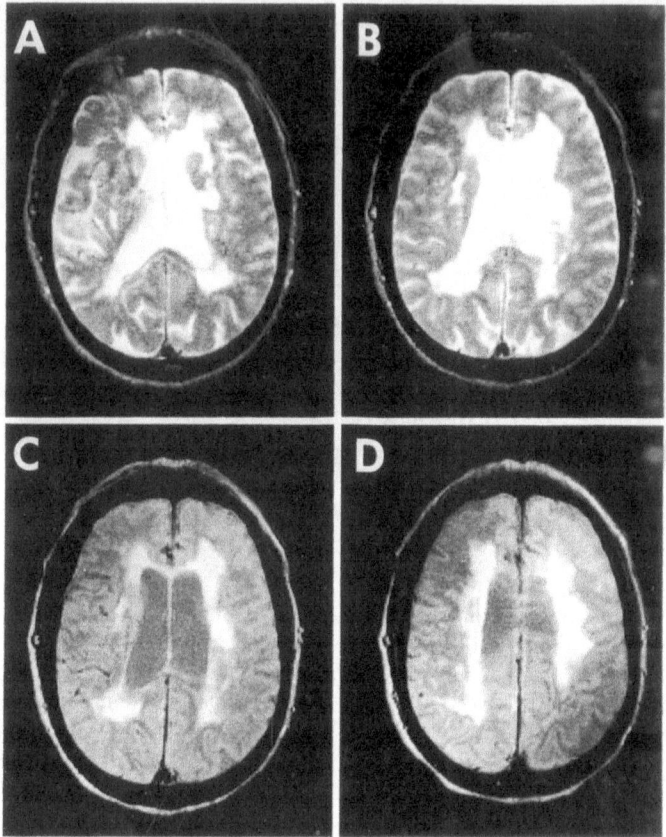

**Fig. 2.** Case 3: MRI of the brain revealed diffuse atrophy of the cerebral hemispheres and diffuse periventricular and subcortical white matter changes on T2 weighted (**A,B**) and spin echo images (**C,D**)

## Discussion

The 128 patients included in this study were diagnosed with PSP based on previously established clinical criteria (Jankovic et al., 1990). The average age of our patients was somewhat older than in previous reports; our patients averaged 69.2 years, compared to the previous reports of 62.0 (Maher and Lees, 1986), 62.9 (Golbe et al., 1987), and 63.2 years (Friedman and Jankovic, 1989). The average age was even higher in the vascular group (72.6 years) and advanced age may be a risk factor for cerebrovascular disease (Dyken et al., 1984; Dyken, 1991). The 61.7% male preponderance was similar to that reported in other series (Friedman and Jankovic, 1989; Davis et al., 1988). The percentage of men was higher in the vascular group (73.3%) as compared to the idiopathic group (58.2%). This higher percentage of males in our vascular group may reflect the role of male sex as a risk factor for cerebrovascular disease (Dyken et al., 1984; Dyken, 1991).

Predominance of asymmetric involvement and corticospinal signs in the vascular group could be explained by focal asymmetric vascular disease. Vascular ischemic injury of the corticospinal tracts, as in the subcortical white matter, could account for corticospinal tract signs. Asymmetric signs could arise from asymmetry in vascular involvement. Bilateral plantar responses were reported in a small series of vascular PSP patients by Moses and Zee (1987). The predominant lower body involvement in vascular PSP is similar to another form of vascular parkinsonism which we previously referred to as "lower body parkinsonism" (FitzGerald, 1989). Critchley in his initial report, described rigidity, greater in the lower extremities than in the upper, as a prominent feature of arteriosclerotic parkinsonism (Critchley, 1929). This lower body predominance correlates well with the increased percentage of gait disorder in our vascular group (93.1%) versus idiopathic group (77.1%). The presence of gait disorder has also been reported in Binswanger's subcortical dementia, a disorder of presumed vascular etiology (Kinkel et al., 1985; Thompson and Marsden, 1987). The gait disorder of Binswanger's may be due to diffuse vascular disease through disruption of the interconnecting fiber tracts between the basal ganglia and cerebellum, and the leg areas of the motor cortex (Thompson and Marsden, 1987). Other gait disorders presumably caused by multi-infarct state have been referred to as "gait ignition failure" (Atchison et al., 1993) and "primary progressive freezing gait" (Achiron et al., 1993).

Nygaard et al. (1986) reported an increased incidence of seizures and focal EEG abnormalities in their series of patients with PSP. The incidence of seizures in our series was low, overall, and in both the vascular and idiopathic groups. The EEG abnormalities noted in our series differed from those reported by Nygaard (1986). They consisted predominately of diffuse slowing and focal temporal lobe slowing. There were no epileptiform abnormalities noted in any of our patients.

We wished to seperate all patients with clear evidence of cerebrovascular disease from the idiopathic group and, therefore, we employed strict criteria for the diagnosis of vascular PSP. Thus, inclusion in the vascular group required evidence of multiple strokes by history, evidence of involvement of multiple vascular territories by imaging, and presence of multiple stroke risk factors. Hypertension and smoking were the only stroke risk factors that were statistically more common in the vascular group. Hypertension was present in the majority of patients reported by Moses and Zee (1987) with possible vascular PSP. An increased frequency of hypertension and smoking were noted in PSP patients with possible multi-infarct state reported by Dubinsky and Jankovic (1987). In one epidemiologic study of PSP, Davis et al. (1988) found no evidence of increased prevalence of hypertension or smoking. There was no significant difference between the two groups in regard to frequency of diabetes, hyperlipidemia, or heart disease. This lack of difference, while similar to the findings of Davis et al. (1988) who compared PSP to a control group, was nonetheless surprising. Failure to find a difference between the two groups may have been the result of under reporting of a given risk factor in the clinical records reviewed for this study. Since the original purpose of the patient evaluations at the time the data was gathered was not to identify vascular risk factors, these may not have been systematically sought, and so their presence may be underestimated in the data gathered. This may be particularly true of hyperlipidemia, since serum lipids were not routinely measured, and since some patients may have been on diet therapy alone, with no drug therapy. Similarly, closer inquiry may have revealed more patients with coronary, or other heart disease.

We used a vascular score which rated historical, clinical, and neuroimaging data to define a subgroup of PSP patients with underlying vascular disease as the cause. A similar method was used by Hachinski to identify patients with multi-infarct dementia (Hachinski et al., 1974, 1975). As modified by Rosen et al. (1980), the Hachinski Ischemia Scale has been used in differentiating patients with vascular dementia from those with Alzheimer's disease, and its validity has been confirmed by pathological studies. Although there are similarities between our scale and the Hachinski Ischemia Scale, we used stricter criteria for the diagnosis of a vascular CNS disorder. The Hachinski scale includes abrupt onset and step-wise deterioration, whereas our vascular scale utilized more specific criteria of onset of PSP symptoms within one month following a clinical stroke. While our scale requires a history of 2 or more strokes, the Hachinski scale requires a history of only 1 stroke. Only one stroke risk factor, hypertension, is included in the Hachinski scale. We have included other risk factors, but require the presence of 2 or more. The Hachinski scale utilizes the presence of emotional incontinence and of somatic symptoms. Since these are also features of idiopathic PSP, we did not include those in the identification of the vascular PSP group. The Hachinski scale includes pseudobulbar and corticospinal signs as vascular features; our vascular group proved to have an increased frequency of such signs, supporting the validity of our vascular criteria.

An important difference between our vascular scale and the Hachinski Ischemia Scale is in the role of neuroimaging studies. Our vascular scale includes neuroimaging evidence of diffuse ischemic changes, while imaging is not a feature of the Hachinski scale. Erkinjunti et al. (1988, 1991) employed a combination of criteria similar to those of Hachinski with use of neuroimaging evidence of multiple infarcts or ischemic white matter changes in identifying cases of multi-infarct dementia. They found an accuracy of 90% when correlating antemortem diagnosis of multi-infarct dementia with subsequent postmortem examination when these combined criteria were utilized (Erkinjuntti and Sulkava, 1991). This experience in multi-infarct dementia suggests that clinical neuroimaging does play a role in identifying patients with PSP due to vascular causes.

The predominant neuroimaging finding in our patients was diffuse subcortical and periventricular white matter change, in particular, increased signal intensity in these areas on T2 weighted MRI scans. These white matter changes have been described both on CT scans (Valentine et al., 1980), and MRI scans (Drayer, 1988a,b), and have been described in vascular dementia, Binswanger's disease, Alzheimer' disease, and normal aging (Drayer, 1988a,b; Hershey et al., 1987). Various mechanisms have been proposed to explain these changes: transependymal fluid leakage, cerebral edema, demyelination, perivascular rarefaction, and discreate lacunar infarction (Valentine et al., 1980; Drayer, 1988a). Drayer (1988a,b) reviewed these changes and their proposed mechanisms and noted that the predominant areas of involvement are in the watershed areas between the deep perforating vessels and the coritcal medullary vessels. He proposed that aging and hypertension, as well as hypoxia and hypoperfusion, lead to arteriolar changes with consequent ischemia, leading to arteriolosclerosis, myelin pallor, and perivascular focal demyelination. He proposed that these mechanisms were the main cause of the white matter changes so often seen on MRI. Our findings of increased periventricular and subcortical white matter changes in our vascular group, associated with increased hypertension and stroke risk factors, are similar to the findings of others (Drayer, 1988a,b; Hershey et al., 1987; Gerard and Weisberg, 1986; Sullivan et al., 1990; Fazekas et al., 1993).

## Conclusion

While the majority of patients with PSP have a neurodegenerative disease of unkown cause, a significant percentage (23.3% in our series) have clinical PSP secondary to cerebrovascular disease. This vascular group is characterized by a higher frequency of pseudobulbar and corticospinal signs, and asymmetric and lower body involvement. PSP patients with such clinical features should have neuroimaging studies and their potential stroke risk factors should be addressed. The extent to which these findings may apply to other patient groups or to the general population of patients with PSP is not clear, since our patient population is drawn from a referral clinic base, in which patients may have more advanced or atypical disease. Whether interventions based on such findings, such as antiplatelet drug therapy or stroke risk factor management, will lead to an improved clinical course in this group of patients remains to be evaluated.

## Acknowledgement

The authors wish to gratefully acknowledge the assistance of Mr. K. Schwartz in technical advice in the data base structure, and in performance of the statistical analysis.

## References

Achiron A, Ziv I, Goren M, et al (1993) Primary progressive freeging gait. Mov Disord 8: 293–297

Atchison PR, Thompson PD, Frackowiak RSJ, Marsden CD (1993) The syndrome of gait ignition failure: a report of six cases. Mov Disord 8: 285–292

Critchley M (1929) Arteriosclerotic parkinsonism. Brain 52: 23–83

Critchley M (1981) Arteriosclerotic pseudo-parkinsonism. In: Rose FC, Capildo R (eds) Research progress in Parkinson's disease. Pitman, London, pp 40–42

Davis PH, Golbe LI, Duvoisin RC, Schoenberg BS (1988) Risk factors for progressive supranuclear palsy. Neurology 38: 1546–1552

Drayer BP (1988a) Imaging of the aging brain, part I. Normal findings. Radiology 166: 785–796

Drayer BP (1988b) Imaging of the aging brain, part II. Pathological conditions. Radiology 166: 797–806

Dubinsky RM, Jankovic J (1987) Progressive supranuclear palsy and a multi-infarct state. Neurology 37: 570–576

Dyken ML (1991) Stroke risk factors. In: Norris JW, Hachinski VC (eds) Prevent of stroke. Springer, New York

Dyken ML, Wolf PA, Barnett, HJM, et al (1984) Risk factors is stroke: a statement for physicians by the subcommitte on risk factors and stroke of the stroke council. Stroke 15: 1105–1111

Eadie MJ, Sutherland JM (1964) Arteriosclerosis in parkinsonism. J Neurol Neurosurg Psychiatry 27: 237–240

Erkinjuntti T, Haltia M, Palo J, et al (1988) Accuracy of the clinical diagnosis of vascular dementia: a prospective clinical and post-mortem neuropathological study. J Neurol Neurosurg Psychiatry 51: 1037–1044

Erkinjuntti T, Sulkava R (1991) Diagnosis of multi-infarct dementia. Alzheimers Dis Assoc Dis 5: 112–121

Fahn S, Elton RL, and Members of the UPDRS Development Committee (1987) Unified Parkinson's disease rating scale. In: Fahn S, Marsden CD, Calne D, Goldstein M (eds) Recent developments in Parkinson's diesease, vol 2. MacMillan Healthcare Information, New Jersey, pp 153–163

Fazekas F, Kleinert R, Offenbacher H, et al (1993) Pathologic correlates of incidental MRI white matter signal hyperinteurities. Neurology 43: 1683–1689

FitzGerald PM, Jankovic J (1989) Lower body parkinsonism: evidence for vascular etiology. Mov Disord 4: 249–260

Friedman A, Kang VJ, Tatemichi TK, Burke RE (1986) A case of parkinsonism following striatal lacunar infarction. J Neurol Neurosurg Psychiatry 49: 1087–1088

Friedman DI, Jankovic J (1989) Progressive supranuclear palsy: a quarter century of progress. In: Appel SH (ed) Current neurology, vol 9, pp 191–218

Gerard G, Weisberg LA (1986) MRI periventricular lesions in adults. Neurology 36: 998–1001

Golbe LI, Davis PH, Schoenberg BS, Duvoisin RC (1987) The natural history and prevalence of progressive supranuclear palsy. Neurology 37 [Suppl] 1: 121

Hachinski VC, Iliff LD, Zilhka E, et al (1975) Cerebral blood flow in dementia. Arch Neurol 32: 632–637

Hachinski VC, Lassen NA, Marshall J (1974) Multi-infarct dementia: a cause of mental deterioration in the elderly. Lancet ii: 207–210

Hershey LA, Modic MT, Greenough G, Jaffe DF (1987) Magnetic resonance imaging in vascular dementia. Neurology 37: 29–36

Jankovic J (1989) Parkinsonism plus syndromes. Mov Disord 4: S95–S119

Jankovic J, Friedman DI, Pirozzolo FJ, McCrary JA (1990) Progressive supranuclear palsy: motor, neurobehavioral, and neuroophthalmic findings. Adv Neurol 53: 293–304

Kinkel WR, Jacobs L, Polachini I, Bates V, Heffner RR (1985) Subcortical arterio-sclerotic encephalopathy (Binswanger's disease): computed tomographic, nuclear magnetic resonance, and clinical correlates. Arch Neurol 42: 951–959

Maher ER, Lees AJ (1986) The clinical features and natural history of the Steele-Richardson-Olszewski syndrome (progressive supranuclear palsy). Neurology 36: 1005–1008

Moses H, Zee D (1987) Multi-infarct PSP. Neurology 37: 1819

Murrow RW, Schweiger GD, Kepes JJ, Koller WC (1990) Parkinsonism due to a basal ganglia lacunar state: clinicopathological correlation. Neurology 40: 897–900

Nygaard TG, Duvoisin RD, Manocha M, Chokroverty S (1986) Epileptic seizures in progressive supranuclear palsy. Neurology 36 [Suppl] 1: 341

Parkes JD, Marsden CD, Rees JE, et al (1974) Parkinson's disease, cerebral arterio-sclerosis, and senile dementia. Q J Med 43: 49–61

Rosen WG, Terry RD, Fuld PA, et al (1980) Pathological verification of ischemic score in differentiation of dementias. Ann Neurol 7: 486–488

Steele JC, Richardson JC, Olszewski J (1964) Progressive supranuclear palsy: a hetero-geneous degeneration involving the brain stem, basal ganglia, and cerebellum with vertical gaze and pseudobulbar palsy, nuchal dystonia, and dementia. Arch Neurol 10: 333–359

Sullivan P, Pary R, Telang F, Rifai AH, Zubenko GS (1990) Risk factors for white matter changes detected by magnetic resonance imaging in the elderly. Stroke 21: 1424–1428

Tanner CM, Goetz CG, Klawans HL (1987) Multi-infarct PSP. Neurology 37: 1819

Thompson PD, Marsden CD (1987) Gait disorder of subcortical arteriosclerotic en-cephalopathy: Binswanger's disease. Mov Dis 2: 1–8

Tolosa ES, Santamaria J (1984) Parkinsonism and basal ganglia infarcts. Neurology 34: 1516–1518

Valentine AR, Moseley IF, Kendall BE (1980) White matter abnormality in cerebral atrophy: clinicoradiological correlations. J Neurol Neurosurg Psychiatry 43: 139–142

Authors' address: Dr. J. Jankovic, Department of Neurology, Baylor College of Medicine, 6550 Fannin, Suite #1801, Houston, TX 77030, U.S.A.

# Biochemical aspects

# Cholinergic and peptidergic systems in PSP

## F. Javoy-Agid

INSERM U 289, Hôpital de la Salpêtrière, Paris, France

**Summary.** PSP is associated with a widespread cholinergic deficit likely corresponding to a loss in cholinergic neurons. The cholinergic damage dramatically affects the basal ganglia and specific cell groups of the mesencephalon and pons. This provides an anatomically defined basis for motor and supranuclear oculomotor syndromes characteristic of PSP.

Unlike Alzheimer's disease and Parkinson's disease with dementia, the disease is not associated with a marked cholinergic deficiency in the cerebral cortex.

Various peptides are present at normal concentrations in extrapyramidal and limbic subcortical areas in brains of patients with PSP. Of particular interest, is somatostatin, the levels of which are subnormal in cerebral cortex of patients with dementia of Alzheimer' or Parkinson's disease type.

## Introduction

PSP is a disease in which a large number of brain structures are lesioned, the pattern of neuronal loss is consistent and characteristic of the disease (Steele et al., 1964). Although it is not known what makes these particular neurones die, neurochemical investigations have identified a number of neuronal systems vulnerable to PSP. The early studies were performed by biochemical assays on various structures of the post-mortem human brain from PSP patients compared to appropriate control populations. The difficulty with these studies is that a brain area is a tissue where heterogeneous populations mix together in addition many nuclei of interest are often too small to be dissected with confidence. In the past few years the advent of quantitative immunocytochemical and histochemical techniques with computer assisted image analysis has considerably increased the precision, the sensivity and the resolution of neurochemical analysis.

In PSP, as in idiopathic Parkinson's disease, a hallmark is the degeneration of dopaminergic neurons in the substantia nigra. Loss of nigral neurones has long been evident on classical neuropathological preparations. The dopaminergic nature of these neurones has been evidenced by measures of tyrosine hydroxylase activity (Jellinger et al., 1980), of concentrations of dopamine, its metabolites (Ruberg et al., 1985) and its uptake sites

(Maloteaux et al., 1988; Pierot et al., 1988) in homogenates of striatum, and cell counts in the substantia nigra, on preparations stained by a specific antiserum against tyrosine hydroxylase (Hirsch et al., 1988).

Such measures have lead to conclude that degeneration of dopaminergic neurons is more severe in PSP than in idiopathic Parkinson's disease, and, affects caudate nucleus and putamen to the same degree, unlike Parkinson's disease where the putamen is more severely denervated (see review Agid et al., 1987; and Ruberg et al., 1992). Another striking difference with Parkinson's disease is the absence of decrease in dopamine concentrations in the nucleus accumbens or frontal cortex which are terminal regions of the mesocorticolimbic dopaminergic system (Ruberg et al., 1985; Kish et al., 1985). As in Parkinson's disease, an increase in the metabolism of dopamine in reaction to the lesions is observed (Kish et al., 1985; Ruberg et al., 1985) in the residual dopaminergic nerve terminals which may reflect compensatory upregulation of dopaminergic transmission. Dopamine D1 receptors were unaltered in the striatum of PSP patients (Pierot et al., 1988), as were levels of the D1 associated second messenger DARP-32, measured by in vitro phosphorylation (Raisman et al., 1990), or immunoblot (Girault et al., 1989), indicating that the medium sized spiny neurones in the striatum, thought to be GABAergic, are not affected by the disease. At difference, the number of D2 receptors was found to decrease in the putamen and caudate nucleus both in studies post-mortem (Ruberg et al., 1985) or in vivo as detected by position emission tomography (Baron et al., 1985). This decrease, reflects loss of dopaminoceptive cholinergic neurones in the striatum (see below) and may have something to do with the irresponsiveness of patients to dopaminergic or anticholinergic drugs, intended to restore normal output from the striatum. The loss of these receptors clearly distinguishes PSP from other Parkinson-like syndrome. Decreases in D2 receptor density were also found in a number of other brain areas (nucleus accumbens, substantia innominata and cerebral cortex) (Ruberg et al., 1985).

The decrease in the levels of the various markers of striatal dopaminergic innervation reflects a loss of the neurones as indicated in a quantitative study of the tyrosine hydroxylase-immunostained cell bodies in the mesencephalon of patients with PSP when compared to controls (Hirsch et al., 1988).

Beside, investigations on the dopaminergic neurones, cholinergic and peptidergic systems have stimulated interest, since these systems have been widely studied in degenerating disorders that share certain clinical and pathological features with PSP (i.e. Parkinson's disease, Alzheimer's disease).

## Cholinergic systems in PSP

Progressive supranuclear palsy is characterized by widespread damage to cholinergic systems in the brain. Immunocytochemical studies have shown that the biochemical evidence of cholinergic deficits correspond to a loss of cholinergic neurons in a number of brain regions.

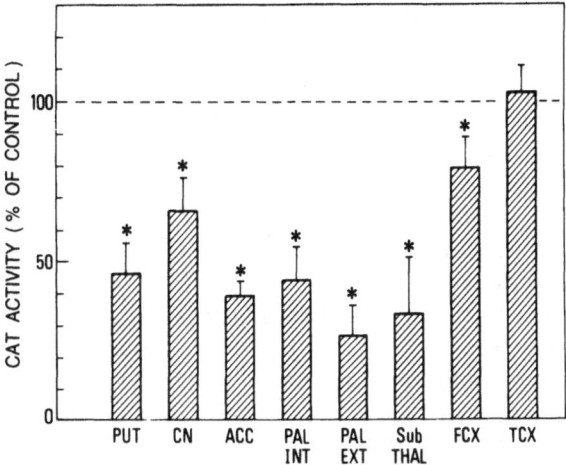

**Fig. 1.** Distribution of choline acetyltransferase (CAT) activity in brains of patients with PSP. Enzyme activity (+/− SEM) is expressed as percentage of control values in respective brain areas. Data were obtained from 9 PSP and 20 to 24 control brains. *p ≤ 0.05, statistically different compared to respective control. *PUT* putamen, *CN* caudate nucleus, *ACC* nucleus accumbens, *PAL INT* internal pallidum, *PAL EXT* external pallidum, *sub THAL* subthalamus, *FCX* frontal cerebral cortex, *TCX* temporal cerebral cortex

In the basal ganglia cholineacetyltransferase (ChAT) activity (an index of cholinergic innervation) is decreased by 40% to 70% in the caudate nucleus, the putamen, the nucleus accumbens, the internal pallidum, and the subthalamic nucleus (Fig. 1) (Agid et al., 1986). However, the number of muscarinic receptors ($^3$H-QNB binding sites) did not differ from control values anywhere in the brain, indicating that cholinergic receptivity had not been altered. The widespread decrease in ChAT activity in these subcortical areas likely reflects neuronal loss. Indeed, quantitative immunohistochemical analysis has shown, that although the neuropil seems to be as immunoreactive in the striatum of PSP brains as in normal subjects, ChAT-stained cell bodies appear to be more sparse (Hirsch et al., 1989). These observations suggest that cholinergic receptor agonists might be used profitably to substitute for the reduced cholinergic transmission due to the loss of cholinergic neurons when treating the motor symptoms of PSP. Finally, it is noteworthy that in the striatum of patients with PSP, loss of cholinergic interneurons is generalized throughout the structure, whereas in Alzheimer's disease the loss is evident only in the ventral striatum, corresponding to the nucleus accumbens (Lehericy et al., 1989). No striatal cholinergic neurons are lost in Parkinson's disease (Hirsch et al., 1989). In PSP, the loss of cholinergic neurons could account for the reduction in density of D2 dopamine receptors (see above).

Cholinergic transmission was expected to be found deficient in the cerebral cortex of PSP patients with neuropsychological disorders, as was the case with intellectually deteriorated patients with Parkinson's or Alzheimer's disease (see review by Ruberg and Agid, 1988). ChAT activity was ob-

F. Javoy-Agid

served to decrease in PSP patients but only to a small degree and in some cortical areas: 20 to 40% in the frontal cortex, the cingulate cortex and in the hippocampus. No change was observed in the temporal cortex (Fig. 1) (Agid et al., 1986; Whitehouse et al., 1988). A dramatic decrease (70%) of ChAT activity was observed, however, in the substantia innominata (Ruberg et al., 1985), which contains cell bodies of cholinergic neurons projecting to the cerebral cortex, but cholinergic afferents from neurons located in the brainstem (Mesulam et al., 1984) as well. The discrepancy between the amplitude of the decrease in ChAT activity in the cerebral cortex and in the substantia innominata suggests that only a small percentage of cholinergic neurons degenerate in the latter structure. This hypothesis was confirmed immunocytochemically since a moderate loss in ChAT-positive cell bodies was observed in the substantia innominata (Brandel et al., 1991) (Fig. 2). Loss of afferent cholinergic projections to the structure would account for the remaining decrease in ChAT activity in the substantia innominata, but their source has not been identified. The cholinergic deficiencies in the cerebral cortex and hippocampus, as well as the substantia innominata, suggest that the innominatocortical and septohippocampal cholinergic pathways are only moderately affected in PSP, which may account for the relatively mild memory impairment observed in these patients. This contrasts with the severe damage of these neurons found in intellectually deteriorated patients with Parkinson's disease or patients with Alzheimer's disease (Ruberg and Agid, 1988; Brandel et al., 1991; Tagliavini et al., 1983).

Among the thalamic nuclei, the mediodorsal nucleus has drawn great attention with respect to functional disorders in patients with PSP. The nucleus is innervated by neurons in several regions of the brain, in particular the forebrain and brainstem. It projects primarily to the prefrontal and premotor cortical areas (for review see Brandel et al., 1990, 1991), modulating cortical activity in response to information from subcortical structures. The cholinergic innervation of the mediodorsal nucleus of the thalamus was studied by ChAT immunocytochemistry on brains from control subjects, patients with PSP, or with Alzheimer's disease (Brandel et al., 1990, 1991). The cholinergic innervation of the mediodorsal nucleus of the thalamus was heterogeneous in control brains, consisting of densely stained patches of neuropil surrounded by less densely stained matrix. In patients with PSP, a prominent loss of innervation was observed in the matrix (75%); loss in the patches was less severe (60%). This loss is likely due to a large extent to the degeneration of neurons in the laterodorsal tegmental nucleus, which reaches 84% in these patients, loss of cholinergic neurons in substantia innominata representing only 33%, the two brain nuclei largely projecting to this thalamic nucleus (Fig. 2). In patients with Alzheimer's disease, ChAT-positive innervation decreased to some extent in the matrix (34%) but more severely in the patches (46%). Loss of innervation was exclusively associated to neuronal loss in the substantia innominata (80%), indicating that his nucleus projects to both matrix and patches but most heavily to the latter (Brandel et al., 1991).

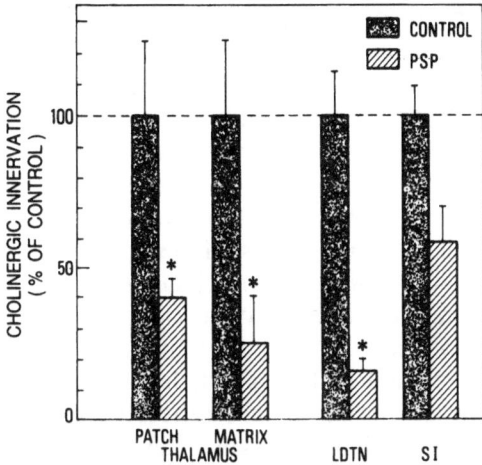

**Fig. 2.** Quantitative estimates of the total number of choline acetyl-transferase-positive neurons in the laterodorsal tegmental nucleus (LDTN) and substantia innominata (SI), and density of choline acetyl-transferase-positive varicosities in the patch and matrix zones of the mediodorsal thalamic nucleus. Data are expressed as percentage of control values. * ≤ 0.05, statistically different compared to respective control

Two conclusions can be drawn from these observations. On the one hand, the differential loss of innervation to matrix and patches of the mediodorsal nucleus of the thalamus in PSP and Alzheimer's disease may differentially affect innervation of the frontal cortex by the thalamus, resulting in different symptomatologies. The predominantly "frontal" type psychopathology of PSP patients (Cambier et al., 1985; Pillon et al., 1986; Dubois et al., 1988), unlike those with Alzheimer's disease, may find part of its explanation here. The principal extrapyramidal circuits also pass through the thalamus: the pallidothalamofrontal pathway to the premotor cortex, the nigrothalamofrontal pathway to the prefrontal cortex, and the dendato-thalamofrontal pathway to the motor cortex. Like the substantia innominata and the laterodorsal tegmental nucleus, the structures of origin of these circuits -the substantia nigra, pallidum, dentate nucleus- are all sites of lesions in PSP patients (Agid et al., 1986). It would be of great interest to investigate these circuits neurochemically as well. On the other hand, from a more general perspective, it is clear that the specific patterns of neuronal death in the neurodegenerative diseases can be powerful tools in the hands of neuroanatomists studying the human brain. Indeed, these natural lesions are models that permit fine differentiation of innervation from distinct and even closely related nuclei. The present example of differential innervation of two cholinergic pathways within a single nucleus is a compelling one (Brandel et al., 1991).

Systematic study of discrete cholinergic nuclei throughout the human brains was notably advanced by the development of immunocytochemical techniques and the availability of an antibody against human ChAT that permitted high-resolution quantitative studies of the many small cholinergic nuclei of the mesencephalon and pontine reticular formation. Groups of neurons that are part of the circuitry thought to play role in the control of eye movement (Carpenter et al., 1970) were obvious candidates for study in PSP patients. In brains from patients with PSP compared to controls, loss of cholinergic neurons was observed in some but not all the mesencephalic nuclei, reaching 69% in the nucleus of Edinger-Westphal, 97% in the rostral interstitial nucleus of the medial longitudinal fasciculus, and 78% in the interstitial nucleus of Cajal. Loss of neurons was also considerable (93%) in the deep layers of the superior colliculus (Juncos et al., 1991). Cell loss in some cranial nerves, known to be cholinergic, has been detected by classic neuropathological techniques (Blumenthal and Miller, 1969). Cholinergic neurons seemed to be spared, however, in cranial nerves III and IV, as well as in the mesencephalic reticular formation and the parabigeminal nucleus. The nucleus of cranial nerve VI seems unaffected as well (Malessa et al., 1991) consistent with a supranuclear source of the oculomotor syndrome in PSP.

The Edinger-Westphal nucleus is thought to transmit visceral parasympathetic input to the oculomotor complex (Carpenter and Sutin, 1983) and may be involved in pupillary abnormalities (Troost and Daroff, 1977). Loss of cholinergic neurons in the deep layers of the superior colliculus may affect coordination of simultaneous neck and gaze-related movements, altering axial tone and causing gaze abnormalities (Steele, 1972; Troost and Daroff, 1977). Loss of cholinergic neurons in the rostral interstitial nucleus, which projects to the third cranial nerve complex, disrupts premotor control of oculomotricity and may lead to impaired vertical eye movements (Büttner-Ennever and Büttner, 1978; Büttner-Ennever et al., 1982). Loss of cholinergic neurons in the interstitial nucleus of Cajal, consistent with histopathological evidence of a lesion in this structure, could, according to animal experiments (Carpenter and Sutin, 1983; Fukushima-Kudo et al., 1987), account for the nuchal dystonia and gait impairments seen in PSP patients.

The lower pontine reticular formation in patients with PSP (Malessa et al., 1991) showed a subnormal number of ChAT-positive cells as well as in the nucleus papillioformis and the nucleus pontis centralis caudalis. Lesion of the former nuclei disrupts communication from the frontal eye field and subcortical nuclei such as the superior colliculus and pretectum with the cerebellum. The latter nucleus, a reticular formation nucleus associated with premotor control of extraocular eye movements, may be related to the alteration of horizontal saccades observed in PSP patients (Pierrot-Deseilligny et al., 1989a,b).

Such quantitative immunocytochemical observations suggest a loss of cholinergic neurons in specific brainstem cell groups, providing an anatomically defined basis for the supranuclear ocular syndrome characteristic of PSP.

Severe (80%) loss of neurons in the nucleus tegmentopedunculopontinus compared to controls, evidenced by acetylcholinesterase histochemistry and biochemical assay of ChAT activity in the nucleus, denervates a number of nuclei of the extrapyramidal motor system (frontal cortex, internal globus

pallidus, substantia nigra pars reticulata, subthalamic nucleus, ventroteg-
mental area, central nucleus of the amygdala) including the thalamus (Edley
and Graybiel, 1983; Hallanger and Wainer, 1988). The loss has been corroborated on other morphological criteria (Zweig et al., 1987; Jellinger,
1988). In addition, the lateral edge of the nucleus seems to be particularly
affected in PSP. In Parkinson's disease these neurons also degenerate, but
the medial part of the nucleus is affected as well. The pedunculopontine
nucleus is affected severely in PSP, less in Parkinson's disease (Hirsch et al.,
1987; Jellinger, 1988), and spared in Alzheimer's disease (Jellinger, 1988).

## Neuropeptide systems in PSP

Peptides have been investigated in neurodegenerative diseases. Such investigations in human brain after autopsy represent one approach to under-
stand peptide function in mature brain by establishing correlations between
dysfunction of peptidergic neurons, pathological evidence of neuronal
damage, and clinical symptoms. In most studies neuropeptides have been
assayed biochemically, although a few immunocytochemical studies are
now helping to approach the significance of the neurochemical deficiencies
observed.

Investigations of brain peptides in PSP are of special interest relative to
data obtained in Parkinson's disease and Alzheimer's disease. Indeed, PSP
shares with parkinsonian syndromes, movement disorders, and in some
cases cognitive deficits characterized essentially by: memory disorders
associated with frontal-lobe deficits, absence of aphasia, apraxia and agnosia.
Further, PSP is associated with cognitive deficits reminiscent to those of
Alzheimer's disease (intellectual deterioration sufficiently severe to interfere with social or occupational functions, memory impairment, impaired
abstract thinking or judgement, aphasia, apraxia or agnosia, personality
changes).

The concentrations of several putative peptide neurotransmitters,
methionine-enkephalin, dynorphin, substance P and cholecystokinin-8,
were normal in basal ganglia (striatum, substantia nigra and pallidum)
(Taquet et al., 1987). Several studies have shown that Parkinson's disease,
but not Alzheimer's disease, is associated with altered levels of neuropeptides in subcortical areas. In particular subnormal, met-enkephalin and
substance P levels have been found in globus pallidus and substantia nigra.
The significance of these neuropeptide deficits remains unknown. It is not
clear whether in Parkinson's disease, they are related to the nigrostriatal
dopamine deficit (indeed i.e. dopamine-methionine-enkephalin interactions
have been suggested in rat brain) or whether they represent and independent feature characteristic of the disease. The absence of significant
alterations of the content in methione-enkephalin, leucine-enkephalin,
Substance-P, cholecystokinin-8 in substantia nigra, caudate nucleus, putamen, pallidum of PSP patients, supports a distinction between PSP and
Parkinson's disease (Fig. 3). Since the reductions in these peptide levels,

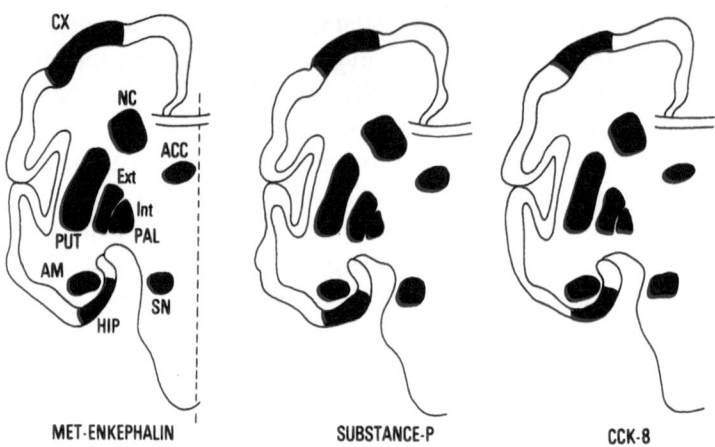

**Fig. 3.** Schematic representation of the distribution of peptides in the brain of PSP patients: black area = normal concentration. *ACC* nucleus accumbens, *AM* amygdala, *CX* frontal cortex, *HIP* hippocampus, *NC* caudate nucleus, *PAL* external and internal pallidum, *PU* putamen, *SN* substantia nigra

particularly in the substantia nigra, in parkinsonian brains were not observed in PSP brains, it is most likely that the neuropeptide deficits associated to Parkinson's disease are probably not secondary to the massive degeneration of nigrostriatal dopaminergic neurons. The differences between PSP and Parkinson's disease regarding peptides levels might also causally be related to the differential alterations of cholinergic neurons. Indeed, the cholinergic deficit in PSP, but not in parkinsonian brains, might induce metabolic alterations in peptidergic systems compensating those associated with changes in dopamine transmission. Little is known so far on the possible influence of cholinergic systems upon enkephalin levels (Chan et al., 1982).

In PSP, the slight reduction in Substance-P levels within the substantia nigra, might be functionally related to the severe deficit in nigrostriatal dopamine neurotransmission. Indeed, chronic treatment of rats with dopamine antagonists induce a marked reduction of nigral Substance-P levels (Hanson et al., 1981; Hong et al., 1978; Oblin et al., 1984).

In parkinsonian brains, a significant depletion of cholecystokinin-8 levels had been shown within the substantia nigra (Studler et al., 1982). It had been proposed that some neurons which in rat substantia nigra colocalize dopamine and the peptide, might also exist in the human substantia nigra, and that their degeneration would cause depletion in the parkinsonian substantia nigra. In PSP, degeneration of nigrostriatal dopaminergic neurons is as extensive as in Parkinson's disease. However, no reduction in cholecytokinin-8 levels was observed in substantia nigra as in any other brain region of PSP brains (Taquet et al., 1987). Therefore, the dopaminergic neurons containing the peptide, if they exist in human, could constitute a limited cell population. Another possibility, is that nigral neurons containing

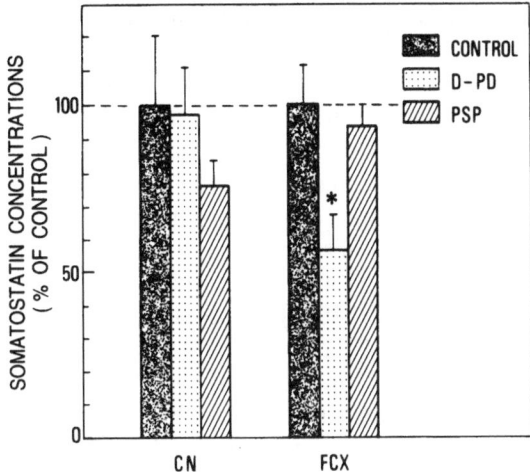

**Fig. 4.** Somatostatin concentrations in the caudate nucleus (CN) and frontal cerebral cortex (FCX) of patients with PSP or in parkinsonian patients with dementia (D-PD). Data are expressed as percentage of control values. *p ≤ 0.05, statistically different compared to respective control

only cholecystokinin-8 might be protected in PSP, unlike in Parkinson's disease.

Finally, the peptide deficits found in the basal ganglia of parkinsonian brains contrast with the absence of changes in peptide levels in extrapyramidal and limbic regions of PSP brains. The data evidence neurochemical distinctions between the two syndromes presenting however a similar severe degeneration of nigrostriatal dopaminergic neurons.

Several peptides have been radioimmunoassayed in cerebral cortex and hippocampus, where they are found within intrinsic neurons (cholecystokinin-8, vasoactive intestinal polypeptide, neuropeptide Y) and/or intrinsic or ascending projections, (substance P, methionine-enkephalin, thyrotropin-releasing-hormone, neurotensin). Their levels did not markedly differ from controls in Alzheimer's disease and demented parkinsonian patients. In subcortical areas, no changes were observed related to cognitive deficits. Thus the corresponding neurons are most likely not implicated in these intellectual disorders.

Among the various neuropeptides, somatostatin is of particular interest, as reduced concentrations of this substance have been observed in the cerebral cortex of patients with senile dementia of the Alzheimer type and parkinsonian patients with intellectual deterioration (see review Ruberg and Agid, 1988). Somatostatin is a peptide widely distributed in the human brain, particularly in the cerebral cortex where it is found in intrinsic neurons. In patients with PSP, somatostatin concentrations were not significantly different from control values (Fig. 4) in either the frontal or the temporal cortex, suggesting that the presumably intrinsic somatostatin-containing neurons are not affected by the disease (Epelbaum et al., 1987). Somatostatin receptors, characterized by binding to membrane preparations, were normal in number in the cerebral cortex of PSP patients (Epelbaum et al., 1988). Somatostatin-mediated functions therefore seem to be normal. In Alzheimer's disease, somatostatin levels are consistently reduced in cortical and hippocampal regions, but not in subcortical areas (see review Ruberg and Agid, 1988). Similar, data have been observed in brains of demented parkinsonian patients as compared to non demented parkinsonians (Epelbaum et al., 1988). In Parkinson's disease, the intensity of the somatostatin deficit was related to the severity of dementia both in cerebral cortex and hippocampus (Epelbaum et al., 1983). Altogether, the data have suggested that intellectual impairment of Parkinson and Alzheimer type is associated with alteration of cortical and hippocampal somatostatin transmission. The role of this neurochemical deficit in cognitive dysfunction remains unknown. At difference, despite the dramatic intellectual deterioration observed in PSP, neocortical and hippocampal somatostatin levels are preserved. The significance of a peptide deficiency is not clear. Does it reflect a neuronal damage or a neuronal loss, or is it an indication of a metabolic change? This important question remains unanswered, and leaves open the idea that somatostatin neurons in PSP may well be preserved anatomically, but physiologically affected. Immunocytochemical and in situ hybridization studies at cellular levels may help elucidate the question.

Finally, PSP is an intriguing clinical paradigm. As other parkinsonian syndromes, the disease is characterized by a severe alteration of the dopaminergic nigrostriatal systems. Unlike Parkinson's disease, PSP is associated with a widespread cholinergic deficit likely corresponding to a loss in cholinergic neurons. However, contrasting both with Alzheimer's disease and Parkinson's disease with dementia, the disease is not associated with a marked cholinergic deficiency in the cerebral cortex. Various peptides are present at normal concentrations in cortical and subcortical areas in brains of patients with PSP. Of particular interest, is somatostatin, since the levels shown to be subnormal in cerebral cortex observed in dementia of Alzheimer's or Parkinson's type, are not affected in PSP.

## Conclusion

The neurochemical study of human autopsy material, in particular in neurodegenerative diseases may serve two purposes, functional and therapeutic. The association of a symptom with a damaged neuronal system may permit analysis of the normal brain circuits involved. As an example, the study of the cholinergic innervation of the thalamus and brainstem nuclei has resulted in a real contribution to human neuroanatomy and advanced our understanding of circuits involved in brain function. Identification of the neurons that die or are damaged in a nucleus known to participate in a physiological function may help improving therapeutics. In PSP the classical treatment of parkinson-like symptoms might be ineffective because the cholinergic relay in the striatum is defective. This cholinergic deficiency must also have repercussions in turn. Thus the search for an effective monotherapy, dopaminergic, cholinergic or other, is probably an illusion. Given the extent of the cholinergic lesions in PSP, anticholinergic drugs to treat parkinsonian symptoms or dystonia should probably be avoided since they may aggravate symptoms due to acetylcholine deprivation.

In summary, while post-mortem neurochemical studies suffer from inherent limitations, the study of neurodegenerative diseases, if one takes advantage of the questions they pose, can provide information on a number of aspects of the disease: diagnostic, anatomical, functional, therapeutic, etiological, as it has been the case with PSP.

## Acknowledgements

The author greatly thanks M. C. Banus for typing the manuscript.

## References

Agid Y, Javoy-Agid F, Ruberg M, Pillon B, Dubois B, Duyckaerts C, Hauw JJ, Baron JC, Scatton B (1986) Progressive supranuclear palsy: anatomoclinical and biochemical considerations. In: Yahr MD, Bergmann KJ (eds) Advances in neurology, vol 45. Raven Press, New York, pp 191–206

Agid Y, Javoy-Agid F, Ruberg M (1987) Biochemistry of neurotransmitters in Parkinson's disease. In: Marsden CD, Fahn S (eds) Movement disorders, vol 2. Butterworth, pp 166–230

Baron JC, Mazière B, Loc'h C, Sgouropoulos P, Bonnet AM, Agid Y (1985) Progressive supranuclear palsy. Loss of striatal dopamine receptors demonstrated in vivo by positron tomography. Lancet i: 1163–1164

Blumenthal H, Miller C (1969) Motor nuclear involvement in progressive supranuclear palsy. Arch Neurol 20: 362–367

Brandel JP, Hirsch EC, Javoy-Agid F (1990) Compartmental ordering of cholinergic innervation in the mediodorsal nucleus of the thalamus in human brain. Brain Res 515: 117–125

Brandel JP, Hirsch EC, Malessa S, Duyckaerts C, Cervera P, Agid Y (1991) Differential vulnerability of cholinergic projections to the mediodorsal nucleus of the thalamus in senile dementia of Alzheimer type and progressive supranuclear palsy. Neuroscience 41: 25–31

Büttner-Ennever JA, Büttner U (1978) A cell group associated with vertical eye movements in the rostral mesencephalic reticular formation of the monkey. Brain Res 151: 31–47

Büttner-Ennever JA, Büttner U (1982) Vertical gaze paralysis and the rostral interstitial nucleus of the medial longitudinal fasciculus. Brain 105: 125–149

Cambier J, Masson M, Viader F, Limodin J, Strube A (1985) Le syndrome frontal de la paralysie supranucléaire progressive. Rev Neurol (Paris) 141: 528–536

Carpenter MB, Sutin J (1983) Human neuroanatomy. Williams & Wilkins, Baltimore, p 428

Carpenter MB, Harbison JW, Peter P (1970) Accessory oculomotor nuclei in the monkey: projections and effects of discrete lesions. J Comp Neurol 140: 131–154

Chau TT, Izazola-Conde C, Dewey WL (1982) Evidence for the existence of peptide and nonpeptide morphine-like materials in mouse brain: effect of an analgesic intracerebroventricular dose of acetylcholine on their levels. J Pharmacol Exp Ther 222: 612–616

Dubois B, Pillon B, Legault F, Agid Y, Lhermitte F (1988) Slowing of cognitive processing in progressive supranuclear palsy. A comparison with Parkinson's disease. Arch Neurol 45: 1194–1199

Edley SM, Graybiel AM (1983) The afferent and efferent connections of the feline nucleus tegmenti pedunculopontinus, pars compacta. J Comp Neurol 217: 187–215

Epelbaum J, Ruberg M, Moyse E, Javoy-Agid F, Dubois B, Agid Y (1983) Somatostatin and dementia in Parkinson's disease. Brain Res 278: 376–379

Epelbaum J, Javoy-Agid F, Hirsch EC, Hauw JJ, Kordon C, Krantic S, Agid Y (1987) Brain somatostatin concentrations do not decrease in progressive supranuclear palsy. J Neurol Neurosurg Psychiatry 50: 1526–1528

Epelbaum J, Javoy-Agid, Enjalbert A, Krantic S, Kordon C, Agid Y (1988) Somatostatin concentrations and binding sites in human frontal cortex are differently affected in Parkinson's disease associated with dementia and in progressive supranuclear palsy. J Neurol Sci 87: 167–174

Fukishima-Kudo J, Fukushima K, Tashiro K (1987) Rigidity and dorsiflexion of the neck in progressive supranuclear palsy and the interstitial nucleus of Cajal. J Neurol Neurosurg Psychiatry 50: 1197–1203

Girault JA, Raisman-Vozari R, Agid Y, Greengard P (1989) Striatal phosphoproteins in Parkinson's disease and progressive supranuclear palsy. Proc Natl Acad Sci USA 86: 2493–2497

Hallanger AE, Wainer BH (1988) Ascending projections from the pedunculontine tegmental nucleus and the adjacent mesopontine tegmentum in the rat. J Comp Neurol 274: 483–515

Hanson GR, Alphs L, Wolf W, Levine R, Lovenberg W (1981) Haloperidol-induced reduction of nigral substance P-like immunoreactivity: a probe for the interactions between dopamine and substance P neuronal systems. J Pharmacol Exp Ther 218: 568–574

Hirsch EC, Graybiel AM, Duyckaerts C, Javoy-Agid F (1987) Neuronal loss in the pedunculupontine tegmental nucleus in Parkinson's disease and in progressive supranuclear palsy. Proc Natl Acad Sci 84: 5976–5980

Hirsch EC, Graybiel AM, Hersh LB, Duyckaerts C, Agid Y (1989) Striosomes and extrastriosomal matrix contain different amounts of immunoreactive choline acetyltransferase in the human striatum. Neurosci Lett 96: 145–150

Hirsch EC, Graybiel AM, Agid Y (1988) Melanized dopaminergic neurons are differentially susceptible to degeneration in Parkinson's disease. Nature: 345–348

Hong JS, Yang HYT, Fratta W, Costa E (1978) Rat striatal methionine-enkephalin content after chronic treatment with cataleptogenic and non cataleptogenic antischizophrenic drugs. J Pharmacol Exp Ther 205: 141–147

Jellinger K, Riederer P, Tomonga M (1980) Progressive supranuclear palsy: clinicopathological and biochemical studies. J Neural Transm [Suppl] 16: 111–128

Jellinger K (1988) The pedunculopontine nucleus in Parkinson's disease, progressive supranuclear palsy and Alzheimer's disease. J Neurol Neurosurg Psychiatry 51: 540–543

Juncos JL, Hirsch EC, Malessa S, Duyckaerts C, Hersh LB, Agid Y (1991) Mesencephalic cholinergic nuclei in Progressive Supranuclear Palsy. Neurology 41: 25–30

Kish SJ, Chang MS, Mirchandani, Shannak K, Hornykiewicz O (1985) Progressive supranuclear palsy: relationship between extrapyramidal disturbances, dementia and brain neurotransmitter markers. Ann Neurol 18: 530–536

Lehéricy S, Hirsch E, Cervera P, Hersh LB, Hauw JJ, Ruberg M, Agid Y (1989) Selective loss of cholinergic neurones in the ventral striatum of patients with Alzheimer's disease. Proc Natl Acad Sci USA 86: 8580–8586

Malessa S, Hirsch EC, Cervera P, Javoy-Agid F, Duyckaerts C, Hauw JJ, Agid Y (1991) Progressive supranuclear palsy: loss of choline acetyltransferase-like-immunoreactive neurons in the pontine reticular formation. Neurology 41: 1593–1597

Mesulam MM, Mufson EJ, Levey AL, Wainer BH (1984) Atlas of cholinergic neurons in the forebrain and upper brainstem of the macaque based on monoclonal choline acetyltranferase immunohistochemistry and acetylcholinesterase histochemistry. Neuroscience 12: 669–686

Monfort JC, Javoy-Agid F, Hauw JJ, Dubois B, Agid Y (1985) Brain glutamate decarboxylase in Parkinson's disease: an index of premortem severity. Brain 108: 301–313

Oblin A, Zivcovic B, Bartholini G (1984) Involvement of the $D_2$ dopamine receptor in the neuroleptic-induced decrease in nigral substance. Eur J Pharmacol 105: 175–177

Pierot L, Desnos C, Blin J, Raisman R, Sherman D, Javoy-Agid F, Ruberg M, Agid Y (1988) D1 and D2 dopamine receptors in patients with Parkinson's disease and progressive supranuclear palsy. J Neurol Sci 86: 291–306

Pierrot-Deseilligny C, Turell E, Penet C, Lebrigand D, Pillon B, Chain F, Agid Y (1989) Increased wave P 300 latency in progressive supranuclear palsy. J Neurol Neurosurg Psychiatry 52: 656–658

Pierrot-Deseilligny C, Rivaud S, Pillon B, Agid Y (1989) Lateral visually-guided saccades in progressive supranuclear palsy. Brain 112: 471–487

Pillon B, Dubois B, Lhermitte F, Agid Y (1986) Heterogeneity of cognitive impairment in progressive supranuclear palsy, Parkinson's disease, and Alzheimer's disease. Neurology 36: 1179–1185

Raisman-Vozari R, Girault JA, Feurstein C, Jenner P, Marsden CD, Agid Y (1990) Lack of changes in striatal DARP-32 levels following nigrostriatal dopaminergic lesions in animals and in parkinsonian syndromes in man. Brain Res 507: 45–50

Ruberg M, Javoy-Agid F, Hirsch E, Scatton B, Lheureux R, Hauw JJ, Duyckaerts C, Gray F, Morel-Maroger A, Rascol A, Serdaru M, Agid Y (1985) Dopaminergic and cholinergic lesions in progressive supranuclear palsy. Ann Neurol 18: 523–529

Ruberg M, Villageois A, Bonnet AM, Pillon B, Rieger F, Agid Y (1987) Acetylcholinesterase and butyrylcholinesterase activity in the cerebrospinal fluid of patients with neurodegenerative diseases involving cholinergic systems. J Neurol Neurosurg Psychiatry 50: 538–543

Ruberg M, Agid Y (1988) Dementia in Parkinson's disease. In: Iversen LL, Iversen SD, Snyder SH (eds) Psychopharmacology of the aging nervous system. Plenum Press, New York, pp 157–206 (Handbook of psychopharmacology, vol 20)

Ruberg M, Hirsch E, Javoy-Agid F (1992) Neurochemistry. In: Litvan I, Agid Y (eds) Progressive supranuclear palsy, vol 5. Oxford University Press, New York, pp 89–109

Steele JC (1972) Progressive supranuclear palsy. Brain 95: 693 704

Steele JC, Richardson JC, Olszewski (1964) Progressive supranuclear palsy. A heterogeneous degeneration involving the brain stem, basal ganglia and cerebellum, with vertical gaze and pseudobulbar palsy, nuclear dystonia and dementia. Arch Neurol 10: 333–359

Studler JM, Javoy-Agid F, Cesselin F, Legrand JC, Agid Y (1982) CCK-8 immunoreactivity distribution in human brain: selective decrease in the substantia nigra from parkinsonian patients. Brain Res 243: 176–179

Tagliavini F, Pilleri G, Gemignani F, Lechi A (1983) Neuronal loss in the basal nucleus of Meynert in progressive supranuclear palsy. Acta Neuropathol (Berl) 61: 157–160

Taquet H, Javoy-Agid F, Mauborgne A, Benoliel JJ, Agid Y, Legrand JC, Hamon M, Cesselin F (1987) Brain neuropeptides in progressive supranuclear palsy. Brain Res 411: 178–182

Troost BT, Daroff RB (1977) The ocular motor defects in progressive supranuclear palsy. Ann Neurol 2: 397–403

Whitehouse PJ, Martino AM, Marcus KA, Zweig RM, Singer HS, Price DL, Kellar KJ (1988) Reductions in acetylcholine and nicotine binding in several degenerative disease. Arch Neurol 22: 18–25

Zweig RM, Whitehouse P, Casanova MF, Walker L, William RJ, Price DL (1988) Loss of pedunculopontine neurons in progressive supranuclear palsy. Ann Neurol 22: 18–25

Author's address: Dr. F. Javoy-Agid, Laboratoire de Medicine Experimentale, Hôpital de la Salpêtrière, F-75634 Paris, France.

# Brain monoamines in progressive supranuclear palsy — comparison with idiopathic Parkinson's disease

## O. Hornykiewicz[1] and K. Shannak[2]

[1] Institute of Biochemical Pharmacology, University of Vienna, Vienna, Austria
[2] Clarke Institute of Psychiatry, University of Toronto, Toronto, Canada

**Summary.** Like idiopathic Parkinson's disease (iPD), Progressive Supra-nuclear Palsy (PSP) is characterized, *inter alia*, by a pronounced non-overlapping loss of dopamine (DA) in caudate, putamen and substantia nigra.

Unlike iPD, in PSP the striatal DA loss is more severe in the caudate than in the putamen; this may contribute to the higher frequency of cognitive deficits in PSP.

In contrast to iPD, in patients with PSP the serotonin (5-HT) levels in the basal ganglia are not significantly reduced, thus resulting in a relative predominance of the inhibitory serotonergic influences on the motor be-haviour in these patients.

It is suggested that combination of levodopa with a 5-HT receptor blocker may substantially improve the (poor) responsiveness of patients with PSP to DA substitution therapy.

## Introduction

The classic symptoms of progressive supranuclear palsy (PSP) include supranuclear ophthalmolegia, pseudobulbar palsy, dysarthria, dystonic extension of the neck and dementia (Steele et al., 1964). In addition to these pathognomonic symptoms, PSP is usually accompanied by several parkinsonian features, including abnormalities of movement and posture such as rigidity, akinesia and postural instability (cf. Brusa et al., 1980), as well as positive, although only limited, response to dopamine (DA) substitution therapy (Jackson et al., 1983). These similarities suggest that changes in some of the brain monoamine, especially DA, systems found in idiopathic Parkinson's disease (iPD) may also be present in PSP brain. In the following discussion, observations on DA, noradrenaline (NA) and serotonin (5-HT) in the postmortem brain of 5 patients with PSP will be summarized and compared with monoamine changes found in iPD brain. The pertinent data are shown in Fig. 1; they are expressed as % of the mean values obtained in 5–8 non-neurological control subjects matched with the PSP subjects for age (63–75 years) and interval between death and freezing of the brain (6–18 hours). For 4 of the 5 PSP patients monoamine data, obtained in a limited number of brain areas, can be found in a previous publication (cases no. 3, 22, 189 and 245, in Kish et al., 1985).

## The nigrostriatal system

As can be seen in Fig. 1, the patients with PSP suffered, as previously reported (Kish et al., 1985; Ruberg et al., 1985) a marked and statistically highly significant loss of DA in caudate nucleus, putamen, and substantia nigra by an average amount of 87–92% (p < 0.001 or less). There was no overlap of the individual DA values between the patients and the controls. The mean levels of NA were markedly reduced in caudate nucleus (−76%) and putamen (−70%), but were within the range of control values in the substantia nigra. On the statistical level, the NA reduction in caudate and putamen attained only weak significance (p < 0.05); in these striatal nuclei there was a small area of overlap between the individual PSP and control values. In contrast to the changes in DA and NA, there was no significant alteration of 5-HT levels in any of the three components of the nigrostriatal system, although on average, the 5-HT levels tended to be mildly to moderately subnormal in these regions (caudate nucleus: −20%; putamen: −18%; substantia nigra: −36%) (see also Kish et al., 1985).

## Non-striatal subcortical regions

Figure 1 also shows the data on monoamine levels in five non-striatal subcortical regions of the PSP brain, namely: lateral globus pallidus, medial globus pallidus, subthalamic nucleus, nucleus accumbens, and the hypothalamus. When compared with the control values, the mean DA levels in the external and internal segments of the globus pallidus were reduced by 71% and 63%, respectively. However, due to overlap of the individual values, neither of these changes was statistically significant. In contrast to the reduction in pallidal DA, the NA and 5-HT levels were distinctly above normal (by 72% to 130%) in both pallidal segments, with the 5-HT elevations being statistically significant. In the subthalamic nucleus the levels of all three monoamines were well within the control range; this was also the case for nucleus accumbens (see also Kish et al., 1985; Ruberg et al., 1985). In the hypothalamus, only mildly, statistically not significantly, reduced monoamine levels were found (see Kish et al., 1985).

## Cerebral cortex

Of the cerebral cortical areas examined in PSP brain (Kish et al., 1985; Ruberg et al., 1985), only the DA level in the parolfactory gyrus (Brodmann area 25) was moderately, but significantly reduced (−63%; p < 0.05) (see

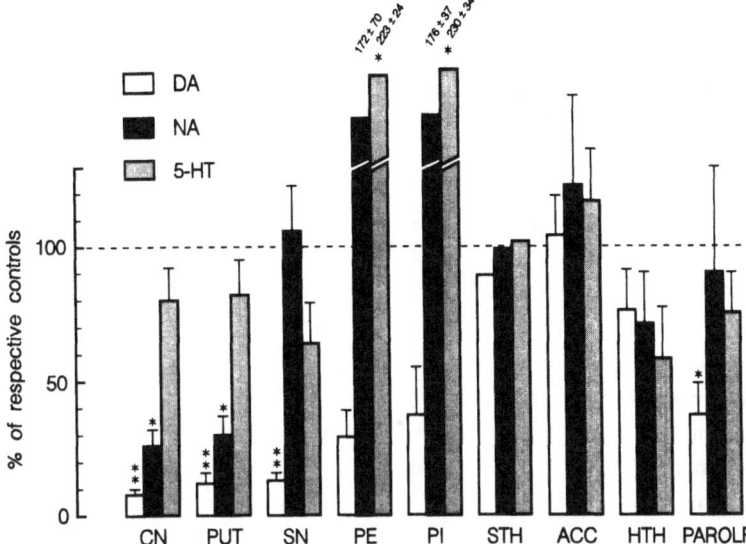

**Fig. 1.** DA, NA and 5-HT in forebrain regions of patients with PSP. Columns are means (bars = SEM), expressed as % of the respective control values (= 100%). Control means ± SEM for DA/NA/5-HT, respectively (in µg/g tissue): caudate nucleus (*CN*): 4.17 ± 0.64/0.06 ± 0.01/0.30 ± 0.06; putamen (*PUT*) 5.10 ± 0.65/0.08 ± 0.03/0.38 ± 0.07; substantia nigra (*SN*) 0.71 ± 0.06/0.12 ± 0.02/0.81 ± 0.20; gl. pallidus, ext. (*PE*) 0.44 ± 0.11/0.05 ± 0.01/0.16 ± 0.08; gl. pallidus, int. (*PI*) 0.08 ± 0.02/0.04 ± 0.01/0.09 ± 0.03; subthalamic nucleus (*STH*) 0.15 ± 0.02/0.14 ± 0.02/0.25 ± 0.03; nucleus accumbens (*ACC*) 2.09 ± 0.49/1.45 ± 0.44/0.43 ± 0.08; hypothalamus (*HTH*) 0.13 ± 0.02/1.29 ± 0.35/0.37 ± 0.06; parolfactory cortex (*PAROLF*) 0.08 ± 0.01/0.10 ± 0.01/0.02 ± 0.03; $^*p < 0.05$; $^{**}p < 0.001$

Fig. 1). In the other cortical areas for which data are available, i.e. frontal pole, parietal (operculum), temporal pole, and occipital pole, the levels of the three monoamines found in individual patients with PSP were highly variable; although in some instances the mean values were considerably below the control means (frontal cortex: DA 50% of control; occipital cortex: DA 75% and 5-HT 57% of control), due to the large scatter of individual (control and PSP) values, the differences were statistically not significant.

## Comparison between PSP and iPD

A synoptic comparison between the monoamine changes in PSP brain and the changes in iPD discloses important similarities, but also significant differences (Table 1).

### Similarities between PSP and iPD

1) The major neurochemical similarity between PSP and iPD is undoubtedly the pronounced and non-overlapping loss of DA in caudate nucleus, pu-

**Table 1.** Synopsis of brain monoamine changes in PSP — comparison with iPD

| Brain region | Dopamine | | Noradrenaline | | Serotonin | |
|---|---|---|---|---|---|---|
| | PSP | iPD | PSP | iPD | PSP | iPD |
| Caudate | ↓ | ↓ | ↓ | ↓ | = | ↓ |
| Putamen | ↓ | ↓ | ↓ | ↓ | = | ↓ |
| Substantia nigra | ↓ | ↓ | = | ↓ | (↓) | ↓ |
| Parolfactory cortex | ↓ | ↓ | = | (↓) | = | (↓) |
| Accumbens | = | ↓ | = | (↓) | = | (↓) |
| Pallidum, ext. | (↓) | ↓ | (↑) | = | ↑ | ↓ |
| Pallidum, int. | (↓) | ↓ | (↑) | = | ↑ | = |
| Subthalamic nucleus | = | ↓ | = | = | = | ↓ |
| Hypothalamus | = | = | = | ↓ | = | (↓) |
| Cortex – frontal | (↓) | ↓ | = | (↓) | = | (↓) |
| 　　　– parietal | (↑) | ? | = | ? | (↓) | ? |
| 　　　– temporal | = | = | = | (↓) | (↑) | (↓) |
| 　　　– occipital | = | ? | = | ? | (↓) | ? |

↓ stat. significant reduction; (↓) (↑) reduction or increase by more than 35% of control values, but statistically not significant; = within control range; ? no data available

tamen and substantia nigra. Since the relationship between nigrostriatal DA loss and parkinsonian symptoms is well established (Bernheimer et al., 1973; Hornykiewicz and Kish, 1986), it appears safe to relate the parkinsonian features frequently seen in patients with PSP to the concomitant marked striatal DA loss. In this respect, PSP behaves like many other degenerative brain disorders accompanied by striatal DA loss (cf. Hornykiewicz et al., 1990). In principle, reduction of striatal DA exceeding the 80% mark (Bernheimer et al., 1973), should produce clinically overt parkinsonian features, esp. brady/akinesia and rigidity, that can be expected to be sensitive to levodopa. In patients with PSP, however, DA substitution treatment brings about only limited improvement of the parkinsonian symptomatology. This may in part be due to the reduction (by 37% to 45%) of striatal (D-2) receptors for DA (Bokobza et al., 1984) indicative of subtle structural changes in the striatum, as well as the known degenerative changes in brain regions that are basically unaffected in classical iPD, such as the globus pallidus, subthalamic nucleus, thalamus and superior colliculi (Steele et al., 1964); these regions are known to be part of the striatal circuitry.

2) Other, less striking similarities between PSP and iPD include statistically significant reductions of NA in caudate nucleus and putamen, and of DA in the parolfactory gyrus (Brodmann area 25). As is the case in iPD, also in PSP no statistically significant changes were observed for DA in the hypothalamus, NA in the subthalamic nucleus and the parolfactory gyrus,

and 5-HT in the hypothalamus and parolfactory gyrus. Finally, the levels of all three monoamines in frontal, temporal, parietal and occipital cortices seem to be only mildly, or not at all, changed in both conditions.

## Differences between PSP and iPD

Against these similarities stand several differences between PSP and iPD. Although some of the differences may be of little functional significance, other differences most likely account for the distinctive clinical and pathophysiological features of each of the two disorders accompanied by severe striatal DA deficiency.

1) The most striking difference between PSP and iPD is the dissimilar inter- and subregional pattern of the striatal DA loss. In iPD the nigral (DA- and melanin-containing) neurones that project to the putamen bear the brunt of the degenerative damage (Hassler, 1938; Gibb, 1991). This results in a markedly more profound loss of DA in the putamen than in the caudate nucleus (Bernheimer et al., 1973). In addition to this essential interregional DA pattern, in iPD each of the two striatal nuclei has its own specific rostro-caudal (and dorso-ventral) pattern of DA loss. Thus, the putamen is more severely affected by DA loss in its caudal, compared with its rostral, portions, whereas in the caudate nucleus this rostro-caudal gradient of DA loss is in the opposite direction (Nyberg et al., 1983; Kish et al., 1988). In contrast to these DA patterns typical for iPD, in our PSP cases the caudate nucleus was as severely, or more, affected by DA loss (−92%) than the putamen (−88%) (see Fig. 1), and there was no indication of a iPD-like subregional striatal pattern of DA loss (unpublished observations). The higher degree of caudate DA loss in PSP (−92%) as compared with iPD (−81%) is interesting because of the more frequent occurrence, in PSP, of signs of cognitive dysfunction, i.e. clinically disabling dementia (Steele et al., 1964). Based on lesion experiments in laboratory animals (Divac et al., 1967; cf. also Divac and Öberg, 1979) and neurophysiological brain studies in monkeys (DeLong et al., 1983; Evarts et al., 1984), the caudate nucleus has been proposed to be involved in cognitive brain functions similar to those subserved by the prefrontal cortex. This notion is supported by data from postmortem brain studies in various striatal DA deficiency conditions accompanied by cognitive disturbances (in addition to PSP) (cf. Hornykiewicz et al., 1990) as well as by recent neuropsychological evidence obtained in patients at different stages of iPD (Owen et al., 1993).

2) Whereas in iPD, the DA levels are significantly reduced throughout the components of the basal ganglia, including the lateral and medial globus pallidus, nucleus accumbens and subthalamic nucleus, in PSP brain the levels of DA in these nuclei were either within control range (nucleus accumbens and subthalamic nucleus) or less markedly, and non-significantly, reduced (globus pallidus). At present it is difficult to assess the functional aspects of this difference between PSP and iPD. The observation that the DA level in nucleus accumbens was not reduced in PSP indicates that, in

contrast to iPD, in PSP the ventral tegmental area, the site of origin of much of the dopaminergic accumbens innervation (Moore and Bloom, 1978), was not involved in the neurodegenerative process. Since the nucleus accumbens is that part of the basal ganglia considered to be an interface between the motor and the limbic systems of the brain (Mogenson and Yim, 1981), studies designed at detecting differences between iPD and PSP regarding interrelations between (basal ganglia) motor and limbic systems should yield interesting insights about the specific functions of this nucleus in the human brain. There exists virtually no information on the origin of the DA fibres innervating the subthalamic nucleus in human brain. Our observation of profoundly reduced nigrostriatal DA levels in both PSP and iPD, with DA in subthalamic nucleus reduced in iPD only, strongly argues in favour of a separate meso-subthalamic DA fibre system in the human brain, distinct from the nigrostriatal DA system. The non-involvement of subthalamic DA and 5-HT in PSP is intriguing in view of recent data on the role of this brain region in production of parkinsonian symptoms in experimental models of iPD (Bergman et al., 1990).

3) In contrast to iPD, in PSP the NA levels in substantia nigra, nucleus accumbens, hypothalamus and globus pallidus were well within, or above, normal limits. Pharmacological experiments in rodents have demonstrated that brain DA and NA are functionally interrelated in a synergistic manner in respect to central control of motor behaviour (Andén and Grabowska, 1976; Lloyd, 1977; Donaldson et al., 1978; Grenhoff and Svensson, 1989; Lategan et al., 1990). Based on these observations, the NA loss frequently seen in iPD brain has been suggested to be among the non-dopaminergic factors possibly aggravating the parkinsonian movement disorder, esp. akinesia (Hornykiewicz, 1976; Narabayashi, 1993). It remains to be seen how the lack of any major NA changes in the extrastriatal nuclei of the PSP basal ganglia may be related to the motor symptomatology and the levodopa response of this condition.

4) The 5-HT levels in PSP brain were within, or significantly above (globus pallidus), normal range, but significantly reduced in iPD in most components of the basal ganglia circuitry analyzed, including caudate nucleus, putamen, substantia nigra, lateral globus pallidus, nucleus accumbens, and subthalamic nucleus. 5-HT has been postulated to exert an inhibitory action on nigrostriatal DA neurone activity (Dray, 1981; Ugedo et al., 1989). This notion is supported by observations in monkeys in which increase in brain 5-HT activity by means of neuronal 5-HT uptake blockade inhibited DA dependent motor behaviours (Korsgaard et al., 1985). It can, therefore, be hypothesized that severe decrease of striatal DA without a concomitant 5-HT reduction may lead to a predominance of inhibitory 5-HT influences, aggravating the parkinsonian symptomatology of patients with PSP. Recently, aggravation of parkinsonian symptoms during administration of the 5-HT re-uptake inhibitor fluoxetine (Jansen, 1993), and a possible amelioration of parkinsonism by the 5-HT$_2$ receptor blocker ritanserin (Meco et al., 1986; Maertens de Noordhout and Delwaide, 1986; Borison et al., 1988) has indeed been reported. This suggests the interesting

possibility that combination of levodopa with a brain 5-HT synthesis or receptor blocking drug may increase the levodopa sensitivity of PSP patients.

## Conclusion

Despite the basic neurochemical similarity between PSP and iPD regarding the profound and non-overlapping DA deficit in the nuclei of the nigrostriatal system, i.e. the caudate nucleus, the putamen and the substantia nigra, neither the interregional nor the subregional patterns of striatal DA loss observed in PSP are comparable with the DA patterns typical for iPD. In this respect, the more severe DA loss in caudate nucleus in PSP compared with iPD may account for the higher frequency of cognitive deficits observed in patients with PSP. In addition, significant neurochemical differences exist between PSP and iPD in respect to the patterns of NA and 5-HT alterations within the various components of the basal ganglia and other subcortical brain regions. These differences, together with the more widespread morphological changes usually present in PSP brain, may account, inter alia, for the poor response of patients with PSP to DA substitution therapy. In addition, the shifting of the ratio DA/5-HT in favour of 5-HT suggests that in PSP brain there exists an imbalance between the locomotor stimulatory dopaminergic mechanisms and the locomotor inhibitory serotonergic influences in favour of inhibition. It is proposed that combination of DA substitution drugs with a 5-HT receptor blocking, or 5-HT synthesis inhibiting, agent may increase the therapeutic responsiveness of PSP patients to DA substitution therapy.

## References

Andén N-E, Grabowska M (1976) Pharmacological evidence for a stimulation of dopamine neurons by noradrenaline neurons in the brain. Eur J Pharmacol 39: 275–282

Bergman H, Wichmann T, DeLong MR (1990) Reversal of experimental Parkinsonism by lesions of the subthalamic nucleus. Science 249: 1436–1438

Bernheimer H, Birkmayer W, Hornykiewicz O, Jellinger K, Seitelberger F (1973) Brain dopamine and the syndromes of Parkinson and Huntington: clinical, morphological and neurochemical correlations. J Neurol Sci 20: 415–455

Bokobza B, Ruberg M, Scatton B, Javoy-Agid F, Agid Y (1984) [³H]Spiperone binding, dopamine and HVA concentrations in Parkinson's disease and supranuclear palsy. Eur J Pharmacol 99: 167–175

Borison RL, Pathiraja AP, Diamond BI (1992) Influence of serotonin on dopaminergically mediated extrapyramidal side-effects. Mov Disord [Suppl] 7: 55

Brusa A, Mancardi GL, Bugiani O (1980) Progressive supranuclear palsy 1979: an overview. Ital J Neurol Sci 4: 205–222

DeLong MR, Georgopoulos AP, Crutcher MD (1983) Cortico-basal ganglia relations and coding of motor performance. Exp Brain Res [Suppl] 7: 29–40

Divac I, Öberg RGE (1979) Current conceptions of neostriatal functions: history and an evaluation. In: Divac I, Öberg RGE (eds) The neostriatum. Pergamon Press, Oxford, pp 215–230

Divac I, Rosvold HE, Szwarcbart MK (1967) Behavioural effects of selective ablation of the caudate nucleus. J Comp Physiol Psychol 63: 184–190

Donaldson IMacG, Dolphin AC, Jenner P, Pycock C, Marsden CD (1978) Rotational behaviour produced in rats by ipsilateral electrolytic lesions of the ascending noradrenergic bundles. Brain Res 138: 487–509

Dray A (1981) Serotonin in the basal ganglia: functions and interactions with other neuronal pathways. J Physiol (Paris) 77: 393–403

Evarts EV, Kimura M, Wurtz RH, Hikosaka O (1984) Behavioural correlates of activity in basal ganglia neurons. Trends Neurosci 7: 447–453

Gibb WRG (1991) Neuropathology of the substantia nigra. Eur Neurol [Suppl 1] 31: 48–59

Grenhoff J, Svensson TH (1989) Clonidine modulates dopamine cell firing in rat ventral tegmental area. Eur J Pharmacol 165: 11–18

Hassler R (1938) Zur Pathologie der Paralysis Agitans und des post-enzephalitischen Parkinsonismus. J Psychol Neurol 48: 387–476

Hornykiewicz O (1976) Neurohumoral interactions and basal ganglia function and dysfunction. In: Yahr MD (ed) The basal ganglia. Raven Press, New York, pp 269–278

Hornykiewicz O, Kish SJ (1986) Biochemical pathophysiology of Parkinson's disease. Adv Neurol 45: 19–34

Hornykiewicz O, Kish SJ, Rajput AH (1990) Neurochemical aspects of Parkinson's disease and the dementing brain disorders: relation to brain ageing. In: Nagatsu T, Fisher A, Yoshida M (eds) Basic, clinical, and therapeutic aspects of Alzheimer's and Parkinson's disease, vol 1. Plenum Press, New York, pp 445–452

Jackson JA, Jankovic J, Ford J (1983) Progressive supranuclear palsy: clinical features and response to treatment in 16 patients. Ann Neurol 13: 273–278

Jansen Steur ENH (1993) Increase of Parkinson disability after fluoxetine medication. Neurology 43: 211–213

Kish SJ, Chang LJ, Mirchandani L, Shannak K, Hornykiewicz O (1985) Progressive supranuclear palsy: relationship between extrapyramidal disturbances, dementia, and brain neurotransmitter markers. Ann Neurol 18: 530–536

Kish SJ, Shannak K, Hornykiewicz O (1988) Uneven pattern of dopamine loss in the striatum of patients with idiopathic Parkinson's disease. N Engl J Med 318: 876–880

Korsgaard S, Gerlach J, Christensson E (1985) Behavioural aspects of serotonin-dopamine interaction in the monkey. Eur J Pharmacol 118: 245–252

Lategan AJ, Marien MR, Colpaert FC (1990) Effects of locus coeruleus lesions on the release of endogenous dopamine in the rat nucleus accumbens and caudate nucleus as determined by intracerebral microdialysis. Brain Res 523: 134–138

Lloyd KG (1977) Neurotransmitter interactions related to central dopamine neurons. In: Youdim MBH, Lovenberg W, Sharman DE, Lagnado TR (eds) Essays in neurochemistry and neuropharmacology. Wiley, Chichester, pp 131–207

Maertens de Noordhout A, Delwaide PJ (1986) Open pilot trial of ritanserin in Parkinsonism. Clin Neuropharmacol 9: 480–484

Meco G, Marini S, Lestingi L, Modarelli F, Agnoli A (1986) Efficacy of ritanserin on tremor and abnormal involuntary movements in Parkinson's disease. Abstract, 15th C.I.N.P. Congress, San Juan, Puerto Rico, p 298

Mogenson GJ, Yim CY (1981) Electrophysiological and neuropharmacological-behavioural studies of the nucleus accumbens: implications for its role as a limbic-motor interface. In: Chronister RB, Defrance JF (eds) The neurobiology of the nucleus accumbens. Haer Institute for Electrophysiological Research, Brunswick, ME, pp 210–229

Moore RY, Bloom FE (1978) Central catecholamine neuron systems: anatomy and physiology of the dopamine systems. Ann Rev Neurosci 1: 129–169

Narabayashi H (1993) Three types of akinesia in the progressive course of Parkinson's disease. Adv Neurol 60: 18–24

Nyberg P, Nordberg A, Wester P, Winblad B (1983) Dopaminergic deficiency is more pronounced in putamen than in nucleus caudatus in Parkinson's disease. Neurochem Pathol 1: 193–202

Owen AM, Beksinska M, James M, Leigh PN, Summers BA, Marsden CD, Quinn NP, Sahakian BJ, Robbins TW (1993) Visuo-spatial memory deficits at different stages of Parkinson's disease. Neuropsychologia (in press)

Ruberg M, Javoy-Agid F, Hirsch E, Scatton B, LHeureux R, Hauw JJ, Duyckaerts C, Gray F, Morel-Maroger A, Rascol A, Serdaru M, Agid Y (1985) Dopaminergic and cholinergic lesions in progressive supranuclear palsy. Ann Neurol 18: 523–529

Steele JC, Richardson JC, Olszewski J (1964) Progressive supranuclear palsy. Arch Neurol 10: 333–359

Ugedo L, Grenhoff J, Svensson TH (1989) Ritanserin, a 5-HT$_2$ receptor antagonist, activates midbrain dopamine neurons by blocking serotonergic inhibition. Psychopharmacology 98: 45–50

Authors' address: Prof. Dr. O. Hornykiewicz, Institute of Biochemical Pharmacology, University of Vienna, Borschkegasse 8a, A-1090 Wien, Austria.

# Alterations of neurotransmitter receptors and neurotransmitter transporters in progressive supranuclear palsy*

## B. Landwehrmeyer[1] and J.M. Palacios[2]

[1] Department of Pathology, Division of Neuropathology, University of Basel, Basel, Switzerland
[2] Laboratorios Almirall, S.A., Research Institute, Barcelona, Spain

**Summary.** Neurotransmitter receptors and neurotransmitter transporters were studied postmortem in the brains of 9 PSP patients by receptor autoradiography. Densities of dopamine uptake sites and neurotensin receptors were significantly reduced in striatum and substantia nigra consistent with a localization of these binding sites on degenerating dopaminergic nigrostriatal projection neurons. The densities of dopamine $D_1$ receptors were unchanged. Dopamine $D_2$ receptors were unaltered when labeled by $[^{125}I]$-Iodosulpride or $[^3H]$-CV 205 502, but appeared to be significantly reduced when labeled by $[^3H]$-spiperone. Levels of $D_2$ mRNA were comparable to control levels, suggesting that only subtypes of Dopamine $D_2$-like receptors may be affected in PSP. Serotonin (5-HT) uptake sites and 5-HT receptors were not altered. The density of muscarinic receptors was reduced in striatum, possibly related to a degeneration of cholinergic striatal interneurons, but increased in internal globus pallidus. $GABA_A/BZ$ receptor binding sites were significantly reduced in both segments of globus pallidus, probably as a consequence of severe degeneration of intrinsic pallidal neurons in PSP. Binding of substance P in striatum tended to be decreased but failed to reach statistical significance. Compared to Parkinson's disease, the densities of more neurotransmitter receptors were altered in PSP. With the exception of increased muscarinic receptor binding sites in medial globus pallidus, the alterations seen in PSP seem to reflect cell loss rather than functional changes.

---

*Abbreviations*: C caudate, DA dopamine, GPe globus pallidus externus, GPi globus pallidus internus, HD Huntington's disease, ISHH in situ hybridization histochemistry, NFT neurofibrillary tangle, NT neurotensin, P putamen, PD Parkinson's disease, PSP Progressive supranuclear palsy, SN substantia nigra, SNc substantia nigra compacta, SNr substantia nigra reticulata, SP substance P, 5-HT 5-hydroxytryptamine, serotonin

## Introduction

In 1964, Steele, Richardson, and Olszewski recognized and described "progressive supranuclear palsy" as a distinct clinical and pathological entity. Clinically, patients suffering from PSP present usually in their sixth decade with visual disturbances, unsteady gait with frequent falls, axial rigidity, pseudobulbar palsy, parkinsonian signs and a certain degree of dementia (Steele et al., 1964; Behrman et al., 1969; Steele, 1972; Maher et al., 1986; Agid et al., 1986; Golbe and Davies, 1988). Pathological features in PSP include neuronal loss, gliosis and cytoskeletal abnormalities in form of neurofibrillary threads and tangles in several brainstem nuclei, basal ganglia (especially globus pallidus), diencephalon (subthalamic nucleus) and cerebellar nuclei (Steele et al., 1964; Steele, 1972; Ishino et al., 1975; Jellinger et al., 1980; Probst et al., 1988). The cerebral cortex is relatively spared, although tangles and threads appear to be consistently present in primary motor cortex (Brodmann area 4) and hippocampal and parahip-pocampal regions (Hauw et al., 1990; Braak et al., 1992; Hof et al., 1992). The extensive pathological changes in brainstem, diencephalon and basal ganglia suggested an involvement of multiple neuronal systems. In a serie of studies summarized here (Chinaglia et al., 1990, 1992, submitted; un-published studies) we analyzed pre- and postsynaptic markers of the dopaminergic and serotoninergic system and studied muscarinic, GABA/BZ and some neuropeptide receptors by in vitro receptor autoradiography on postmortem brain tissues of PSP patients.

## Material and methods

### Case material

9 patients with neuropathologically confirmed PSP were studied (Table 1). 5 out of 9 patients (cases 5 to 9) had a clinical diagnosis of "probable PSP" based on parkinsonian features and paralysis of vertical gaze developing in the course of their disease. Three of these patients were thought to be demented (see Table 2 for clinical details). Of the four patients clinically not recognized as PSP one (case 1) was thought to suffer from severe PD with Alzheimer component, one (case 2) from striatonigral degeneration and one (case 4) from olivopontocerebellar atrophy. Case 3 presented atypically with onset of symptoms at 31 ys, lack of vertical gaze palsy, prominent resting tremor, a positive family history for PD and blepharospasm for 14 years and were thought to suffer from postencephalitic Parkinsonism. Response to L-DOPA therapy was unsatisfactory in all patients. All patients except case 3 had a relatively rapidly pro-gressing course of 3 to 10 years. Neuropathological examination demonstrated cell loss and gliosis in SN of all cases as well as in other areas, e.g. subthalamic nucleus (STN), globus pallidus (GP) and tectum (no details were available for case 8). Neurofibrillary tangles (NFT) were abundant in affected areas, whereas no Lewy bodies and only occasional senile plaques were observed. Two cases (2 and 5) showed abundant axonal spheroids in substantia nigra pars reticulata (SNpr) and were therefore regarded as neuropathologically atypical cases. The clinically atypical case 3, however, was found to have pathological changes consistent with PSP.

**Table 1.** Characteristics of PSP patients

| Case | Sex | Age | PMI | Cause of death |
|------|-----|-----|------|----------------|
| 1 | m | 74 | 12.5 | Bronchopneumonia |
| 2 | m | 72 | n.d. | Bronchopneumonia |
| 3 | m | 67 | 24.5 | Cardiac failure |
| 4 | m | 70 | 15 | n.d. |
| 5 | m | 79 | 28.5 | Bronchopneumonia |
| 6 | f | 69 | n.d. | n.d. |
| 7 | f | 71 | 18 | Bronchopneumonia |
| 8 | m | 72 | 41 | Cardiac failure |
| 9 | m | 68 | 16.5 | Respiratory failure |

*f* female; *m* male; *n.d.* not documented; *PMI* post mortem intervall

**Table 2.** Clinical features in PSP patients

| Case | 1 | 2 | 3 | 4 | 5 | 6 | 7 | 8 | 9 |
|------|---|---|---|---|---|---|---|---|---|
| Bradykinaesia | + | + | + | + | + | + | + | + | + |
| Postural instability (frequent falls) | n.d. | + | n.d. | + | ++ | + | + | ++ | ++ |
| Rigidity (axial) | + | + | + | + | + | + | + | + | + |
| Tremor at rest | n.d. | (+) | + | − | + | − | − | − | − |
| SNOP | n.d. | + | n.d. | + | + | + | + | + | + |
| Akinetic dysarthria/dysphonia | + | n.d. | + | + | + | + | + | + | + |
| Dysphagia | + | + | n.d. | + | + | n.d. | + | + | + |
| Dementia | ++ | ++ | n.d. | n.d. | + | n.d. | n.d. | + | + |
| Pyramidal signs | − | + | − | − | + | − | + | − | − |
| Other | | | BS | | | | | | |

*BS* blepharospasm; *n.d.* not documented; *SNOP* supranuclear ocular palsy

### Autoradiographic studies

The ligands used in the present studies are summarized in Table 3. For autoradiographic studies 10 μm thick tissue sections were cut with a cryostat-microtome (Leitz 1720 from Leitz, Wetzlar, F.R.G.), thaw-mounted onto gelatine-coated glass slides and stored at −20°C until use. For ligand binding tissue sections were brought to room temperature, preincubated, incubated and washed according to the incubation protocols listed in Table 4. After the washing period, tissue sections were dipped in ice-cold distilled water and dried under a stream of cold air. Autoradiograms were generated by apposing the labeled tissue together with autoradiographic [$^3$H]- or [$^{125}$I]-micro-scales (Amersham) to [$^3$H]-Hyperfilm (Amersham) at 4°C. Films were analyzed using a computerized image analysis system (MCID, Imaging Research Inc., St. Catharines, Ontario, Canada). The statistical significance of the effect of pathology on the densities of binding sites was determined by oneway analysis of variance (ANOVA) followed by a Dunnett's multiple comparison test (alpha error 5%).

**Table 3.** Neurotransmitter and neuropeptide binding sites studied in PSP

| Receptors or uptake sites | Subtypes | Ligands |
|---|---|---|
| Dopamine uptake sites | | [³H]Mazindol |
| Dopamine | $D_1$ | [³H]-SCH 23390 |
| | $D_2$ | [³H]Spiperone |
| | | [³H]CV 205-502 |
| | | [¹²⁵I]Iodosulpride |
| Serotonin uptake site | | [³H]Citalopram |
| Serotonin | $5\text{-HT}_{1C}$ | [³H]Mesulergine |
| | $5\text{-HT}_{1D}$ | [³H]5-Hydroxytryptamine |
| Acetylcholine | M | [³H]N-methyl-Scopolamine |
| Gamma-amino-butyric acid | A | [³H]Flunitrazepam |
| Substance P | | [¹²⁵I]-BH-SP |
| Neurotensin | | [¹²⁵I-Tyr3]Neurotensin |

## In situ hybridization histochemistry study

Oligonucleotides complementary to the base sequences 19–69 and 1019–1064 of the human dopamine $D_2$ receptor (Grandy et al., 1989) were synthesized on a 380B Applied Biosystems DNA synthesizer, purified on a 20% polyacrylamide/8M urea preparative sequencing gel, labeled using [³²P]dATP and terminal deoxynu-cleotidyltransferase and hybridized to tissue sections as described (Mengod et al., 1992). Autoradiograms were generated by apposing the labeled slide mounted tissue sections to β-max films (Amersham) for 3 weeks at −70°C. Optical densities of film signals were read using a computerized image analysis system (MCID, Imaging Research Inc., St. Catharines, Ontario, Canada) and converted into artificial units (AU) with the help of co-exposed standards. Values are used for relative comparison only.

## Results

### Dopamine uptake sites labeled with [³H]-mazindol

In PSP 82% and 83% of dopamine uptake sites were lost in caudate and putamen, respectively (Fig. 1C, Table 5). In nucleus accumbens the density of mazindol binding sites was decreased by 66% (Fig. 1C, Table 5). In PD a comparable decrease of dopamine uptake sites was observed (Fig. 1B; Chinaglia et al., 1992).

### Dopamine D₁ receptors labeled with [³H]-SCH 23390

$D_1$-like receptors labeled with [³H]-SCH 23390 were similarly distributed in PSP and control brains, showing high densities in a patchy distribution in striatum and intermediate levels in medial globus pallidus. The densities of [³H]-SCH 23390 were not significantly different between PSP patients, con-trols or PD patients (Cortés et al., 1989) in any region examined (Table 5;

**Table 4.** Conditions of ligand binding studies

| Ligand | Conc. | Nonspecific | Buffer | Preincubation | Incubation | Washing |
|---|---|---|---|---|---|---|
| [$^3$H]-mazindol[1] (15.8 Ci/mmol) | 4 nM | 1 μM mazindol | 50 mM Tris HCl, pH 7.4 120 mM NaCl; 5 mM KCl | 5 min at 4°C | 40 min at 4°C 300 nM desipramine | 2 × 1 min in ice cold-buffer |
| [$^3$H]-SCH 23390[2] (60.4 Ci/mMol) | 1 nM | 1 μM cis-flupentixo | 150 mM Tris-HCl, pH 7.4 120 mM NaCl, 5 mM KCl, 2 mM CaCl₂, 1 mM MgCl₂ | 20 min at RT | 60 min at RT | 1 × 5 min in ice cold-buffer |
| [$^3$H]-spiperone[3] (69.6 Ci/mMol) | 0.4 nM | 1 μM(+)butaclamol | 170 mMTris-HCl, pH 7.4 120 mM NaCl, 5 mM KCl 2 mM CaCl₂, 1 mM MgCl₂ | 15 min at RT | 30 min at RT | 2 × 5 min in ice cold-buffer |
| [$^3$H]-CV 205 502[4] (86.3 Ci/mMol) | 1 nM | 1 μM(+)butaclamol | 170 mM Tris-HCl, pH 7.5 | 30 min at RT | 90 min at RT | 2 × 1 min in ice cold-buffer |
| [$^{125}$I]-Iodosulpride[5] (1,950 Ci/mmol) | 0.1 nM | 1 μM sulpiride | 50 mM Tris-HCl, pH 7.4 120 mM NaCl, 5 mM KCl 2 mM CaCl₂, 1 mM MgCl₂ | 15 min at RT | 30 min at RT | 2 × 2 min in ice cold-buffer |
| [$^3$H]-citalopram[6] (68.5 Ci/mmol) | 1 nM | 1 μM imipramine | 50 mM Tris-HCl, pH 7.4 120 mM NaCl, 5 mM KCl | 15 min at RT | 60 min at RT | 2 × 10 min in ice-cold-buffer |
| [$^3$H]-NMS[7] (85 Ci/mmol) | 1 nM | 10 μM atropine | 0.3 M PBS, pH 7.4 | — | 60 min at RT 100 μM carbachol 300 nM pirenzepine | 1 × 10 min in ice-cold-buffer |
| [$^3$H]-flunitrazepam[8] (79.6 Ci/mmol) | 2 nM | 2 μM diazepam | 170 nM Tris-HCl, pH 7.4 | 40 min at RT | 60 min at 4°C 300 nM CL 218872 | 2 × 1 min in ice-cold-buffer |
| [$^{125}$I]-Bolton–Hunter SP[9] (2,000 Ci/mmol) | 50 pM | 1 μM SP | 50 mM Tris-HCl, pH 7.4 200 mg/l BSA, 2 mg/l chymostatin 4 mg/l leupeptin, 40 mg/l bacitracin, 5 mM MnCl₂ | 15 min at RT | 60 min at RT | 4 × 30 sec in ice-cold Tris |
| [$^{125}$I]-neurotensin[10] (2,200 Ci/mmol) | 0.1 nM | 0.5 μM NT | 50 mM Tris-HCl, pH 7.4 200 mg/l BSA, 20 μM bacitracin, 5 mM MnCl₂ | — | 60 min at 4°C | 4 × 2 min in ice-cold-buffer |

[1] Compare Javitch et al., 1985; [2] Compare Cortés et al., 1989; [3] Compare Palacios et al., 1981; [4] Compare Charuchinda et al., 1987; [5] Compare Martres et al., 1985; [6] Compare D'Amato et al., 1987; [7] Compare Wamsley et al., 1980; Cortés et al., 1987; [8] Compare Zezula et al., 1988; [9] Compare Dietl et al., 1989; [10] Compare Uhl and Snyder, 1977; Palacios and Kuhar, 1981

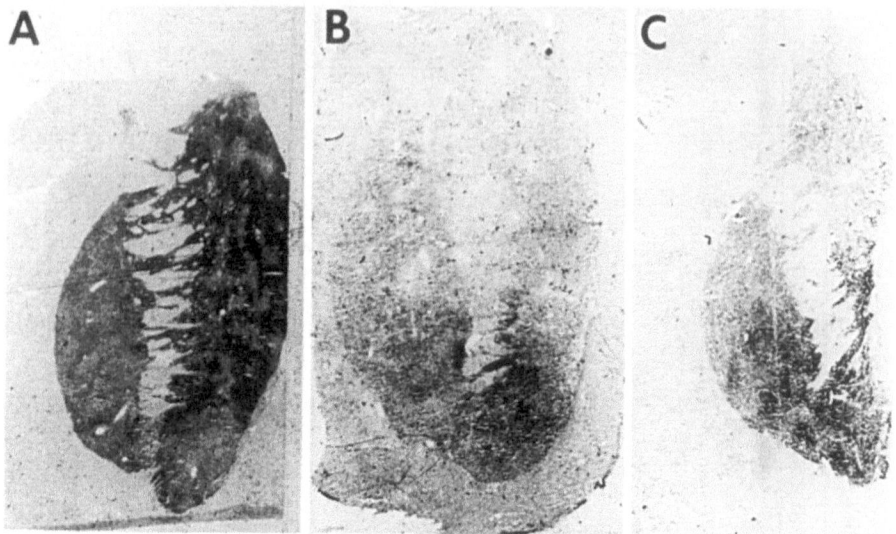

**Fig. 1.** Dopamine uptake sites labeled by [$^3$H-]mazindol in the anterior basal ganglia of a normal control (**A**), a PD (**B**) and PSP (**C**) patient. The marked loss of dopamine uptake sites is comparable in PSP and PD

Landwehrmeyer, Chinaglia, Alvarez, Probst, Mengod, and Palacios, in preparation).

### *Dopamine D$_2$ receptors labeled with [$^{125}$I]-Iodosulpride, [$^3$H]-CV 205 502 or [$^3$H]-spiperone*

The distribution of all three ligands were qualitatively very similar in controls, PD and PSP, showing high labeling in caudate and putamen with a ventrally increasing gradient (for [$^{125}$I]-Iodosulpride and [$^3$H]-CV 205 502) and low levels in lateral globus pallidus. Quantitatively there was a decrease in [$^3$H]-spiperone binding by 50% and 40% in accumbens and putamen/caudate, respectively (Fig. 2, Table 5). However, binding of the D$_2$ receptor agonist [$^3$H]-CV 205 502 and the antagonist [$^{125}$I]-Iodosulpride in PSP was not significantly different from controls and tended to be higher than control values in experiments with [$^{125}$I]-Iodosulpride (Fig. 2, Table 5; Landwehrmeyer, Chinaglia, Alvarez, Probst, Mengod, and Palacios, in preparation).

### *Dopamine D$_2$ mRNA in PSP*

Regional levels of D$_2$ mRNA in PSP were in no striatal subregion significantly different from control values (Fig. 3; Landwehrmeyer, Chinaglia, Alvarez, Probst, Mengod, and Palacios, in preparation).

**Table 5.** Alterations of binding sites in basal ganglia of PSP — quantitative data

| Ligand | Acc C | Acc PSP | C C | C PSP | P C | P PSP | SN C | SN PSP | PN C | PN PSP |
|---|---|---|---|---|---|---|---|---|---|---|
| [3H]-mazindol[1] | 941 ± 63 | 327 ± 71* | 1,020 ± 53 | 144 ± 36* | 1,087 ± 62 | 157 ± 44* | 356 ± 31 | n.a. | | |
| [3H]-SCH 23390[2] | 291 ± 25 | 249 ± 28 | 252 ± 27 | 266 ± 38 | 322 ± 34 | 309 ± 30 | 74 ± 16 | 61 ± 8 | | |
| [3H]-spiperone[2] | 337 ± 19 | 169 ± 17* | 314 ± 23 | 158 ± 17* | 330 ± 14 | 230 ± 17* | 56 ± 7 | 22 ± 11* | | |
| [3H]-CV 205 502[2] | 221 ± 28 | 175 ± 19 | 152 ± 15 | 148 ± 16 | 161 ± 18 | 182 ± 12 | 51 ± 7 | 23 ± 3* | | |
| [125I]-Iodosulpride[2] | 12.7 ± 2.6 | 18.2 ± 4.6 | 14.9 ± 1.8 | 16.6 ± 3.2 | 18.2 ± 3.9 | 16.7 ± 2.2 | 4.4 ± 0.9 | 2.4 | | |
| [3H]-citalopram[3] | 128 ± 13 | 110 ± 21 | 89 ± 22 | 63 ± 13 | 144 ± 15 | 118 ± 16 | 439 ± 147 | 350 | | |
| [3H]-NMS[4] | 2,621 ± 108 | 1,805 ± 158* | 2,233 ± 96 | 1,725 ± 118* | 2,384 ± 81 | 1,805 ± 158* | 289 ± 26 | 306 ± 49 | | |
| [3H]flunitrazepam[5] | 672 ± 65 | 528 ± 54 | 389 ± 24 | 326 ± 28 | 416 ± 30 | 361 ± 17 | 239 ± 13 | n.a. | | |
| [125I]-BH-SP[6] | 6.0 ± 1.6 | 4.9 ± 1.0 | 5.6 ± 1.4 | 3.2 ± 0.8 | 6.6 ± 1.3 | 2.6 ± 0.5 | n.a. | n.a. | | |
| [125I]neurotensin[7] | 4.7 ± 0.5 | 3.6 ± 0.5 | 2.4 ± 0.3 | 1.8 ± 0.3 | 2.8 ± 0.2 | 1.7 ± 0.2* | 8.3 ± 0.6 | 3.2 ± 0.6* | 8.1 ± 0.5 | 7.4 ± 1.1 |

| Ligand | GPe C | GPe PSP | GPi C | GPi PSP | PD C | PD PSP |
|---|---|---|---|---|---|---|
| [3H]-NMS[4] | 266 ± 26 | 329 ± 31 | 292 ± 27 | 417 ± 64* | 299 ± 22 | 436 ± 65* |
| [3H]flunitrazepam[5] | 221 ± 14 | 108 ± 42* | 179 ± 12 | 73 ± 27* | 211 ± 19 | 169 ± 14 |

[1] Data from Chinaglia et al., 1992; [2] Data from Landwehrmeyer, Alvarez, Chinaglia, Probst, Mengod and Palacios, in preparation; [3] Data from Chinaglia et al., 1993; [4] Data from Vilaró, Landwehrmeyer, Chinaglia, Alvarez, Probst and Palacios, in preparation; [5] Data from Palacios, Chinaglia, Landwehrmeyer, Alvarez, and Probst; [6] Data from Palacios et al., unpublished; [7] Data from Chinaglia et al., 1990; (fmol/mg protein; mean ± SEM)

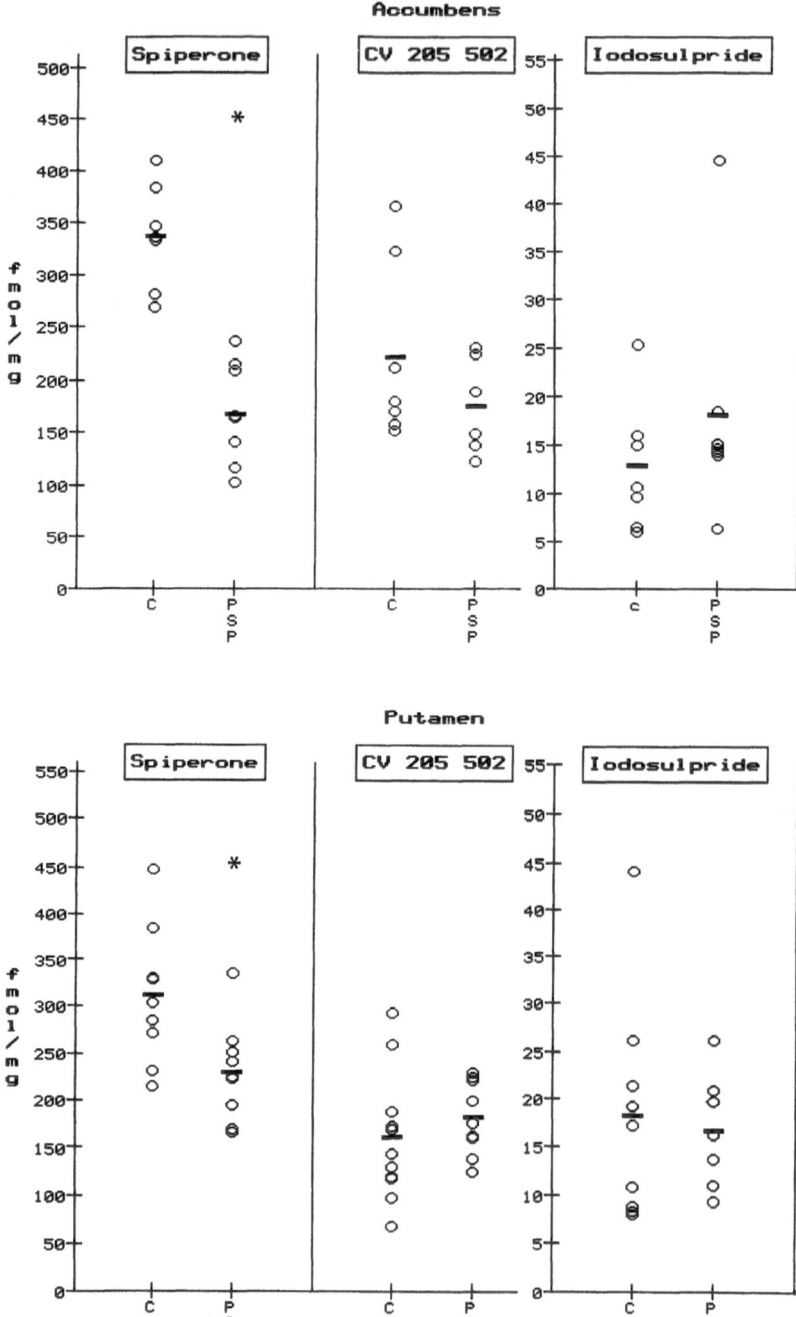

**Fig. 2.** Scattergram showing densities of dopamine "$D_2$" receptors labeled with three different putative $D_2$-selective ligands ([$^3$H]-spiperone, [$^3$H]-CV 205 502, [$^{125}$I]-Iodosulpride) in controls (*C*) with no history of neurological or psychiatric disease and PSP patients. Values for n. accumbens and putamen (middle third) are shown.
*p < 0.05

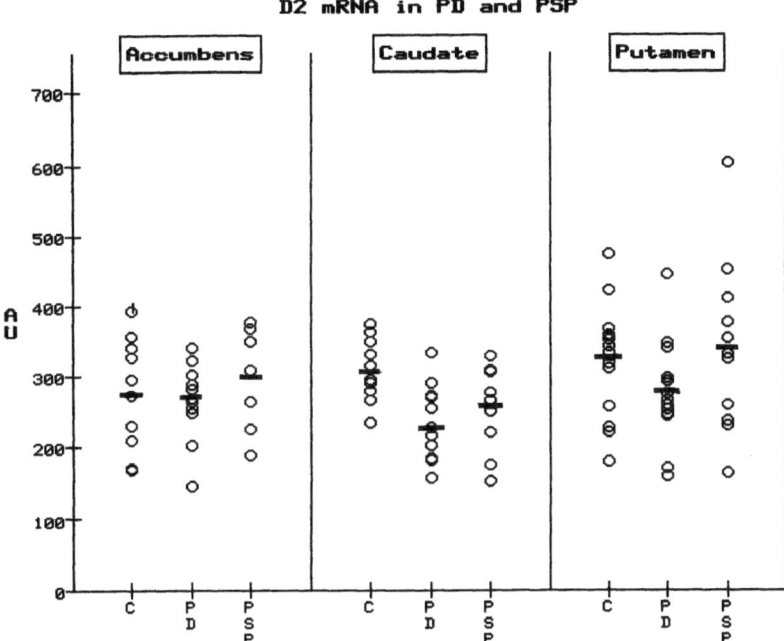

**Fig. 3.** Scattergram showing the relative regional levels of dopamine $D_2$ mRNA determined by in situ hybridization histochemistry in n. accumbens, caudate head and putamen of control (*C*), PD and PSP patients. Values are expressed in artificial units (*AU*) derived from optical densities of film autoradiograms

*5-HT uptake sites labeled with [³H]-citalopram*

In basal ganglia of PSP [³H]-citalopram binding sites were relatively well preserved (Fig. 4B, Table 5). In contrast, in PD patients 5-HT uptake sites were significantly reduced to 30–60% of control values (Fig. 4C; Chinaglia et al., submitted). In cortical regions, however, a significant decrease of serotonin transporters was found in both PD and PSP (Chinaglia et al., submitted).

*5-HT₁C and 5-HT₁D receptors labeled with [³H]-mesulergine and [³H]-5-HT*

In preliminary studies no significant alterations were found in striatum of PSP patients (data not shown).

*Muscarinic receptors labeled with [³H]-N-methyl-scopolamine (NMS)*

Using [³H]-NMS (a compound labeling all muscarinic receptor subtypes with comparable affinity) we observed a reduced density of [³H]-NMS binding sites in striatum of PSP but not of PD (−30%; Fig. 5, Table 5). In

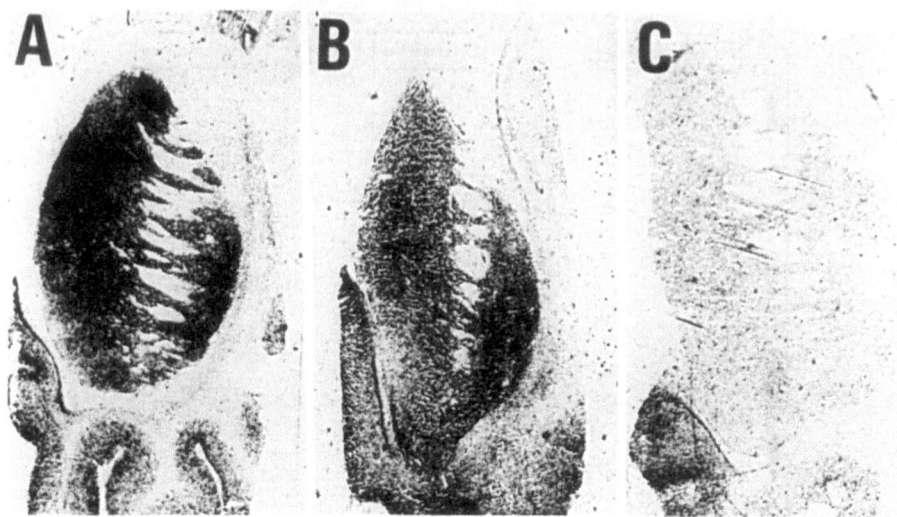

**Fig. 4.** 5-HT uptake sites labeled with [³H]-citalopram in anterior basal ganglia of a control patient (**A**), a PSP patient (**B**) and a PD patient (**C**). 5-HT uptake sites in striatum are relatively well preserved in PSP

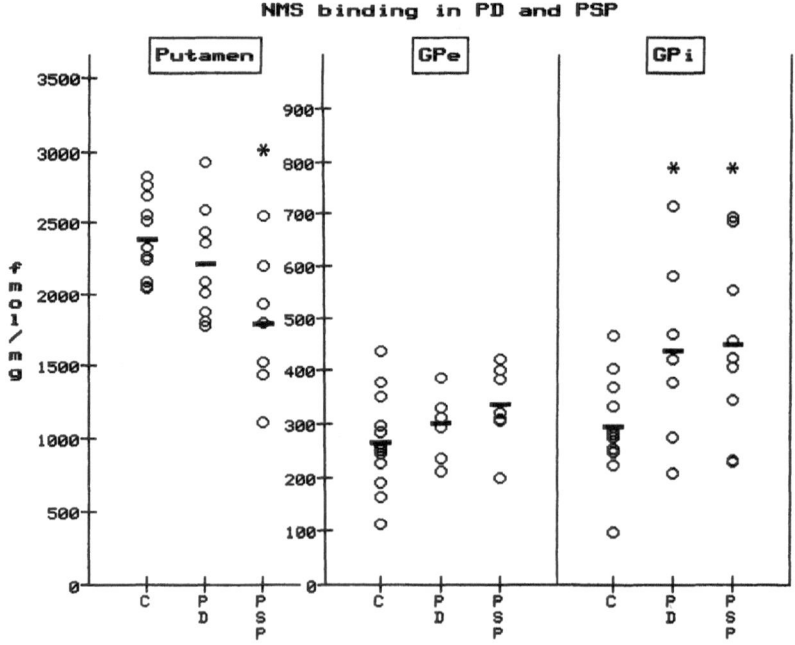

**Fig. 5.** Scattergram showing alterations of [³H]-NMS binding densities in putamen and globus pallidus of PD and PSP patients. *p < 0.05

addition, there was an increase of [³H]-NMS binding in GPi of about 40% in both PD and PSP (Fig. 5, Table 5; Vilaró, Landwehrmeyer, Chinaglia, Probst, and Palacios, in preparation).

### GABA$_A$/BZ receptors labeled with [³H]-flunitrazepam

GABA$_A$/BZ receptors as labeled with [³H]-flunitrazepam were significantly reduced in GP of PSP (−51% in GPe, −60% in GPi; Table 5). In contrast, binding in striatum of PSP was unchanged (Table 5). No differences from control values were observed in PD (Table 5; Palacios et al., unpublished).

### Neurotensin receptor binding using [¹²⁵I]-neurotensin

There was an extensive loss of [¹²⁵I]-NT binding to SNc in PSP (−62%; Table 5) and in PD (Chinaglia et al., 1990); binding to medial paranigral nucleus, however, was not significantly reduced in PSP (−9%; Table 5). In striatum we observed a trend to reduced densities in all striatal subregions and a significant decrease by 43% in putamen of PSP (Table 5; Chinaglia et al., 1990).

### Substance P binding sites labeled with [¹²⁵I]-BHSP

There was a trend to decreased densities of SP receptors in PSP, most notable in putamen (Table 5). In no striatal subregion statistical significance was reached. In PD the densities of SP receptors were not altered (Palacios et al., in preparation).

## Discussion

### Dopaminergic system

In PSP (like in PD) the dopaminergic mesostriatal system appears to be severely affected as indicated by a marked decrease in striatal dopamine (DA) concentrations (−75% to −86%; Kish et al., 1985; Ruberg et al., 1985), a loss of dopaminergic neurons in lateral SN (Steele et al., 1964) and a severe reduction of dopamine uptake sites located presynaptically on dopaminergic afferents (Chinaglia et al., 1992). In contrast, the mesolimbic dopaminergic system projecting to n. accumbens is thought — (unlike in PD) — not to be affected in PSP (Agid et al., 1986). Normal DA concentrations in n. accumbens (−11% or +22%; Ruberg et al., 1985; Kish et al., 1985) and preserved neurotensin binding (Chinaglia et al., 1990) to nucleus paranigralis, the putative site of origin of mesolimbic dopaminergic projections, support the concept of an unaffected mesolimbic system in

PSP. However, the finding of significantly reduced dopamine uptake sites in n. accumbens in PSP suggests that the mesolimbic dopaminergic system is altered to some degree, at least in the sense that dopaminergic projection neurons to n. accumbens are no longer able to express DA-uptake sites on their terminals. It remains to be established whether the reduced density of DA-uptake sites reflects a defect of axonal transport as suggested by prominent axonal cytoskeletal abnormalities in PSP (Probst et al., 1988) or changes in synthesis or turnover of DA-uptake sites in mesolimbic projection neurons.

The at best limited benefit of treatment with L-DOPA and dopamine agonists in PSP raises the question whether (in contrast to idiopathic Lewy body Parkinson's disease) the therapeutic inefficacy of dopaminergic drugs may be explained by altered densities of postsynaptic dopamine receptors in PSP (Agid et al., 1986). Our studies indicate that (as in idiopathic Lewy body Parkinson's disease) the densities of dopamine $D_1$ receptors are unchanged in PSP. This agrees well with previous membrane binding studies (Pierot et al., 1988) and is further supported by the finding that levels of dopamine $D_1$ receptor mRNA do not differ significantly between control and PSP populations (data not shown; Landwehrmeyer, Chinaglia, Alvarez, Probst, Mengod, and Palacios, in preparation). Concerning dopamine $D_2$ receptors, in contrast, there are conflicting data, indicating unchanged or decreased density of dopamine $D_2$ receptors depending on the radioligand used (Bokobza et al., 1984; Ruberg et al., 1985; Pierot et al., 1988; this study). Our study confirms the previously reported loss of $[^3H]$-spiperone binding sites (Bokobza et al., 1984; Ruberg et al., 1985; Pierot et al., 1988). However, dopamine $D_2$ receptors labeled with the putative dopamine $D_2$ selective compounds $[^{125}I]$-Iodosulpride and $[^3H]$-CV 205 502 did not differ from control values. This finding raises the question whether there is in fact a real reduction in density of dopamine $D_2$ receptors that accounts for the marked loss of $[^3H]$-spiperone binding in PSP. Spiperone is known to label $5HT_{2/1C}$ receptors and a nonreceptor site, the spirodecanone site (Palacios et al., 1981), apart from dopamine $D_2$ receptors. In addition, all available radioligands label more than one of the recently discovered $D_2$ receptor subtypes (Landwehrmeyer et al., 1993; Palacios et al., 1993). We therefore compared the relative levels of dopamine $D_2$ receptor mRNA by ISHH in PSP and control patients. Levels of $D_2$ mRNA in PSP did in no striatal subregion significantly differ from control levels (Landwehrmeyer, Chinaglia, Alvarez, Probst, Mengod, and Palacios, in preparation). Although the marked loss of $[^3H]$-spiperone binding sites remains to be explained, these data favor the view that the regional levels of neither dopamine $D_1$ nor postsynaptic dopamine $D_2$ receptors are significantly altered in PSP.

*Serotoninergic system*

Concentrations of 5-HT (Kish et al., 1985), 5-HT uptake sites (Chinaglia et al., 1993) and serotonin receptors in basal ganglia — as far as studied to date — are unchanged in PSP, suggesting a relatively well preserved 5-HT system in basal ganglia. The sparing of serotoninergic terminals in the basal ganglia of PSP patients might have some therapeutical implications, as the stimulation of $5\text{-HT}_1$ receptors results in a hyperlocomotion in globus pallidus lesioned rats by apparently dopamine independent mechanisms (Oberlander et al., 1986). Amitriptyline has been shown to result in some improvement of motor disability in PSP patients that can be striking in occasional patients (Newman, 1985). Although the mechanism of this action of amitriptyline is not understood in detail, it is likely that the preservation of striatal serotoninergic terminals in PSP is critical for the symptomatic effect of amitriptyline.

*Cholinergic system*

In addition to marked changes in cholinergic neurons in the nucleus basalis of Meynert (Tagliavini, 1983) and brainstem (e.g. Hirsch et al., 1987; Jellinger, 1988; Zweig et al., 1987) a significant decrease of CAT activity in basal ganglia has been reported (Ruberg et al., 1985). In contrast, Kish et al. (1985) found no significant changes in striatal CAT activity, suggesting some variability in the degree of affection of cholinergic striatal interneurons in PSP. In addition, there is morphological evidence for a loss or degeneration of cholinergic striatal interneurons (Hirsch et al., 1989; Oyanagi et al., 1991). The alteration of cholinergic interneurons may in part explain the relative infrequent occurrence of tremor in PSP by restoring a dopaminergic/cholinergic balance at a lower level. If cholinergic interneurons degenerate in PSP one should expect to see changes in muscarinic receptors in striatum. In our study we observed a small, but significant reduction in the density of muscarinic receptors labeled by $[^3\text{H}]$-NMS in caudate, putamen and n. accumbens of PSP. In a previous membrane binding study using $[^3\text{H}]$-QNB as a ligand, no loss of muscarinic binding sites has been observed (Ruberg et al., 1985). It is tempting to speculate that the apparent loss of $[^3\text{H}]$-NMS binding sites in striatum of PSP may reflect a loss of muscarinic receptors of the m2 subtype thought to be expressed on cholinergic interneurons (Vilaró et al., 1991). Further studies are under way to determine which subtype(s) of muscarinic receptors are affected (Vilaró, Landwehrmeyer, Chinaglia, Probst, and Palacios, in preparation). In addition, the trend to reduced densities of substance P receptors in PSP, but not in PD or Huntington's disease, is compatible with an expression of a population of SP receptors on cholinergic striatal interneurons as demonstrated in rat brain (Gerfen, 1991). In addition, we observed an increased density of muscarinic receptors in GPi in both PD and PSP. Observations in Huntington's disease (HD) suggest that the muscarinic receptors found in GP are located on afferents from striatal projection neurons (Vilaró, Landwehrmeyer, Chinaglia, Probst, and Palacios, in preparation). This implies an increased expression of a muscarinic receptor subtype(s) on a subpopulation of striatal neurons projecting to GPi (Vilaró, Landwehrmeyer,

Chinaglia, Probst, and Palacios, in preparation). The reasons for the up-regulation of muscarinic receptors observed are unknown and might differ between PD and PSP. The functional significance of the increased binding in GPi of both PD and PSP is not known either at present and will require further studies.

### GABA-ergic system

Levels of glutamic acid decarboxylase (GAD) activity in PSP were not significantly different from control levels in a number of brain areas, but are notably reduced in GP ($-60\%$; Agid et al., 1986). As described in our study, densities of $GABA_A/BZ$ receptors in GP were also significantly reduced. We suggest therefore that the loss of $GABA_A/BZ$ receptors in GP reflects a loss of intrinsic GABA-ergic pallidal neurons. This interpretation is supported by studies in HD where the marked loss of GABA-ergic striatopallidal afferents did not result in a decrease of $GABA_A/BZ$ receptors in the region of GP. Instead, an increased density of $GABA_A/BZ$ receptors was observed in GP of HD (Fig. 6), probably reflecting denervation hypersensitivity of intrinsic pallidal neurons.

### Conclusion

Alterations of neurotransmitter receptors and neurotransmitter uptake sites studied in PSP are summarized in Table 6. Like in PD, significant reductions of DA-uptake sites, and NT-receptors and increased densities of muscarinic receptors in GPi can be observed in PSP. However, unlike in PD, $GABA_A$ receptors (in GP) and [$^3$H]-NMS- (and possibly SP) binding sites (in striatum) are lost in PSP, whereas striatal serotonin uptake sites are relatively preserved. In our view only the increase in [$^3$H]-NMS binding in GPi appears to be a functional change. The majority of alterations seen in basal ganglia of PSP, however, appear to reflect loss or degenerative dysfunction of either dopaminergic mesostriatal and mesolimbic neurons, of striatal cholinergic interneurons or of intrinsic GABA-ergic neurons in GP.

### Acknowledgements

The authors thank Profs. A. Probst and J. Ulrich, Department of Pathology, Division of Neuropathology, University of Basel and Prof. E. Bird, Brain Tissue Resource Center, McLean Hospital Belmont, MA, USA for kindly supplying specimen of PSP patients. The latter institution was in part supported by grant MH/NS 31862. We also want to express our sincere gratitude to Drs. Alvarez, Chinaglia, and Mengod for contribution to the work summarized in this paper. Part of the work was performed at Sandoz Pharma Ltd., Basel, Switzerland; this support is gratefully acknowledged. B.L. was supported by a grant of the Deutsche Forschungsgemeinschaft, Bonn, Federal Republic of Germany.

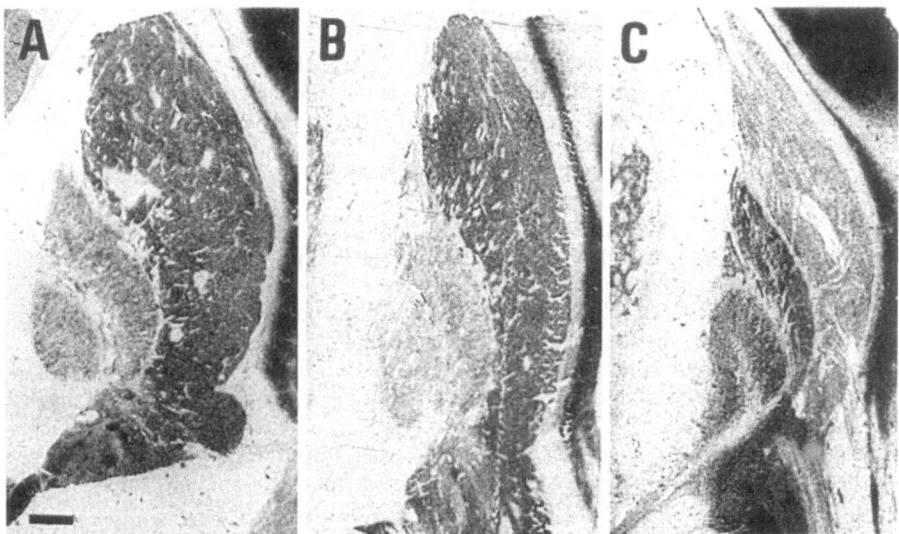

**Fig. 6.** GABA$_A$/BZ receptors labeled with [$^3$H]-flunitrazepam in posterior basal ganglia of a control patient (**A**), a PSP patient (**B**) and a patient with Huntington's disease (**C**). Compared to controls, densities of [$^3$H]-flunitrazepam binding sites are decreased in both segments of globus pallidus of PSP patients (**B**), but increased in Huntington's disease (**C**). Note in addition the loss of GABA$_A$/BZ receptors in the putamen of Huntington's disease; densities were within normal range in putamen of PSP patients

**Table 6.** Alterations of binding sites in PSP — summary

| Ligand | Caud | Put | Acc | GPe | GPi | SNc | SNr |
|---|---|---|---|---|---|---|---|
| [$^3$H]Mazindol | ↓ | ↓ | ↓ | low | low | low | low |
| [$^3$H]SCH 23390 | = | = | = | low | = | = | = |
| [$^3$H]Spiperone | ↓ | ↓ | ↓ | ↓ | low | ↓ | low |
| [$^{125}$I]Iodosulpride | = | = | = | = | = | na | na |
| [$^3$H]CV 205 502 | = | = | = | = | = | ↓ | low |
| [$^3$H]Citalopram | = | = | = | = | = | na | na |
| [$^3$H]NMS | ↓ | ↓ | ↓ | = | ↑ | = | = |
| [$^3$H]Flunitrazepam | = | = | = | ↓ | ↓ | na | na |
| [$^{125}$I]Neurotensin | = | ↓ | = | = | = | ↓ | = |
| [$^{125}$I]BH-Substance P | = | = | = | = | = | = | = |

*low* very low specific binding; *na* not available

## References

Agid Y, Javoy-Agid F, Ruberg M, Pillon B, Dubois B, Duyckaerts C, Hauw J-J, Baron J-C, Scatton B (1986) Progressive supranuclear palsy: anatomoclinical and biochemical considerations. In: Yahr MD, Bergman KJ (eds) Advances in neurology, vol 45. Raven Press, New York

Behrman S, Carroll JD, Janota I, Matthews WB (1969) Progressive supranuclear palsy. Brain 92: 663–678

Bokobza B, Ruberg M, Scatton B, Javoy-Agid F, Agid Y (1984) [$^3$H]-spiperone binding, dopamine and HVA concentrations in Parkinson's disease and supranuclear palsy. Eur J Pharmacol 99: 167–175

Braak H, Jellinger K, Braak E, Bohl J (1992) Allocortical neurofibrillary changes in progressive supranuclear palsy. Acta Neuropathol 84: 478–483

Charuchinda C, Supavilai P, Karobath M, Palacios JM (1987) Dopamine D2 receptors in the rat brain: autoradiographic visualization using a high-affinity selective agonist ligand. J Neurosci 7: 1352–1360

Chinaglia G, Probst A, Palacios JM (1990) Neurotensin receptors in Parkinson's disease and progressive supranuclear palsy: an autoradiographic study in basal ganglia. Neuroscience 39: 351–360

Chinaglia G, Alvarez FJ, Probst A, Palacios JM (1992) Mesostriatal and mesolimbic dopamine uptake binding sites are reduced in Parkinson's disease and progressive supranuclear palsy: a quantitative autoradiographic study using [$^3$H]-mazindol. Neuroscience 49: 317–327

Chinaglia G, Landwehrmeyer B, Probst A, Palacios JM (1993) Serotoninergic terminals are differentially affected in Parkinson's disease and progressive supranuclear palsy: an autoradiographic study with [$^3$H]-citalopram. Neuroscience 54: 691–699

Cortés R, Probst A, Palacios JM (1987) Quantitative light microscopic autoradiographic localization of cholinergic muscarinic receptors in the human brain: forebrain. Neuroscience 20: 65–107

Cortés R, Camps M, Gueye B, Probst A, Palacios JM (1989) Dopamine receptors in human brain: autoradiographic distribution of D1 and D2 sites in Parkinson syndrome of different etiology. Brain Res 483: 30–38

D'Amato RJ, Largent BL, Snowman AM, Snyder SH (1987) Selective labeling of serotonin uptake sites in rat brain by [$^3$H]citalopram contrasted to labeling of multiple sites by [$^3$H]imipramine. J Pharmacol Exp Ther 242: 364–371

Dietl MM, Sanchez M, Probst A, Palacios JM (1989) Substance P receptors in the human spinal cord: decrease in amyotrophic lateral sclerosis. Brain Res 483: 39–49

Gerfen CR (1991) Substance P (neurokinin-1) receptor mRNA is selectively expressed in cholinergic neurons in the striatum and basal forebrain. Brain Res 556: 165–170

Golbe LI, Davies PH (1988) Progressive supranuclear palsy. In: Jankovic J, Tolosa E (eds) Parkinson's disease and movement disorders. Urban & Schwarzenberg, Baltimore, pp 121–130

Grandy DK, Marchionni MA, Makam H, Stofko RE, Alfano M, Frothingham L, Fischer JB, Burke-Howie KJ, Bunzow JR, Server AC, Civelli O (1989) Cloning of the cDNA and gene for a human D2 dopamine receptor. Proc Natl Acad Sci USA 86: 9762–9766

Hauw J-J, Verny M, Delaère P, Cervera P, He Y, Duyckaerts C (1990) Constant neurofibrillary changes in the neocortex in progressive supranuclear palsy: basic differences with Alzheimer's disease. Neurosci Lett 119: 182–186

Hirsch EC, Graybiel AM, Duyckaerts C, Javoy-Agid F (1987) Neuronal loss in the pedunculopontine tegmental nucleus in Parkinson disease and in progressive supranuclear palsy. Proc Natl Acad Sci USA 84: 5976–5980

Hirsch EC, Graybiel AM, Hersh LB, Duyckaerts C, Agid Y (1989) Striosomes and extrastriosomal matrix contain different amounts of immunoreactive choline acetyltransferase in the human striatum. Neurosci Lett 96: 145–150

Hof PR, Delacourte A, Bouras C (1992) Distribution of cortical neurofibrillary tangles in progressive supranuclear palsy: a quantitative analysis of six cases. Acta Neuropathol 84: 45–51

Ishino H, Ikeda H, Otsuki S (1975) Contribution to clinical pathology of progressive supranuclear palsy (subcortical argyrophilic dystrophy). On the distribution of neurofibrillary tangles in the basal ganglia and brain-stem and its clinical significance. J Neurol Sci 24: 471–481

Javitch JA, Strittmatter SM, Snyder SH (1985) Differential visualization of dopamine and norepinephrine uptake sites in rat brain using [$^3$H]mazindol autoradiography. J Neurosci 5: 1513–1521

Jellinger K, Riederer P, Tomonaga M (1980) Progressive supranuclear palsy: clinico-pathological and biochemical studies. J Neural Transm [Suppl] 16: 111–128

Jellinger K (1988) The pedunculopontine nucleus in Parkinson's disease, progressive supranuclear palsy and Alzheimer's disease. J Neurol Neurosurg Psychiatry 51: 540–543

Kish SJ, Chang LJ, Mirchandani L, Shannak K, Hornykiewicz O (1985) Progressive supranuclear palsy: relationship between extrapyramidal disturbances, dementia, and brain neurotransmitter markers. Ann Neurol 18: 530–536

Landwehrmeyer B, Mengod G, Palacios JM (1993) Differential visualization of dopamine D2 and D3 receptor sites in rat brain. A comparative study using in situ hybridization histochemistry and ligand binding autoradiography. Eur J Neurosci 5: 145–153

Maher ER, Lees AJ (1986) The clinical features and natural history of the Steele-Richardson-Olszewski syndrome (progressive supranuclear palsy). Neurology 36: 1005–1008

Martres M-P, Bouthenet M-L, Sales N, Sokoloff P, Schwartz J-C (1985) Widespread distribution of brain dopamine receptors evidenced with [$^{125}$I]Iodosulpride, a highly selective ligand. Science 228: 752–755

Mengod G, Vilaró MT, Landwehrmeyer GB, Martinez-Mir MI, Niznik HB, Sunahara RK, Seeman P, O'Dowd BF, Probst A, Palacios JM (1992) Visualization of dopamine D1, D2 and D3 receptor mRNAs in human and rat brain. Neurochem Int 20: 33S–43S

Newman GC (1985) Treatment of progressive supranuclear palsy with tricyclic antidepressants. Neurology 35: 1189–1193

Oberlander C, Blaquière B, Pujol J-F (1986) Distinct functions for dopamine and serotonin in locomotor behavior: evidence using the 5-HT$_1$ agonist RU 24969 in globus pallidus-lesioned rats. Neurosci Lett 67: 113–118

Oyanagi K, Takahashi H, Wakabayashi K, Ikuta F (1991) Large neurons in the neostriatum in Alzheimer's disease and progressive supranuclear palsy: a topographic, histologic and ultrastructural investigation. Brain Res 544: 221–226

Palacios JM, Niehoff DL, Kuhar MJ (1981) [$^3$H]spiperone binding sites in brain: autoradiographic localization of multiple receptors. Brain Res 213: 277–289

Palacios JM, Landwehrmeyer B, Mengod G (1993) Brain dopamine receptors: characterization, distribution and alteration in disease. In: Jankovic J, Tolosa E (eds) Parkinson's disease and movement disorders, 2nd edn. Williams & Wilkins, Baltimore, pp 35–54

Pierot L, Desnos C, Blin J, Raisman R, Scherman D, Javoy-Agid F, Ruberg M, Agid Y (1988) D1 and D2-type dopamine receptors in patients with Parkinson's disease and progressive supranuclear palsy. J Neurol Sci 86: 291–306

Probst A, Langui D, Lautenschlager C, Ulrich J, Brion JP, Anderton BH (1988) Progressive supranuclear palsy: extensive neuropil threads in addition to neurofibrillary tangles. Very similar antigenicity of subcortical neuronal pathology in progressive supranuclear palsy and Alzheimer's disease. Acta Neuropathol 77: 61–68

Ruberg M, Javoy-Agid F, Hirsch E, Scatton B, Heureux RL, Hauw J-J, Duyckaerts Ch, Gray F, Morel-Maroger A, Rascol A, Serdaru M, Agid Y (1985) Dopaminergic and cholinergic lesions in progressive supranuclear palsy. Ann Neurol 18: 523–529

Steele JC, Richardson JC, Olzewski J (1964) Progressive supranuclear palsy. Arch Neurol 10: 333–359

Steele JC (1972) Progressive supranuclear palsy. Brain 95: 693–704

Tagliavini F, Pilleri G, Gemignani F, Lechi A (1983) Neuronal loss in the basal nucleus of Meynert in progressive supranuclear palsy. Acta Neuropathol 61: 157–160

Uhl GR, Snyder SH (1977) Neurotensin receptor binding, regional and subcellular distributions favour transmitter role. Eur J Pharmacol 41: 89–91

Vilaró MT, Wiederhold KH, Palacios JM, Mengod G (1991) Muscarinic M2 receptor mRNA expression and receptor binding in cholinergic and noncholinergic cells in the rat brain: a correlative study using in situ hybridization and receptor autoradiography. Neuroscience 47: 367–393

Wamsley JK, Zarbin MA, Birdsall JM, Kuhar MJ (1980) Muscarinic cholinergic receptors: autoradiographic localization of high and low affinity agonist binding sites. Brain Res 200: 1–12

Zezula J, Cortés R, Probst A, Palacios JM (1988) Benzodiazepine receptor sites in the human brain: autoradiographic mapping. Neuroscience 25: 771–795

Zweig RM, Whitehouse PJ, Casanova MF, Walker LC, Jankel WR, Price DL (1987) Loss of pedunculopontine neurons in progressive supranuclear palsy. Ann Neurol 22: 18–25

Authors' address: Dr. J. M. Palacios, Laboratorios Almirall, Cardoner 68, Barcelona 08024, Spain.

# Changes in aminergic receptors in a PSP postmortem brain: correlation with pathological findings

**J. Pascual**[1,2], **J. Figols**[3], **B. Grijalba**[1], **A. M. González**[1], **E. del Olmo**[1], **J. Berciano**[2], and **A. Pazos**[1]

[1] Department of Physiology and Pharmacology, Unit of Pharmacology,
[2] Service of Neurology, and [3] Section of Neuropathology, University Hospital
"Marqués de Valdecilla", University of Cantabria, Santander, Spain

**Summary.** The state of different aminergic receptors was assessed, by quantitative autoradiography in tissue sections, in several representative brain regions from a typical progressive supranuclear palsy (PSP) patient and from 9 matched brains. The densities of muscarinic receptors were within control limits in most of the brain areas of this PSP brain. Serotonin$_1$ receptors were clearly reduced only in areas with very relevant neuropathological damage, such as locus niger and globus pallidus. The density of D$_1$ dopamine receptors in the caudate-putamen and frontal cortex of the patient was within control limits. By contrast, nigral D$_1$ and striatal D$_2$ dopamine receptors were dramatically reduced in the patient as compared to controls. Finally, alpha$_2$-adrenoceptors were clearly reduced in all the examined areas of this PSP patient as compared to control group. Both the potential role of these receptor changes in the pathophysiology of the clinical features of PSP and their correlation with the neuropathological findings of this PSP patient are discussed.

## Introduction

Progressive supranuclear palsy (PSP) is an uncommon neurodegenerative disease representing about 4% of patients with parkinsonism (Jankovic, 1984). The clinico-pathological features of PSP were already described more than two decades ago (Steele et al., 1964; Steele, 1972). PSP patients exhibit a characteristic clinical picture including loss of voluntary vertical gaze, dystonic extension of the neck and subcortical dementia as well as rigidity and akinesia with very poor response to levodopa or dopaminergic agonists (Klawans and Ringel, 1971). Neuropathologically, there are degenerative changes in the brainstem, diencephalic and cerebellar nuclear masses. In these subcortical areas neurofibrillary tangles, granulovacuolar degeneration, nerve cell loss and gliosis are observed (Steele, 1975; Barr, 1986).

In contrast to the detailed knowledge of the clinical and pathological features of this entity, neurochemical data in PSP are not numerous, in particular those on the state of receptors for classical neurotransmitters. In this work, we show the changes that in some of the most relevant aminergic receptors we have found in a typical PSP patient by using quantitative autoradiography. The results concerning dopamine and noradrenergic receptors have already been reported in part (Pascual et al., 1992a, 1993). The correlation between the neuropathological findings and the neuro-chemical data of this PSP patient is discussed.

## Patient report

This man was first seen in our hospital in 1983 when he was 64 years old. His past medical history was unremarkable. He described an unsteady gait, blurred vision not corrected by glasses and subtle mental changes. On examination he was mentally slowed, apathetic and moderately depressed. In addition, lack of voluntary vertical gaze with normal reflex movements, slurred speech and a rigid-akinetic syndrome were observed. Computed tomography (CT) showed slight brain atrophy. Despite clorimipramine and increasing doses of levodopa plus carbidopa, his symptoms pro-gressed and two years later axial rigidity with neck extension, dysphagia, osteotendinous hyperreflexia, complete loss of ocular voluntary movements, and urinary incontinence ensued. During 1984 he took for several months 50 mg of clorimipramine at night. Another CT, performed three years after the beginning of the clinical picture, disclosed progression of the brain atrophy, especially marked in the midbrain. The patient was maintained on levodopa (up to 1 g daily) plus carbidopa without a significant response. Amitriptyline in a dose of 25–50 mg daily was added with no benefit in 1986. During his last three weeks of life the patient was admitted to our hospital receiving only levodopa plus carbidopa and trihexyphenidyl with no apparent benefit. He died sud-denly in July 1986 while sleeping.

## Methods

At necropsy, the brain was promptly removed, one half being fixed in 10% formalin for neuropathological examination. The other half was cut into blocks, quickly frozen and stored at $-70°C$ until it was used in binding assays.

Autoradiographic labelling of aminergic receptors was performed in multiple brain regions of this PSP patient (age, 67 years; postmortem delay, 24 hours) as well as in 9 male patients with no history of neuropsychiatric disease (average age ± standard deviation [SD], 64 ± 13; postmortem delay ± SD, 20 ± 10).

Autoradiographic labelling of different aminergic receptors (muscarinic cholinergic, serotonin$_1$, dopamine $D_1$ and $D_2$ and alpha$_2$-adrenoceptors) was performed incubating 10-μm-thick tissue sections with specific radioligands following the already described protocols (see Table 1). Adjacent sections were incubated in the presence of different cold compounds (see Table 1) to define nonspecific binding.

Autoradiograms were generated by apposing the slide-mounted tissue sections to tritium-sensitive films ($^3$H-Ultrofilm, Leica, Nussloch, Germany) at 4°C for the time required in every case. The autoradiograms were analyzed and quantified by using a computer-assisted microdensitometer (Microm IP, Microm España, Barcelona, Spain). Appropriate standards (Amersham [Buckinghamshire, UK] tritiated microscales), exposed together with the tissues, allowed the transformation of densitometric readings into receptor densities (fmol/mg protein) (Unnerstall et al., 1982).

**Table 1.** Protocols followed for the labeling of aminergic receptors

| Receptor | $^3$H-ligand | Cold compound | Conditions Authors, year |
|---|---|---|---|
| Muscarinic | N-methyl-scopolamine | Atropine | Cortés et al., 1984 |
| Serotonin$_1$ | 5-HT | 5-HT | Pazos et al., 1987 |
| D$_1$ | SCH 23390 | Dopamine | Cortés et al., 1989b |
| D$_2$ | Spiperone | Haloperidol | Camps et al., 1989 |
| Alpha$_2$ | Bromoxidine | Phentolamine | Pazos et al., 1988 |

## Results

### Neuropathological studies

Macroscopical examination showed generalized brain atrophy most prominent in the midbrain, with an enlarged sylvian aqueduct and third ventricle. Microscopical findings were diagnostic for PSP (Steele, 1975; Barr, 1986). There were prominent neurofibrillary tangles in the globus pallidus, subthalamic nucleus, substantia nigra (Fig. 1), periacueductal gray matter and locus ceruleus (Fig. 1). In these brain regions as well as in others such as the substantia innominata, dentate nucleus of the cerebellum, and red nucleus, we observed relevant but variably intense gliosis and neuronal loss, with central chromatolysis or granulovacuolar degeneration in the remaining neurons. Slight to moderate gliosis with neither definitive neuronal loss nor neurofibrillary degeneration was observed in the caudate-putamen and neocortex.

### Neurochemical studies

These results appear in Table 2 and are illustrated in Figs. 2 and 3. The densities of cholinergic muscarinic receptors, as labelled with $^3$H-N-methylscopolamine, were within normal limits in most of the examined brain areas of this PSP patient. Only in the striatum was there a slight reduction in the density of muscarinic receptors ($-23\%$ in the putamen and 18% in the caudate) as compared to controls.

Serotonin$_1$ receptors in the PSP brain were clearly reduced only in specific areas, such as locus niger and globus pallidus, slightly reduced over the striatum and preserved along frontal cortex, hippocampus and cerebellar cortex.

The density of dopamine D$_1$ receptors in the caudate-putamen and the frontal cortex of the patient was within normal limits as compared to the control group. By contrast, D$_1$ receptors were clearly reduced in the substantia nigra of the patient with PSP in comparison to the control subjects.

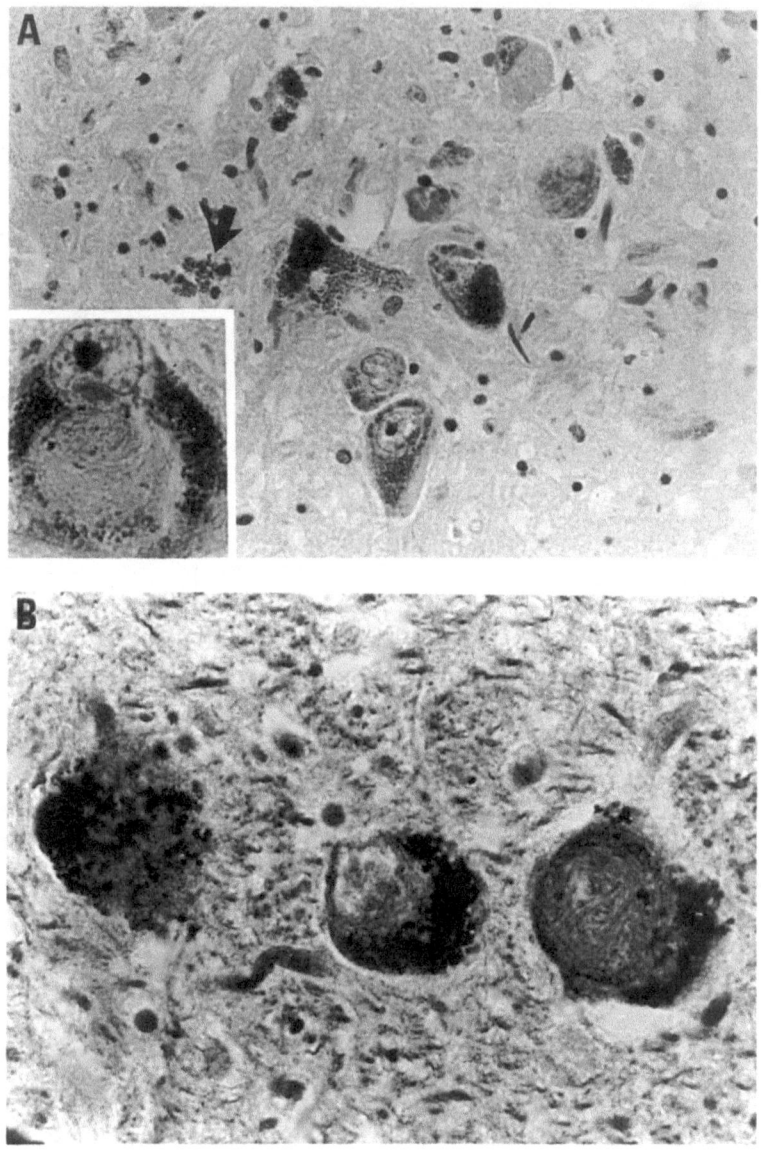

Fig. 1. A Zona compacta of the locus niger showing loss of neurons, gliosis, scattered melanin (arrow) and balloned neurons (HE ×600). The insert illustrates a pigmented neuron with neurofibrillary degeneration (HE ×1,500, before % reduction). B Photograph of locus ceruleus neurons which exhibit typical neurofibrillary degeneration (Bielchowsky ×1,000)

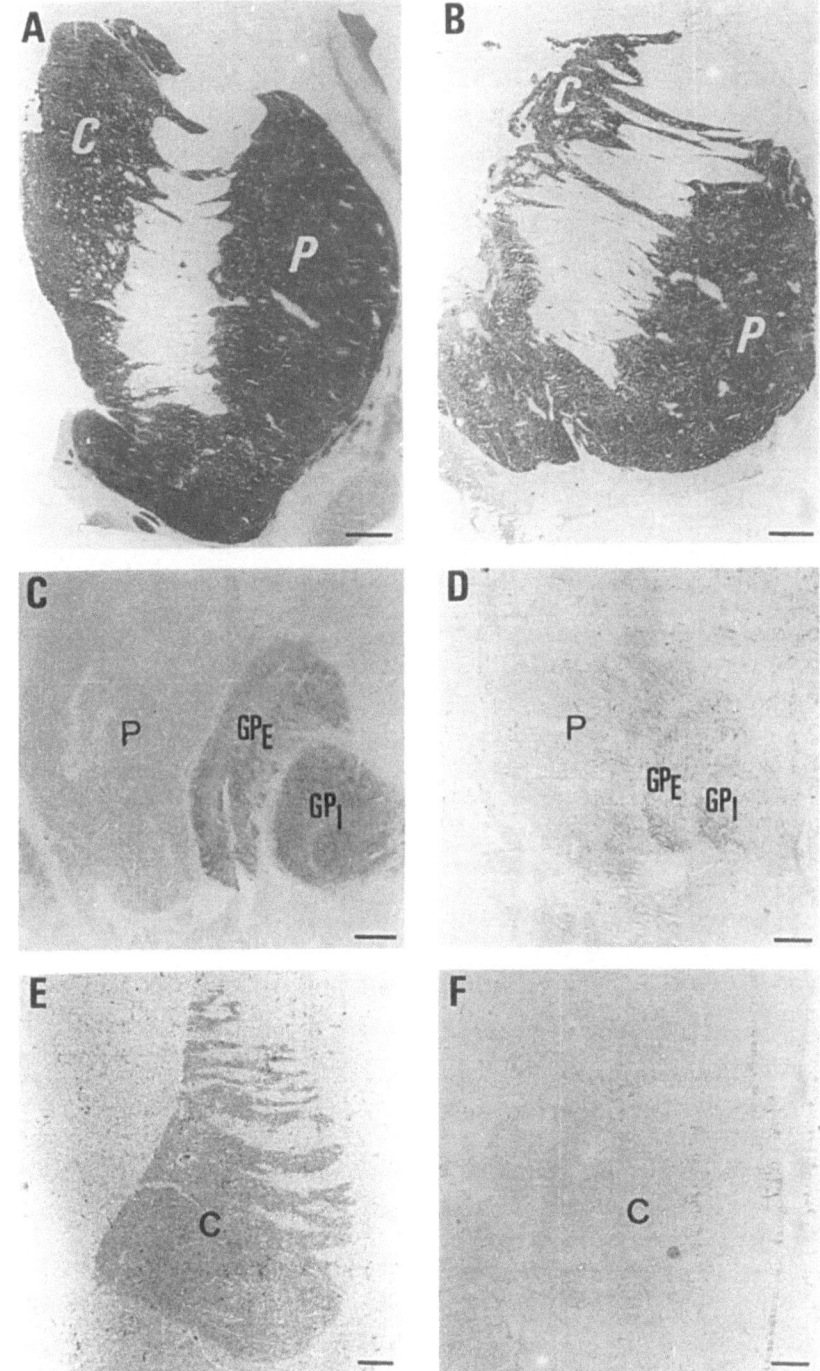

**Fig. 2.** Autoradiograph of aminergic receptors at different striatal levels of controls (left) vs. PSP patient (right). Note the very slight reduction in muscarinic receptors in PSP striatum (**B**) as compared to a control (**A**). Regarding serotonin$_1$ receptors a selective reduction was found for the external and internal globus pallidus ($GP_E$, $GP_I$, respectively) but not in the putamen (*P*) of this PSP case (**D**) as compared to a control (**C**). Finally, note the dramatic loss of dopamine D$_2$ receptors in caudate (*C*) and putamen (*P*) of this PSP brain (**F**) in comparison with a control case (**E**). Bar = 2 mm

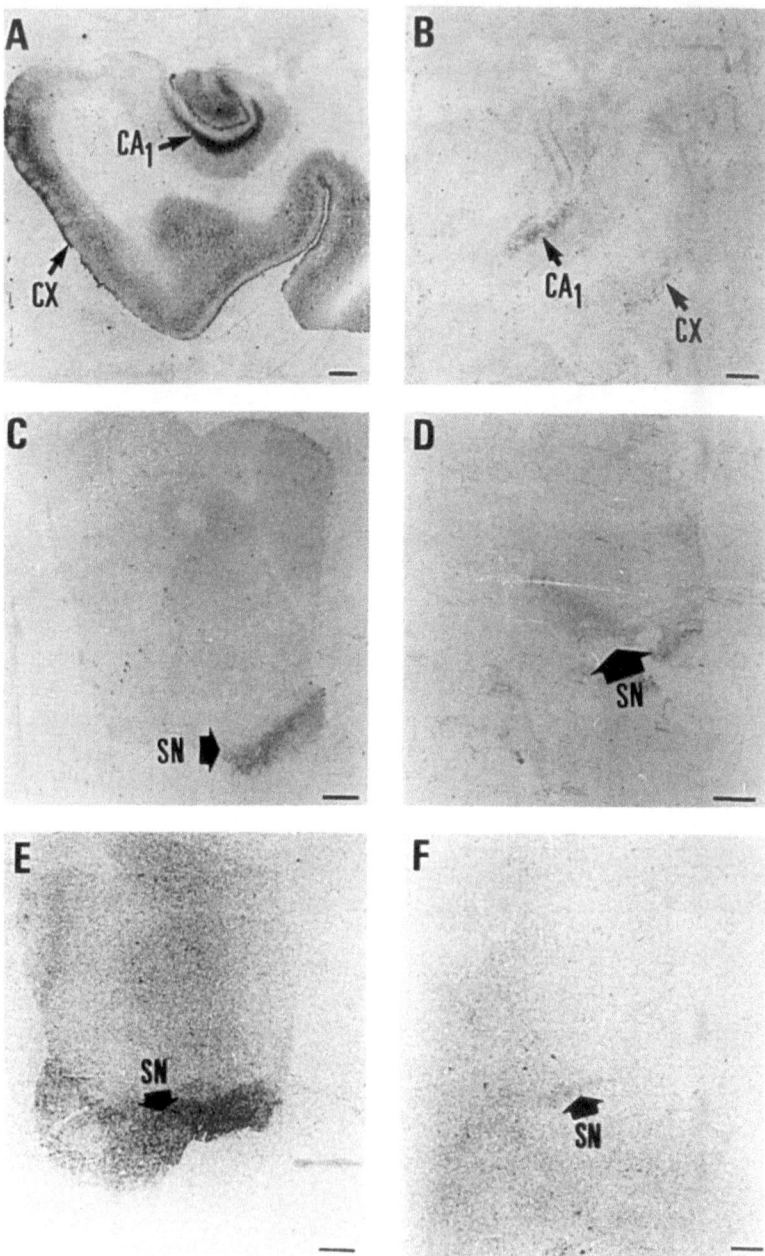

**Fig. 3.** Autoradiographs of aminergic receptors at hippocampal (**A,B**) and nigral (**C,D,E,F**) levels. Note the dramatic loss of alpha$_2$-adrenoceptors over the hippocampus and entorhinal cortex (*CX*) of this PSP brain (**B**) as compared to a control brain (**A**). In the substantia nigra (*SN*) of this PSP patient, there is a very relevant loss of D$_1$ dopamine (**D**) and serotonin receptors (**F**) as compared to controls (**C,E**). Bar = 2 mm.
*CA$_1$* stratum lacunoso-moleculare of the CA$_1$ field of the hippocampus

**Table 2.** Density (fmol/mg prot) of aminergic receptors in PSP brain vs control brains (mean ± SD)

| Area | Muscarinic | | | Serotonin$_1$ | | | Dopamine$_1$ | | | Dopamine$_2$ | | | Alpha$_2$ | | |
|---|---|---|---|---|---|---|---|---|---|---|---|---|---|---|---|
| | Controls | PSP | % | Controls | PSP | % | Controls | PSP | % | Controls | PSP | % | Controls | PSP | % |
| Frontal cortex | | | | | | | | | | | | | | | |
| External layers | 459 ± 29[a] | 526[a] | +15 | 659 ± 156[a] | 666[a] | +1 | 109 ± 14[b] | 122[b] | +12 | | | | 945 ± 200[c] | 63[c] | −51 |
| Internal layers | 445 ± 22[d] | 478[d] | +7 | 357 ± 108[d] | 422[d] | +15 | | | | | | | 665 ± 189[e] | 275[e] | −58 |
| Hippocampus | | | | | | | | | | | | | | | |
| CA$_1$ field | 545 ± 58[f] | 599[f] | +10 | 1,134 ± 203[f] | 1,364[f] | +17 | | | | | | | 1,098 ± 273[g] | 440[g] | −60 |
| | 466 ± 41[h] | 455[h] | −1 | | | | | | | | | | 910 ± 131[i] | 253[i] | −72 |
| Dentate gyrus | 712 ± 60 | 586 | −18 | 414 ± 144 | 561 | +26 | | | | 324 ± 58 | <30 | <91 | 120 ± 34 | 74 | −38 |
| | | | | 279 ± 44 | 202 | −28 | | | | 429 ± 69 | <20 | <95 | 121 ± 37 | 75 | −38 |
| Caudate | 843 ± 75 | 647 | −23 | 620 ± 212 | 224 | −64 | 317 ± 78 | 325 | +3 | | | | | | |
| Putamen | 85 ± 39 | 67 | −21 | 471 ± 76 | 214 | −55 | 275 ± 69 | 339 | +23 | | | | | | |
| G. pallidus, ext. | 160 ± 68 | 166 | −4 | | | | | | | | | | | | |
| G. pallidus, int. | 325 ± 38 | 308 | −5 | | | | | | | | | | | | |
| Subst. innominata | | | | 562 ± 195 | 278 | −51 | 151 ± 40 | 73 | −52 | | | | 222 ± 134 | 121 | −46 |
| Locus niger | | | | | | | | | | | | | | | |
| Cerebellum, cortex | | | | 65 ± 10[j] | 60[j] | −8 | | | | | | | 750 ± 136[k] | 350[k] | −53 |

[a] layers I–III, [b] layers I–VI, [c] layer, [d] layers III–VI, [e] layer III, [f] stratum pyramidalis I, [g] stratum lacunoso-moleculare, [h] molecular layer, [i] stratum granularis, [j] molecular layer, [k] granularis layer

The density of dopamine $D_2$ receptors was almost negligible in the PSP striatum as compared to control levels. In the human brain, $D_2$ receptors are concentrated over the striatum, their very low density ($<30-40$ fmol/mg of protein) in the substantia nigra not allowing comparison between the patient and control subjects (Camps et al., 1989). Muscarinic and alpha$_2$-receptors are also present in negligible quantities in the substantia nigra (Cortés et al., 1984; Pascual et al., 1992b).

The density of alpha$_2$-adrenoceptors of this PSP case was markedly reduced in all the brain areas examined as compared to the control group. In fact, the percentage of this reduction ranged from 38% in the striatum (an area with low densities of alpha$_2$-receptors) to 72% in the dentate gyrus of the hippocampus.

## Discussion

In this work we show the autoradiographic distribution of some of the most relevant aminergic receptors in a typical PSP patient. The detailed study of this patient allows a good correlation between his neurochemical and clinico-pathological data, this being of interest for a better knowledge of the pathophysiology and management of this devastating disorder. It must be emphasized here, however, that all the neurochemical findings occurring in our typical PSP patient will not neccessarily be present in all PSP cases. In fact, in PSP the neuropathological damage appearing in several nuclei quantitatively varies from case to case (Kish et al., 1985; Ruberg et al., 1985; Jellinger, 1987) and, some of the receptor changes found in PSP cases may partly be secondary to the administration of neuropsychiatric medications.

The state of the cholinergic neurotransmission system in PSP has not been fully clarified from previous studies. While Ruberg and co-workers found a decrease in choline acetyltransferase (CAT) activity in cerebral cortex, striatum and substantia innominata of 9 PSP brains (Ruberg et al., 1985), Kish's group found no significant changes in the cerebral cortex or basal ganglia, even in 2 of the 5 PSP patients with severe dementia (Kish et al., 1985). Therefore, complete clarification of the state of the presynaptic cholinergic component must await further studies. The slight reduction in muscarinic receptors found here over the caudate-putamen can be due to the loss of large neurons selectively appearing in the striatum in PSP (Oyanagi et al., 1988). However, from Ruberg and colleagues' work and from our results it seems that the predominantly postsynaptic muscarinic receptors are spared in PSP. Taken together, all these data suggest that a deficit in the cholinergic neurotransmission system is not a crucial factor for the dementia appearing in this entity, though there may be a slight to moderate presynaptic cholinergic deficit in some PSP brains.

In the brain of our patient with PSP serotonin$_1$ receptors, as labelled with $^3$H-5HT, were selectively reduced in subcortical areas such as globus pallidus and locus niger, and preserved in cortical areas, such as frontal cortex or hippocampus. These data show that the loss of serotonin$_1$ receptors in PSP depends directly on the degree of local degenerative neuropathological changes and affects preferably to the $D_1$ subtype (Hoyer et al., 1988; Waeber et al., 1988). In animals, some behavioural motor responses are mediated by serotonin$_1$ receptors and motor activity can be affected by serotonergic lesions because nigrostriatal dopaminergic transmission is under inhibitory control of ascending serotonergic systems. However, there is at present no clear connection between parkinsonism and serotonergic deficits. By contrast, there is some evidence suggesting that this serotonergic deficiency may play a part in depressive and cognitive symptoms (Pazos and Palacios, 1985; Agid et al., 1987).

The PSP patient studied here exhibited a dramatic loss of striatal $D_2$ dopamine receptors. Previous studies have shown $D_2$ dopamine receptors to be decreased in PSP caudate-putamen both in vitro, by using membrane binding procedures in postmortem brains, and in vivo, by using SPECT and PET (see Table 3). Dopamine $D_2$ receptors are thought to be located preferably on postsynaptic striatal spiny neurons, this fact explaining the normal densities of this dopamine receptor subtype in *pure* presynaptic parkinsonisms (Guttmann et al., 1986; Pierot et al., 1988; Cortés et al., 1989a; Pascual et al., 1991). Although $D_2$ down-regulation secondary to chronic levodopa intake might partly explain the loss of $D_2$ receptors seen in this study, our data, surely representing the end stage of this disease, show again that the loss of $D_2$ dopamine receptors is the most plausible explanation for the absence of response to dopaminergic drugs in PSP.

In agreement with Pierot and co-workers' data obtained in membrane homogenates (Pierot et al., 1988), $D_1$ dopamine receptors are preserved in striatal sections of our PSP case. In the human species, the preservation of $D_1$ receptors in Parkinson's disease (PD) and PSP, despite the marked nigrostriatal degeneration taking place in these degenerative disorders (Pimoule et al., 1985; Raisman et al., 1985; Rinne et al., 1985; Pierot et al., 1988; Cortés et al., 1989) lends weight to a postsynaptic localisation of dopamine $D_1$ receptors. Furthermore, the observation that $D_1$ dopamine receptors are not modified in PSP patients suggests that they are not located on the same striatal neurons as $D_2$ receptors. The neuronal degeneration occuring in PSP striatum seems to be selective for large neurons where $D_2$ receptors are believed to be located (Oyanagi, 1988). Thus, $D_1$ receptors could be located on one of the several kinds of intrinsic striatal neurons that do not degenerate in this entity.

The striatum is the main site of action of the dopamine formed from levodopa. In addition, evidence has accumulated to suggest that dopamine is also released by the dendrites of dopamine neurons in the substantia nigra and $D_1$ dopamine receptors in this region of the brain appear to play an important role in the actions of levodopa (Robertson, 1992). Nigral $D_1$ dopamine receptors have been studied for the first time in PSP in this case. Although the loss of nigral $D_1$ receptors in this PSP patient could be explained by the observed nigral degeneration, the preservation of nigral $D_1$

J. Pascual et al.

**Table 3.** Summary of reports analyzing striatal $D_2$ dopamine receptors in PSP brains

| Tissue | Method | Ligand | Results | | Author, year |
|---|---|---|---|---|---|
| | | | Caudate | Putamen | |
| Postmortem brain | | | | | |
| | Membrane homogenates | $^3$H-spiperone | −30% | −37% | Bokobza et al., 1984 |
| | Membrane homogenates | $^3$H-spiperone | −48% | −42% | Ruberg et al., 1985 |
| | Membrane homogenates | $^3$H-spiperone | −32% | −36% | Pierot et al., 1988 |
| | Tissue sections | $^3$H-spiperone | −91% | −95% | Pascual et al., 1992 |
| Living brain | PET | $^{76}$Br-spiperone | −24% | | Baron et al., 1985 |
| | PET | $^{18}$F-fluoroethylspiperone | −17% | | Wienhard et al., 1990 |
| | PET | $^{11}$C-raclopride | −32% | −12% | Brooks et al., 1992 |
| | SPECT | $^{123}$I-iodobenzamide | Significant reduction (semiquantitative) | | Schwarz et al., 1992 |

receptors in patients with PD may indicate that the loss of $D_1$ nigral receptors in PSP is probably secondary to the degeneration of nigral terminals of gamma-aminobutyric acid (GABA)ergic or substance P-containing neurons projecting from the striatum (Robertson, 1992). Therefore, the two main neurochemical differences between PD and PSP seem to be the loss of striatal $D_2$ receptors and nigral $D_1$ receptors in PSP, both explaining the poor response to dopaminergic treatments in PSP parkinsonism. On the other hand, the preservation of postsynaptic dopamine $D_1$ receptors indicates that the clinical effects of agonist drugs acting selectively on $D_1$ receptors are worth testing in PSP.

This is the first work analyzing the density of alpha$_2$-receptors in PSP showing a generalized loss of alpha$_2$ receptors in this PSP case as compared with a matched control group. Although, theoretically, the antidepressant drugs taken by our patient could give rise to an alpha$_2$-adrenoceptor down-regulation (Charney et al., 1983), these drugs were given to our patient for short periods and had been dropped at least three weeks before death. Since alpha$_2$-adrenoceptors are thought to be located mostly on presynaptic terminals arising from the locus ceruleus (Starke et al., 1989), the most plausible explanation for this dramatic loss of alpha$_2$-receptors is the degeneration of locus ceruleus observed here. Although the locus ceruleus neuronal loss was not massive, the remaining neurons exhibited extensive degenerative changes which could be accompanied by nerve terminal loss as part of a "dying-back" phenomenon.

In Alzheimer's disease (AD) and PD several clinical manifestations such as depression, the severity of dementia and freezing episodes have been thought to be connected with abnormalities in the noradrenergic system (Agid et al., 1987). Locus ceruleus degeneration is fairly constant in PD (Jellinger, 1987) and AD (Bondareff et al., 1982). While postsynaptic adrenergic receptors are essentially preserved in PD and AD, a selective decrease in alpha$_2$-receptors in the frontal cortex has been described in these conditions (Cash et al., 1984; Meana et al., 1992; Pascual et al., 1992c), a fact also suggesting that ceruleo-cortical pathway degeneration is the main reason for alpha$_2$-receptor loss. Some of the clinical manifestations of PSP, such as aspects of subcortical dementia and akinesia, have been tentatively attributed to a possible adrenergic deficit and noradrenergic supplementation has recently been shown to improve the motility and balance in several PSP patients (Ghika et al., 1991; Matsuo et al., 1991). However, locus ceruleus degeneration in PSP, though frequent, seems to be less constant and severe than in PD or AD (Mann et al., 1984; Kish et al., 1985; Ruberg et al., 1985; Jellinger, 1987) and it would not be surprising if the behaviour of alpha$_2$-receptors in further PSP brains varies from case to case depending on the grade of locus ceruleus degeneration appearing in each patient.

Our findings suggest that alpha$_2$-receptor loss occurs in at least some PSP patients, this fact potentially explaining some of the features of this devastating disorder. Moreover, our data support further studies of the clinical effects of noradrenergic drugs in PSP.

## Acknowledgments

We want to thank Mr. J. Hawkins for his stylistic revision of this manuscript.

## References

Agid Y, Javoy-Agid F, Ruberg M (1987) Biochemistry of neurotransmitters in Parkinson's disease. In: Marsden CD, Fahn S (eds) Movement disorders, vol 2. Butterworths, London, pp 166–230

Baron JC, Mazière B, Loc'h C, Sgouropoulos P, Bonnet AM, Agid Y (1985) Progressive supranuclear palsy: loss of striatal dopamine receptors demonstrated in vivo by positron tomography. Lancet i: 1163–1164

Barr AN (1986) Progressive supranuclear palsy. In: Vinken PJ, Bruyn GW, Klawans HL (eds) Handbook of clinical neurology, vol 49. Extrapyramidal disorders. Elsevier, Amsterdam, pp 239–254

Bokobza B, Ruberg M, Scaton B, Javoy-Agid F, Agid Y (1984) [$^3$H]spiperone binding, dopamine and HVA concentrations in Parkinson's disease and supranuclear palsy. Eur J Pharmacol 99: 167–175

Bondareff W, Mountjoy CQ, Roth M (1981) Selective loss of neurons of origin of adrenergic projection to cerebral cortex (nucleus locus coeruleus) in senile dementia. Lancet i: 783–784

Brooks DJ, Ibanez V, Sawle GV, Playford ED, Quinn N, Nathias CJ, Lees AJ, Marsden CD, Bannister R, Frackowiak SJ (1992) Striatal $D_2$ receptor status in patients with Parkinson's disease, striatonigral degeneration, and progressive supranuclear palsy, measured with $^{11}$C-raclopride and positron emission tomography. Ann Neurol 31: 184–192

Camps M, Cortés R, Gueye B, Probst A, Palacios JM (1989) Dopamine receptors in human brain: autoradiographic distribution of $D_2$ sites. Neuroscience 28: 275–290

Cash R, Ruberg M, Raisman R, Agid Y (1984) Adrenergic receptors in Parkinson's disease. Brain Res 322: 269–275

Charney DS, Heninger GR, Sternberg DE (1983) Alpha$_2$-adrenergic receptor sensitivity and the mechanism of action of antidepressant therapy: the effect of long-term amitriptyline treatment. Br J Psychiatry 142: 265–275

Cortés R, Camps M, Gueye B, Probst A, Palacios JM (1989a) Dopamine receptors in human brain: autoradiographic distribution of $D_1$ and $D_2$ sites in Parkinson syndrome of different etiology. Brain Res 483: 30–38

Cortés R, Gueye B, Pazos A, Probst A, Palacios JM (1989b) Dopamine receptors in human brain: autoradiographic distribution of $D_1$ sites. Neuroscience 28: 263–273

Cortés R, Probst A, Palacios JM (1984) Quantitative light microscopic autoradiographic localization of cholinergic muscarinic receptors in the human brain: brainstem. Neuroscience 12: 1003–1026

Ghika AJ, Tennis M, Schoenfield D, Growdon JH (1991) Idazoxan improves motor function in progressive supranuclear palsy. Neurology 41 [Suppl 1]: 173

Guttmann M, Seeman P, Reynolds GP, Riederer P, Jellinger K, Tourtellote WW (1986) Dopamine $D_2$ receptor density remains constant in treated Parkinson's disease. Ann Neurol 19: 487–492

Hoyer D, Waeber C, Pazos A, Probst A, Palacios JM (1988) Identification of a 5-HT$_1$ recognition site in human brain membranes different from 5-HT$_{1A}$, 5-HT$_{1B}$ and 5-HT$_{1C}$ sites. Neurosci Lett 85: 357–362

Jankovic J (1984) Progressive supranuclear palsy: clinical and pharmacological update. Neurol Clin 2: 473–486

Jellinger K (1987) The pathology of parkinsonism. In: Marsden CD, Fahn S (eds) Movement disorders, vol 2. Butterworths, London, pp 124–165

Kish SJ, Chang LJ, Mirchandani L, Shannak K, Hornykiewicz O (1985) Progressive supranuclear palsy: relationship between extrapyramidal disturbances, dementia, and brain neurotransmitter markers. Ann Neurol 18: 530–536

Klawans HL, Ringel SP (1978) Observations on the efficacy of L-dopa in progressive supranuclear palsy. J Can Sci Neurol 5: 167–173

Mann DMA, Yates PO, Hawkes J (1983) The pathology of the human locus coeruleus. Clin Neuropathol 2: 1–7

Matsuo H, Takashima H, Kishikawa M, Kinoshita I, Mori M, Tsujihata M, Nagataki S (1991) Pure akinesia: an atypical manifestation of progressive supranuclear palsy. J Neurol Neurosurg Psychiatry 54: 397–400

Meana JJ, Barturen F, Asier M, García-Sevilla JA, Fontán A, Zarranz JJ (1992) Decreased density of presynaptic alpha$_2$-adrenoceptors in postmortem brains of patients with Alzheimer's disease. J Neurochem 58: 1896–1904

Oyanagi K, Takahashi H, Wakabayashi K, Ikuta F (1988) Selective decrease of large neurons in the neostriatum in progressive supranuclear palsy. Brain Res 458: 218–223

Pascual J, Berciano J, Grijalba B, del Olmo E, González AM, Figols J, Pazos A (1992a) Dopamine D1 and D2 receptors in progressive supranuclear palsy: an autoradiographic study. Ann Neurol 32: 703–707

Pascual J, Berciano J, González AM, Grijalba B, Figols J, Pazos A (1993) Autoradiographic demonstration of loss of alpha$_2$-adrenoceptors in progressive supranuclear palsy. Preliminary report. J Neurol Sci 114: 165–169

Pascual J, del Arco C, González AM, Pazos A (1992b) Quantitative light microscopic autoradiographic localization of alpha$_2$-adrenoceptors in the human brain. Brain Res 585: 116–127

Pascual J, Grijalba B, García-Sevilla J, Zarranz JJ, Pazos A (1992c) Loss of high-affinity alpha$_2$-adrenoceptors in Alzheimer's disease: an autoradiographic study in frontal cortex and hippocampus. Neurosci Lett 142: 36–40

Pascual J, Pazos A, del Olmo E, Figols J, Leno C, Berciano J (1991) Presynaptic parkinsonism in olivopontocerebellar atrophy: clinical, pathological and neurochemical evidence. Ann Neurol 30: 425–428

Pazos A, González AM, Pascual J, Meana JJ, Barturen F, García-Sevilla JA (1988) Alpha$_2$-adrenoceptors in human forebrain: autoradiographic visualization and biochemical parameters using tha agonist $^3$H-UK 14304. Brain Res 475: 361–365

Pazos A, Probst A, Palacios JM (1987) Serotonin receptors in the human brain-III. Autoradiographic mapping of serotonin-1 receptors. Neuroscience 21: 97–122

Pazos A, Palacios JM (1985) Quantitative autoradiographic mapping of serotonin receptors in the rat brain. I. Serotonin-1 receptors. Brain Res 346: 205–230

Pierot L, Desnos C, Blin J, Raisman R, Scherman D, Javoy-Agid F, Ruberg M, Agid Y (1988) D1 and D2-type dopamine receptors in patients with Parkinson's disease and progressive supranuclear palsy. J Neurol Sci 86: 291–306

Pimoule C, Schoemacker H, Reynolds GP, Langer SZ (1985) $^3$H-SCH 23390 labeled D$_1$ receptors are unchanged in schizophrenia and Parkinson's disease. Eur J Pharmacol 114: 235–237

Raisman R, Cash Ruberg M (1985) Binding of $^3$H-SCH 23390 to D$_1$ receptors in the putamen of control and parkinsonian subjects. Eur J Pharmacol 113: 467–468

Rinne JO, Rinne JK, Laakso K, Lönnberg P, Rinne UK (1985) Dopamine D$_1$ receptors in the parkinsonian brain. Brain Res 359: 306–310

Robertson HA (1992) Dopamine receptor interactions: some implications for the treatment of Parkinson's disease. Trends Neurosci 15: 201–206

Ruberg M, Javoy-Agid F, Hirsch E, Scatton B, LHeureux R, Hauw JJ, Duyckaerts C, Gray F, Morel-Maroger A, Rascol A, Serdaru M, Agid Y (1985) Dopaminergic and cholinergic lesions in progressive supranuclear palsy. Ann Neurol 18: 523–529

Schwarz J, Tastsch K, Arnold G, Gasser T, Trenkwalder C, Kirsch CM, Oertel WH (1992) $^{123}$I-iodobenzamide-SPECT predicts dopaminergic responsiveness in patients with de novo parkinsonism. Neurology 42: 556–561

Starke K, Gothert M, Kilbinger H (1989) Modulation of neurotransmitter release by presynaptic autoreceptors. Physiol Rev 69: 864–989

Steele JC (1972) Progressive supranuclear palsy. Brain 95: 693–704

Steele JC (1975) Progressive supranuclear palsy. In: Vinken PJ, Bruyn GW (eds) Handbook of clinical neurology, vol 22. System disorders and atrophies. North-Holland, Amsterdam, pp 217–230

Steele JC, Richardson JC, Olzsewski J (1964) Progressive supranuclear palsy: a heterogeneous degeneration involving the brainstem, basal ganglia and cerebellum with vertical gaze and pseudobulbar palsy, nuchal dystonia and dementia. Arch Neurol 10: 333–359

Unnerstall JR, Niehoff DL, Kuhar MJ, Palacios JM (1982) Quantitative receptor autoradiography using [$^3$H]-Ultrofilm. J Neurosci Methods 6: 59–73

Waeber C, Dietl MM, Hoyer D, Probst A, Palacios JM (1988) Visualization of a novel serotonin recognition site (5-HT$_{1D}$) in the human brain by autoradiography. Neurosci Lett 88: 11–16

Wienhard K, Coenen HH, Pawlik G, Rudolf J, Laufer P, Jovkar S, Stocklin G (1990) PET studies of dopamine receptor distribution using [$^{18}$F]fluoroethylspiperone: findings in disorders related to the dopaminergic system. J Neural Transm 81: 195–213

Authors' address: Dr. J. Pascual, Neurology Service, University Hospital "Marqués de Valdecilla", 39008 Santander, Spain.

# Epidemiology and treatment

# The epidemiology of PSP

## L. I. Golbe

Department of Neurology, University of Medicine and Dentistry of New Jersey —
Robert Wood Johnson Medical School, New Brunswick, New Jersey, U.S.A.

**Summary.** The age-adjusted prevalence of PSP as measured in central New Jersey is 1.5 cases per million population, about 1% of that of Parkinson's disease. Its incidence is 3–4 new cases per million population per year, similar to that of such better-known illnesses as myasthenia gravis, the hereditary ataxias as a group and Tourette syndrome. Median actuarially adjusted survival after symptom onset is 5.9–6.9 years. PSP appears to favor no geographical, racial, ethnic or occupational group, though there is anecdotal evidence for hydrocarbon exposure as a candidate etiologic factor. No familial cases of typical PSP have been proven. The one formal case-control study failed to implicate any particular causal agent and the rural predilection of PD appears to be absent in PSP. Better diagnostic methods, more multi-center organization, additional case-control studies and new etiologic hypotheses are needed in the epidemiological investigation of PSP.

Even as clues to the causes of other neurodegenerative conditions accumulate encouragingly, the cause of PSP remains a mystery. The few attempts to date to elucidate the cause through epidemiologic means have fared poorly. There has emerged no correlation with geography, race, toxin exposure, family history, personality type, head trauma or occupation. No immunologic, infectious, toxic or genetic pathogenetic mechanism has suggested itself. This article will review the work to date, not least for the purpose of stimulating new epidemiologic efforts toward finding the cause of PSP.

## Descriptive epidemiology

### Prevalence and incidence

The aforementioned paucity of research on PSP is partly the result of its perceived rarity. The International Classification of Diseases, 9th revision (ICD-9) groups PSP with olivopontocerebellar atrophy, striatonigral degeneration, Shy-Drager syndrome and Hallervorden-Spatz disease under a

**Table 1.** Crude and age-adjusted prevalence of PSP in Middlesex and Somerset Counties, New Jersey, 1986 (total population of area: 799,022)

| Sex | Number of patients ascertained | Age-adjusted prevalence ratio |
|---|---|---|
| Men | 6 | 1.53/100,000 |
| Women | 5 | 1.23/100,000 |

Note: Among all persons over age 65, the prevalence ratio is 7.00/100,000. Age-adjusted prevalence ratios (Data from Goble et al., 1988)

single code, 333.0. This makes epidemiologic study and identification of PSP difficult, even when cases have been diagnosed accurately, and further hinders such study. Unfortunately, a late draft of ICD-10 continues this nosologic lumping.

Among referral series of patients with parkinsonian disorders, PSP comprises approximately 4–6 percent (Jackson et al., 1983). Such populations, however, are biased in favor of PSP because referring neurologists are often aware of the research interest of the tertiary academic neurologist, the source of our published data. Furthermore, the poor medication response of patients with PSP drives their referral to the academic center, while patients with PD, which is far more amenable to treatment, often remain unreferred.

We measured the population prevalence of PSP in two counties in central New Jersey in 1986 (Golbe et al., 1988). We contacted all neurologists and chronic care facilities in and near the area, which had a population of 800,000, inquiring as to their knowledge of patients with PSP living in the two counties. We assumed that any patient carrying a diagnosis of PSP would have been seen at some point by a neurologist, as non-neurologists very rarely feel confident in making that diagnosis. Another convenient but unfortunate circumstance — the relatively short survival of patients with PSP — minimized the possibility that a neurologist who diagnosed a living patient would have since left practice in the area. We verified all alleged cases by personal examination.

Six men and five women were ascertained (Table 1). The prevalence ratio, age-adjusted to the 1980 U.S. population, was 1.39 per 100,000, 1.53 per 100,000 for men and 1.23 per 100,000 for women. This is about one percent of the prevalence of Parkinson's disease (PD) in Western countries, which approximates 100 to 170 per 100,000.

This figure for PSP must be considered a minimal prevalence, as many patients being followed by a family physician with diagnosis of PD undoubtedly had not displayed atypical features that might have prompted a referral. Even among alert neurologists familiar with PSP, the diagnosis may appear to be PD for several years. We may quantify this problem as follows:

Maher and Lees (1986) found that the median delay from symptom onset to diagnosis is three years and the median actuarially adjusted survival after symptom onset was 5.9 years. Dividing provides a ratio of living to diagnosed cases of 1.96:1. A similar ratio may be calculated from the data of Golbe et al. (1988), in which median delay to diagnosis was 5 years and median actuarially adjusted survival was 9.7 years, giving a ratio of 1.94:1. There are therefore approximately twice as many living patients who will eventually receive a diagnosis of PSP as there are living patients who have already received that diagnosis (Table 2). Undoubtedly, the true ratio is higher, as some additional patients with PSP never receive an accurate diagnosis, even within the referral areas of the centers that conducted the two studies cited.

Allocation of research resources in neurology, being based in part on prevalence, emphasizes such public health problems as stroke, Alzheimer's disease and multiple sclerosis. However, the number of new cases of PSP per unit population per unit time, the incidence, is probably similar to that of some other well-known and well-researched neurologic entities.

The incidence of PSP has not been measured directly in a large population, but it can be estimated from the prevalence, the number of cases alive in a defined population at a defined point in time. Using the data of Golbe et al. (1988), wherein the measured prevalence was about 14 living diagnosed cases per million population, and multiplying by 1.96, the factor needed to correct for still-undiagnosed cases, produces a "true" prevalence of about 28 per million. If the median survival is 9.7 years, (Golbe et al., 1988) and that figure and the prevalence do not change over the short term, then the frequency with which new cases arise, the incidence, would be about 2.9 cases per million population per year. Using the survival duration from the data of Maher and Lees (1986) produces an incidence of 4.7 per million per year (Table 3).

The incidence of PSP has been measured directly in two small populations as secondary aspects of two studies (Table 3). Mastaglia et al. (1973) found eight new cases over two years' observation of an Australian metropolitan area with a population of one million. The crude incidence was 4.0 per million per year. Rajput et al. (1984) reviewed the Mayo Clinic database for cases of PSP arising among the local population of approximately 50,000 over a 13-year period, finding two cases. The crude prevalence there becomes 3.1 per million per year. Both of the observed rates from Minnesota and Australia are reassuringly similar to the calculated rates from Great Britain and New Jersey.

The incidence of PSP, about 3–4 per million per year, lies very close to that of myasthenia gravis, Huntington's disease, the hereditary ataxias as a group, polymyositis, and Tourette syndrome (Table 4). It is fair to say that these conditions have received far more attention than has PSP. The reason undoubtedly is the difficulty distinguishing PSP from PD, a circumstance

**Table 2.** Underdiagnosis of PSP

|  | Maher and Lees, 1986 | Golbe et al., 1988 |
|---|---|---|
| Delay from symptom onset to diagnosis: |  |  |
| Mean (sd) | 3.6 years (1.9) | 4.7 years (3.3) |
| Median | 3 years | 5 years |
| Median adjusted survival after onset: | 5.9 years | 9.7 years |
| Ratio of medians: |  |  |
| Living symptomatic cases: living diagnosed cases | 1.96:1 | 1.94:1 |

**Table 3.** Incidence of PSP (newly diagnosed cases per unit time)

| Source | Period of observation (years) | Number of cases newly diagnosed | Population base | Incidence (per million per year) |
|---|---|---|---|---|
| Mastaglia et al., 1973 | 2 | 8 | 1,000,000 | 4.0 |
| Rajput et al., 1984 | 13 | 2 | 50,000 | 3.1 |
| Maher and Lees, 1986 | Incidence calculated indirectly |  |  | 2.9 |
| Golbe et al., 1988 | See text for details |  |  | 4.7 |

**Table 4.** Incidence of PSP (new cases/million pop./year): comparison with other neurological disorders

| Disease | Incidence |
|---|---|
| Parkinson's disease | 20 |
| Amyotrophic lateral sclerosis | 4–18 (mean 10) |
| Guillain-Barre syndrome | 10 |
| Intracranial abscess | 10 |
| All muscular dystrophies | 7 |
| Polymyositis | 5 |
| Tourette syndrome | 5 |
| Syringomyelia | 4 |
| All hereditary ataxias | 4 |
| Huntington's disease | 4 |
| Myasthenia gravis | 4 |
| PSP | 3–4 |
| Charcot-Marie-Tooth disease | 2 |
| Wilson's disease | 1 |

that delayed its identification until the 1960's and which hinders its recognition today.

## Onset age

Adding to the difficulty in recognition of PSP is its late-life onset. In the series of 50 patients (41 alive and examined by the authors, 9 deceased) surveyed by Golbe et al. (1988), mean age at symptom onset was 62.9 years (sd 6.4). Thirty percent had onset before age 60 and 14 percent after age 69. Nearly identical results were reported by Maher and Lees (1986) in whose record review of 52 patients (14 alive, 33 deceased, 5 status unknown) mean onset occurred at age 62.0 (sd 6.9). The onset age of PD in most referral series is 60 to 61, but in one excellent community-based study was 65.3 (sd 12.6) (Mutch et al., 1986). A truly community-based series of patients with PSP could well reveal a older onset age than did the cases that reached neurologists' attention in the reported epidemiologic series.

Maher and Lees (1986) found no correlation between onset age and duration of survival or between onset age and delay to diagnosis. In PD, by contrast, younger onset correlates with a more benign clinical course (Golbe, 1991).

## Survival

The duration of survival after symptom onset in PSP appears to be a few years shorter than it was for PD before the advent of levodopa. Among the PD referral population of Hoehn and Yahr reported in 1967, mean survival was 9.4 years. The 33 deceased patients from the PSP referral series of Maher and Lees (1986) had a median actuarially adjusted survival of 5.9 years, with a range of 1.2–10.3 years. Among the 15 deceased patients of Golbe et al. (1988), the median duration was 6.9 years, with the range 2–17 years.

### Analytical epidemiology

Referral populations may be biased with regard to many patient characteristics. In view of the absence of any true community-based series of patients with PSP, we must rely on the referral series to obtain demographic and environmental clues to the etiology of PSP. Case-control studies, one of the most powerful tools of the analytical epidemiologist, will also therefore rely on referral populations. Nevertheless, one may surmise which characteristics may influence the likelihood of referral to an academic center (income, education, age, etc.) and discount the data accordingly.

## Ethnic and geographic correlates

There has never been a formal study of PSP prevalence in a sufficiently large black population. The informal observation of the author and of many others in the United States is that the frequency of blacks among referral populations with PSP reflects the frequency of that race among patients referred for all illnesses to that medical center. Among the PD population, on the other hand, the black race appears to be far underrepresented (Kessler et al., 1972).

The prevalence of PSP has never been formally measured (or informally reported) among an Asian or Asian-American population. There have been reports of patients with PSP from most parts of the developed and developing world and from many ethnic groups (Hynd et al., 1982).

## Gender

The original clinicopathologic description of Steele et al. (1964) and several reports during the ensuing decade reported a strong male preponderance approximating 8:1. However, recent reports have found a more equal ratio, still with a moderate male preponderance. The two-county survey of Golbe et al. (1988) revealed six men and five women. A 1985 review tallied 302 reported cases, 60 percent of whom were male. The referral series of Golbe et al. (1988) was 58 percent male while that of Maher and Lees was 58 percent female.

We may conclude from these data that the cause of PSP is unlikely to be an occupational exposure that correlates closely with male sex.

## A case-control study

A more efficient means to discover an occupational, environmental, historical or familial exposure as a disease risk factor is the case-control questionnaire study. Only one such study has been performed for PSP, however, because of the difficulty in assembling a sufficiently large group of patients. For a risk factor with low prevalence (e.g., a history of boxing or of working with hydrocarbons), large case and control groups are necessary for a scientifically important difference to reach statistical significance. This difficulty can be mitigated slightly by matching multiple controls to each case. Another difficulty with case-control studies is their reliance on a formal questionnaire based on existing etiologic hypotheses. An unduly long and inclusive questionnaire will fatigue the respondent and degrade the validity of the data. However, if the cause of the disease is not among the questions, the most meticulously executed study of the most motivated respondents will not reveal it.

Undaunted, Davis et al. (1988) matched two controls from a suburban university hospital's inpatients to each of 50 personally examined outpatients with PSP from throughout New Jersey. To reduce ascertainment bias, controls were disqualified if the present hospital admission was for an illness known to have a risk factor examined in the questionnaire. (Otherwise, controls would have appeared to have a greater frequency of atherosclerosis, smoking, etc. than patients.) All questionnaires were administered in person. Because many of the patients but few of the controls were demented or severely bradyphrenic, and because nine of the patients were deceased, the questionnaire was given only to surrogate respondents, usually spouses or children, for all cases and controls. Deceased controls were matched to the nine deceased patients.

One statistically significant finding was that patients with PSP, relative to controls, were 3.1 times as likely to have completed high school and 2.9 times as likely to have completed college. The explanation for these odds ratios may simply be a bias related to demographic differences between the population living near the university hospital (the source of most of the controls) and that of the state of New Jersey in general (the source of the PSP patients). Alternatively, those with less education may be more likely to pursue occupations or recreations that present exposure to an etiologic toxicant. This scientifically interesting possibility deserves further scrutiny.

The odds ratio for having lived as an adult over age 40 in a locality of <10,000 population was 2.4 (p < 0.05). However, odds ratios for having lived in rural areas or in small towns as a child or young adult, or of having lived on a farm were not statistically significant. The isolated "significant" finding is therefore unlikely to be scientifically significant and may represent a multiple comparisons error, wherein random variation alone will cause one of every 20 comparisons to satisfy an alpha of 0.05. Still, this result is intriguing in light of the ample evidence that rural living confers increased risk of PD (Tanner and Langston, 1990).

Of the total of 85 questions, those that gave negative results concerned living overseas, occupations implicated in other neurologic illnesses, potential exposure to occupational toxins, smoking, alcohol, caffeine, contact sports, head trauma, type A personality, early menopause, estrogen supplementation, multiparity, various medical conditions, surgical history, psychiatric history, animal exposure, maternal age, birth order, and family neurologic history.

### The rural question

To better investigate the hint of greater rural experience in PSP, we have obtained preliminary data comparing the PSP with PD with regard to residential history. The ideal method would have been to compare PSP with normals, but we were unable to obtain a population of normals without the geographic bias inherent in the referral pattern that brought the PSP patients to our attention. Using neighbors as controls would constitute "overmatching" and could conceal a difference if one existed.

We therefore used patients with PD presently living in New Jersey as "controls" on the assumption that their referral pattern was similar to that of PSP patients presently living in New Jersey. We found that *urban* residence 40 and 45 years before disease onset was more common in PSP (p < 0.05) and that a similar but nonsignificant such effect occurred in other years before onset.

These data suggest that the cause of PSP differs from that of PD, but say nothing about PSP relative to healthy controls. The element of the rural environment that confers increased risk for developing PD has not been identified.

The same study found that none of the 21 counties of New Jersey included a disproportionate number of cases of PSP. This is negative evidence for the hypothesis that PSP is the result of common-source infectious or of post-infectious or toxic etiology.

It should be noted that an attempt to transmit PSP to chimpanzees by intracerebral inoculation with brain tissue from human PSP patients was unsuccessful (Steele, 1972). However, the details of that experiment have never been published.

### Suggestive case reports

#### *An occupational cluster?*

McCrank and Rabheru (1989) have reported four patients with PSP with unusual exposure to organic solvents. They were a housewife who used household pesticides (which use organic solvents as a vehicle) against insects in an "obsessional" way, a banker who in his retirement sprayed large quantities of pesticides in his rose garden, and two lithographers. McCrank subsequently (1990) reported greater than usual exposure to organic solvents for 12 of his 13 patients with PSP.

This finding, while not accompanied by control data, is intriguing in light of the description of parkinsonism in a young man (Tetrud et al., 1990) and in rodents (Pezzoli et al., 1990) after hydrocarbon exposure. Also suggestive, though not statistically significant, is the odds ratio of 3.7 for "auto repair" as an occupation in the large case-control study cited above (Davis et al., 1988) (six of 50 cases and four of 100 controls reported that occupation). However, the odds ratios for exposure to "chemicals," "pesticides at home," "pesticides on a farm" and "insect extermination" did not exceed 1.0. One patient and one control reported pesticide exposure "at work" (odds ratio 2.0, not significant).

#### *A genetic link?*

Mata et al. (1983) reported three siblings who in their 20's developed a condition suggestive of PSP, but with more pyramidal signs and dementia.

Light microscopy showed neurofibrillary tangles in many areas that are involved in PSP, plus the hippocampus. Electron microscopy was not performed. The very early onset age and the distribution of tangles would be highly atypical for PSP and the authors concluded that their patients had a previously undescribed illness. This author agrees.

Ohara et al. (1992) reported two siblings with consanguineous parents and an illness with onset at ages 58 and 60 that closely resembled PSP. The neurofibrillary tangles and their immunohistologic properties were similar to those of PSP by electron microscopy. However, their heavy predilection for the limbic system and the absence of neuronal loss or tangles from the red nuclei, superior colliculus, interstitial nucleus of Cajal and pontine nuclei distinguished the pathology from that of PSP. Again, this is probably a hitherto-undescribed entity rather than "hereditary PSP."

A case of "familial PSP" often cited in reviews is that mentioned by David et al. (1968). The mother of one of their PSP cases was said by other relatives to have had "slowing of voluntary movement over the last few years of her life, progressing to total immobility and invalidism," expiring at age 57. Dr. David has informed me that no further information was, or is, available. Neither of us feels justified in calling this "familial PSP."

One may compare this to the prevalence of secondary cases in PD, where one typical recent survey (Vieregge et al., 1992) found PD in 9.3% of first- and second-degree relatives of PD patients and in 1.0% of relatives of controls. It therefore appears that the hereditary contribution to the cause of PSP is minimal to absent.

One caveat is in order. If the penetrance of a PSP gene is low or if the disease is caused by the combined action of multiple genes with a "threshold effect," then a familial predilection could be masked for a disease as rare as PSP. Even if the prevalence of familial secondary cases is, for example, 1,000 times that among the general population, the prevalence among relatives of PSP patients would still be only 1 or 2 percent. This could easily be missed without personal examination of all elderly relatives by a suspicious and experienced neurologist. Such a study has not been reported.

### The future

As physicians become more aware of PSP, additional properly diagnosed patients will be available for epidemiologic research at one center or in one surveyable geographic area. Before then, multi-institutional studies may be a partial answer to the relative paucity of patients who can be surveyed for etiologic factors.

Formal diagnostic criteria, a staging system, a clinical rating scale and a PSP database are all being put into place under the auspices of the Society for Progressive Supranuclear Palsy. These will facilitate data sharing, descriptive epidemiology and access to patients for analytical epidemiology and other forms of research.

There has been only one case-control etiologic survey, so the potential of this technique, despite its limitations, has not been exhausted for PSP. Newly-generated hypotheses deserve to be tested in this fashion.

PSP can avail itself of advances in the epidemiology of closely allied and more intensively studied diseases, PD and Alzheimer's disease. It may also benefit from investigations into the cause of the ALS-parkinson-dementia complex of Guam, discussed by Steele elsewhere in this volume. Finally, every neurologist can and should take detailed and directed histories of unusual events and exposures from their patients with PSP in search of one able to relinquish the secret.

## References

David NJ, Mackey EA, Smith JL (1968) Further observations in progressive supranuclear palsy. Neurology 18: 349–356

Davis PH, Golbe LI, Duvoisin RC, Schoenberg BS (1988) Risk factors for progressive supranuclear palsy. Neurology 38: 1546–1552

Golbe LI, Davis PH, Schoenberg BS, Duvoisin RC (1988) Prevalence and natural history of progressive supranuclear palsy. Neurology 38: 1031–1034

Golbe LI (1991) Young-onset Parkinson's disease: a clinical review. Neurology 41: 168–173

Hoehn MM, Yahr MD (1967) Parkinsonism: onset, progression and mortality. Neurology 17: 427–442

Hynd GW, Pirozzolo FJ, Maletta GJ (1982) Progressive supranuclear palsy. Int J Neurosci 16: 87–98

Jackson JA, Jankovic J, Ford J (1983) Progressive supranuclear palsy: clinical features and response to treatment in 16 patients. Ann Neurol 13: 273–278

Kessler II (1972) Epidemiologic study of Parkinson's disease. Am J Epidemiol 96: 242–254

Maher ER, Lees AJ (1986) The clinical features and natural history of the Steele-Richardson-Olszewski syndrome (progressive supranuclear palsy). Neurology 36: 1005–1008

Mastaglia FL, Grainger K, Kee F, Sadka M, Lefroy R (1973) Progressive supranuclear palsy (the Steele-Richardson-Olszewski syndrome): clinical and electrophysiological observations in eleven cases. Proc Aust Assoc Neurol 10: 35–44

Mata M, Dorovini-Zis K, Wilson M, Young AB (1983) New form of familial Parkinson-dementia syndrome: clinical and pathologic findings. Neurology 33: 1439–1443

McCrank E, Rabheru K (1989) Four cases of progressive supranuclear palsy in patients exposed to organic solvents. Can J Psychiatry 34: 934–935

McCrank E (1990) PSP risk factors. Neurology 40: 1637

Mutch WJ, Dingwall-Fordyce I, Downie AW, Paterson JG, Roy SK (1986) Parkinson's disease in a Scottish city. Br Med J 292: 534–536

Ohara S, Kondo K, Morita H, Maruyama K, Ikeda S, Yanagisawa N (1992) Progressive supranuclear palsy-like syndrome in two siblings of a consanguineous marriage. Neurology 42: 1009–1014

Pezzoli G, Ricciardi S, Masotta C, Mariani CB, Carenzi A (1990) n-Hexane induces parkinsonism in rodents. Brain Res 531: 355–357

Rajput AH, Offord KP, Beard CM, Kurland LT (1984) Epidemiology of parkinsonism: incidence, classification, and mortality. Ann Neurol 16: 278–282

Steele JC, Richardson JC, Olszewski J (1964) Progressive supranuclear palsy: a heterogeneous degeneration involving the brain stem, basal ganglia and cerebellum,

with vertical gaze and pseudobulbar palsy, nuchal dystonia and dementia. Arch Neurol 10: 333–359
Steele JC (1972) Progressive supranuclear palsy. Brain 95: 693–704
Tanner CM, Langston JW (1990) Do environmental toxins cause Parkinson's disease? A critical review. Neurology 40 [Suppl] 3: 17–30
Tetrud JW, Langston JW, Irwin I, Snow B (1990) Acute and persistent parkinsonism associated with ingestion of petroleum product mixture. Ann Neurol 28: 296
Vieregge P, Glaese A, Ulm G, Kompf D (1992) Familial Parkinson's disease. Mov Disord 7 [Suppl] 1: 23

Author's address: Dr. L. I. Golbe, Department of Neurology, UMDNJ-Robert Wood Johnson Medical School, New Brunswick, NJ 08903, U.S.A.

# Cholinergic approaches to the treatment of progressive supranuclear palsy

## I. Litvan

Neuroepidemiology Branch, National Institute of Neurological Disorders and Stroke, National Institutes of Health, Bethesda, Maryland, U.S.A.

**Summary.** In spite of the severe loss of cholinergic neurons in the brains of patients with progressive supranuclear palsy (PSP), marginal or null benefits are seen in clinical trials after the administration of physostigmine, a cholinesterase inhibitor, or RS-86, a cholinergic agonist. The possible role of cholinergic therapy in PSP is reevaluated.

It has been shown that several cholinergic regions are affected in progressive supranuclear palsy (PSP): the basal forebrain, basal ganglia, mediodorsal thalamic nuclei, midbrain and pontine areas (Brandel et al., 1991; Hirsch et al., 1987; Juncos et al., 1991; Kish et al., 1965; Malessa et al., 1991; Ruberg et al., 1985; Tagliavini et al., 1983). Lesions in particular structures may result in specific neurobehavioral deficits; e.g. lesions in the basal forebrain and mediodorsal thalamus may lead to the memory deficits seen in PSP. Involvement of the mesencephalic and pontine nuclei (rostral interstitial nucleus of the medial longitudinal fasciculus, superior colliculus and laterodorsal tegmental nucleus) may result in supranuclear ophthalmoplegia. The affected striatum and pedunculopontine nuclei may contribute to the motor deficits exhibited by PSP patients. These brain-behavior correlates prompt the question: Could cholinergic stimulation improve the neurobehavioral abnormalities seen in PSP patients?

The rationale for cholinoceptive treatment of the memory and cognitive disorders found in these patients is as follows: in PSP, the mediodorsal and basal forebrain cholinergic nuclei are affected, but the cortical neurons (Hauw et al., 1990) and the noradrenergic, dopaminergic and serotoninergic subcorticocortical pathways are relatively spared (Agid et al., 1986; Ruberg et al., 1992). This differentiates PSP from Alzheimer's disease (AD) which has severe cortical pathology, and from Parkinson's disease (PD) which also has severe subcortical pathology. Therefore, the relative selectivity of the cholinergic lesions found in PSP makes it an ideal condition for treatment with cholinergic agents.

In contrast to patients with PD, motor function in PSP patients does not change after treatment with oral or intravenous physostigmine (Duvoisin, 1967; Litvan et al., 1989, in preparation). This differential response may be attributed to degeneration of cholinergic striatal interneurons and the pedunculopontine nuclei. Thus, in addition to the nigral degeneration, also seen in PD, the degeneration of these cholinergic neurons might contribute to the interruption of the nigrostriatocortical pathways and explain why PSP patients do not respond to dopaminergic therapy. Degeneration of the pedunculopontine nuclei also seen to a lesser degree in PD patients might explain the failure of dopaminergic agents to improve the balance disorder often seen in these patients (Hirsch et al., 1987; Jellinger, 1988; Koller et al., 1989).

To date, only two cholinergic agents have been used to treat PSP. Foster et al. (1989) evaluated the effects of RS-86, an M1–M2 muscarinic agonist, on motor and cognitive function in 10 PSP patients. No effects were found with regard to either cognitive tasks or motor functions. However, there were changes in two measures of REM sleep which are typically sensitive to cholinergic treatment (increase in total, and percentage of, REM sleep and decrease in REM latency).

Litvan et al. (1989) administered oral physostigmine to 8 PSP patients at an "optimal" dose while evaluating a wide range of cognitive and motor abilities. The "optimal dose" was selected by titrating physostigmine on consecutive days from 0.5 mg to 2 mg every 2 hr, 6 times a day, using the total number of words recalled on the Selective Reminding Test as the dependent variable (Buschke et al., 1974). In this test, patients had to recall a list of 12 words. Before each trial, the subject is reminded of the words that he/she did not recall on the previous trial. Both short- and long-term explicit memory can be measured with this test. The group mean dose curve (Fig. 1) indicates that we successfully found an optimal dose for each patient in the open study. The effects of physostigmine were marginal (in the double-blind, controlled phase) on two measures of secondary memory: long term memory (the ability of the subject to recall a word on two consecutive trials without being reminded) and consistent long term memory (the capacity to recall a word consistently through the end of the test) (Fig. 2). Similar results were found on two indices of forgetting from the Brown-Peterson paradigm (Brown, 1958). In this paradigm, patients were given three words, and after interference-filled intervals of different durations (3–36 sec), they were asked to recall the originally presented words. Only two patients showed any consistent (improvement in all 4 outcome measures) positive effects (Fig. 2). The pattern of results was similar to those found with physostigmine treatment of AD (Thal et al., 1983).

In another experiment, Kertzman et al. (1990) evaluated spatial attention in the same group of PSP patients. The task used was designed by Posner who with Rafal had originally shown that PSP patients had difficulty with spatial attention (Rafal et al., 1988). In this task, the subject fixates on a spotlight and presses a button as soon as a target light appears. A cue stimulus precedes the target and cues the subject to look toward the same or opposite spatial location of the subsequent target (valid or invalid condition, respectively) (Fig. 3). The subjects' reaction times were measured from target onset to button press for all the conditions. The validity effect

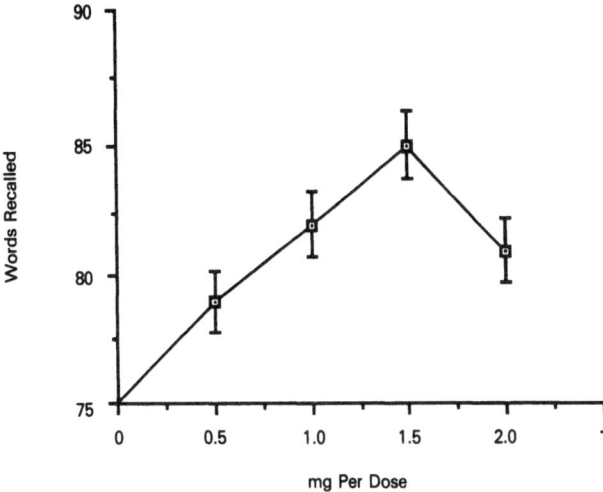

**Fig. 1.** Group mean words recalled in the selective reminding test at different doses of physostigmine

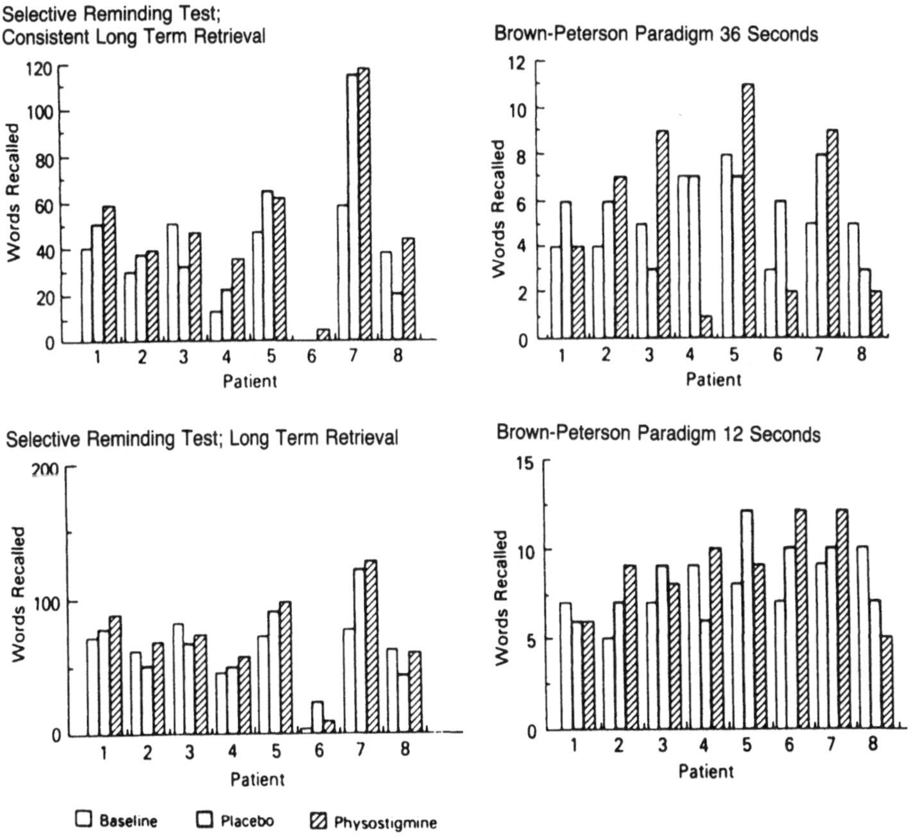

Selective Reminding Test;
Consistent Long Term Retrieval

Brown-Peterson Paradigm 36 Seconds

Selective Reminding Test; Long Term Retrieval

Brown-Peterson Paradigm 12 Seconds

☐ Baseline    ☐ Placebo    ▨ Physostigmine

**Fig. 2.** Case-by-case analysis of the effects of physostigmine and placebo on verbal memory. *Left upper figure*: Selective Reminding Test: Consistent Long-Term Retrieval; *left lower figure*: Selective Reminding Test: Long-Term Retrieval; *right upper figure*: Brown-Peterson Paradigm at 36 seconds and *right lower figure*: Brown-Peterson Paradigm at 12 seconds (reprinted with permission from Ann Neurol)

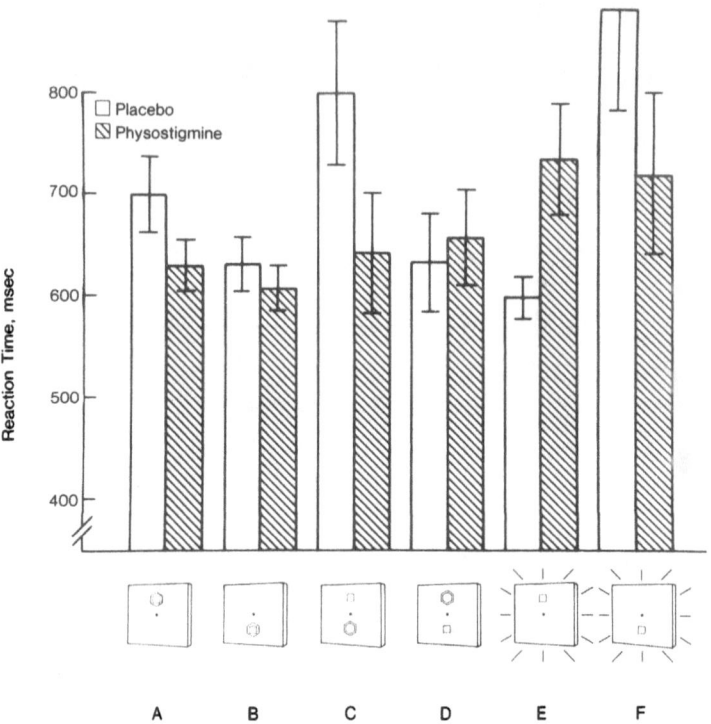

**Fig. 3.** PSP patients and controls performance on a measure of vertical shift of attention during placebo and physostigmine administration. In the schema below, the hexagon represents the cue and the square the subsequent target. *A* and *B* are validly cued trials, *C* and *D* are invalidly ones, and *E* and *F* are diffusely cued trials, to the upper and lower fields respectively. There was a significant improvement of the validity effect after physostigmine treatment. Reprinted with permission from Arch Neurol

(which is the difference in reaction times and accuracy between the invalid and valid conditions) is a measure of visual attention. Physostigmine treatment induced a faster response to invalidly cued targets (Fig. 3). The greater the duration of the disease, the smaller was the effect obtained. On the other hand, physostigmine treatment did not improve any of several cognitive measures of frontal lobe function of these PSP patients (Grafman et al., 1990). Neither motor performance, assessed with the modified Columbia scale, nor extraocular movements which were recorded with a magnetic coil, showed any changes.

In summary, physostigmine marginally improved memory and attention without any changes in motor or oculomotor function in PSP patients. RS-86 did not modify any motor or cognitive abnormalities in these patients despite its action on REM sleep.

Why did RS-86 fail to improve motor and behavioral response in PSP patients? Perhaps the cholinergic role in cognitive and motor functions in this disease is negligible, or alternatively, RS-86 is a weak agent since it acts on both M2 and M1 receptors. The stimulation of M2 autoreceptors may decrease acetylcholine (ACh) release so that it may be necessary to use a selective M1 agonist to realistically evaluate the role of cholinergic agonists (Marchi et al., 1991; Potter, 1987). The M1 muscarinic receptors not only are the most important cortical muscarinic receptors but they are also the most important striatal ones (Weiner et al., 1990). They are fully coexpressed with the D2 receptor (which is the dopaminergic receptor affected in PSP) (Pierot et al., 1988; Weiner et al., 1990).

Why was physostigmine not as effective as we expected? It may be that insufficient drug arrived to the brain because of low bioavailability. Indeed, cerebrospinal fluid acetylcholinesterase (AChE) activity did not change after orally administered physostigmine (Atack et al., 1991). In order to evaluate this possibility, we recently administered physostigmine intravenously at higher doses (Litvan et al., in preparation). No changes in memory or motor parameters were seen after 30-min. infusions of 0.5 to 1.5 mg of physostigmine. It still remains to be seen if this lack of response to physostigmine infusion could be attributed to a failure to achieve steady-state and/or sufficient brain ACh levels to produce a behavioral response. Actually, physostigmine has been classified as a first generation cholinesterase inhibitor due to its lack of selective AChE inhibition and failure to achieve steady-state CNS acetylcholine levels (Giacobini, 1991). More potent cholinesterase inhibitors or M1 muscarinic agonists now becoming available may be more beneficial than those previously administered (Giacobini, 1991).

We recently assessed the role of cholinergic blockade in patients with PSP. Preliminary results from these studies using measures of memory and motor performance suggest that PSP patients are sensitive to cholinergic blockade at doses that have no effect in normal controls (Litvan et al., in preparation). PSP patients memory and gait functions significantly worsened after muscarinic blockade with a low dose of scopolamine (0.25 mg). Moreover, in contrast to AD and PD patients, PSP patients also became stuporous at the higher doses. This difference in behavior may be attributable to the degeneration of cholinergic neurons in the upper brainstem in this disorder. Thus, PSP appears associated not only with degeneration of cholinergic neurons but also with substantial cholinergic hypofunction. Surprisingly, in a recent review of uncontrolled single cases, cholinergic blockers were reported to improve the parkinsonian features of certain patients (Litvan and Chase, 1992). Given the evidence of marked cholinergic depletion in PSP, the use of oral cholinergic blockers in patients with this disorder should ordinarily be avoided (exceptions to this would include patients with dystonia who may respond well to anticholinergic agents).

The cholinergic replacement therapies remain a viable alternative treatment in PSP and it is likely that as more specific agents become available for phase I trials, they will be used in studies with PSP patients in an attempt to ameliorate their symptoms.

## Acknowledgments

I am grateful to the PSP patients who generously participated in our studies and to D. Schoenberg for her editorial assistance.

## References

Agid Y, Javoy-Agid F, Ruberg M, Pillon B, Dubois B, Duyckaerts C, Hauw JJ, Baron JC, Scatton B (1986) Progressive supranuclear palsy: anatomoclinical and biochemical considerations. In: Yahr MD, Bergmann KJ (eds) Advances in neurology, vol 45. Raven Press, New York, pp 191–206

Atack JR, Litvan I, Thal LJ, May C, Rapaport SI, Chase TN (1991) Cerebrospinal fluid acetylcholinesterase in progressive supranuclear palsy: reduced activity relative to normal subjects and lack of inhibition by oral physostigmine. J Neurol Neurosurg Psychiatry 54: 832–835

Brandel JP, Hirsch EC, Malessa S, Duyckaerts C, Cervera P, Agid Y (1991) Differential vulnerability of cholinergic projections to the mediodorsal nucleus of the thalamus in senile dementia of Alzheimer type and progressive supranuclear palsy. Neuroscience 41: 25–31

Brown J (1958) Some test of decay theory of immediate memory. Q J Exp Psychol 39: 15–22

Buschke H, Altman Fuld P (1974) Evaluating storage, retention, and retrieval in disordered memory and learning. Neurology 24: 1019–1025

Duvoisin RC (1967) Cholinergic-anticholinergic antagonism in parkinsonism. Arch Neurol 17: 124–136

Duvoisin R (1992) Clinical diagnosis. In: Litvan I, Agid Y (eds) Progressive supranuclear palsy: clinical and research approaches. Oxford University Press, pp 15–33

Foster NL, Aldrich MS, Bluemlein MS, White RF, Berent S (1989) Failure of cholinergic agonist RS-86 to improve cognition and movement in PSP despite effects on sleep. Neurology 39: 257–261

Giacobini E (1991) The second generation of cholinesterase inhibitors: pharmacological aspects. In: Becker R, Giacobini E (eds) Cholinergic basis for Alzheimer therapy. Birkhäuser, Boston, pp 247–262

Grafman J, Litvan I, Gomez C, Chase TN (1990) Frontal lobe function in progressive supranuclear palsy. Arch Neurol 47: 553–558

Hauw JJ, Verny M, Delaere P, Cervera P, He Y, Duyckaerts C (1990) Constant neurofibrillary changes in the neocortex in progressive supranuclear palsy. Basic differences with Alzheimer's disease and aging. Neurosci Lett 119: 182–186

Hirsch EC, Graybiel AM, Duyckaerts C, Javoy-Agid F (1987) Neuronal loss in the pedunculopontine tegmental nucleus in Parkinson disease and in progressive supranuclear palsy. Proc Natl Acad Sci USA 84: 5976–5980

Jellinger K (1988) The pedunculopontine nucleus in Parkinson's disease, progressive supranuclear palsy and Alzheimer's disease. J Neurol Neurosurg Psychiatry 51: 540–543

Juncos JL, Hirsch EC, Malessa S, Duyckaerts C, Hersh LB, Agid Y (1991) Mesencephalic cholinergic nuclei in progressive supranuclear palsy. Neurology 41: 25–30

Kish SJ, Chang LJ, Mirchandani L, Shannak K, Hornykiewicz O (1985) Progressive supranuclear palsy: relationship between extrapyramidal disturbances, dementia, and brain neurotransmitter markers. Ann Neurol 18: 530–536

Kertzman C, Robinson DL, Litvan I (1990) Effects of physostigmine on spatial attention in patients with progressive supranuclear palsy. Arch Neurol 47: 1346–1350

Koller WC, Glatt S, Vetere-Overfield B, Hassanein R (1989) Falls and Parkinson's disease. Clin Neuropharmacol 12: 98–105

Litvan I, Gomez C, Atack JR, Gillespie M, Kask A, Mouradian MM, Chase TN (1989) Physostigmine treatment of progressive supranuclear palsy. Ann Neurol 26: 404–407

Litvan I, Chase TN (1992) Traditional and experimental therapeutic approaches. In: Litvan I, Agid Y (eds) Progressive supranuclear palsy: clinical and research approaches. Oxford University Press, New York, pp 254–269

Malessa S, Hirsch EC, Cervera P, Javoy-Agid F, Duyckaerts C, Hauw JJ, Agid Y (1991) Progressive supranuclear palsy: loss of choline acetyltransferase-like immunoreactive neurons in the pontine reticular formation. Neurology 41: 1593–1597

Marchi M, Besana E, Codignola A (1991) Presynaptic cholinergic mechanisms in human cerebral cortex of adult and aged patients. In: Becker R, Giacobini E (eds) Cholinergic basis for Alzheimer therapy. Birkhäuser, Boston, pp 92–97

Pierot L, Desnos C, Blin J, Raisman R, Scherman D, Javoy-Agid F, Ruberg M, Agid Y (1988) D1 and D2-type dopamine receptors in patients with Parkinson's disease and progressive supranuclear palsy. J Neurol Sci 86: 291–306

Potter LT (1987) Muscarinic receptors in the cortex and hippocampus in relation to the treatment of Alzheimer's disease. In: Cohen S, Sokolovsky M (eds) International symposium on muscarinic cholinergic mechanisms. Freud Publishing Ltd, London, pp 294–301

Rafal RD, Posner MI, Friedman JH, Inhoff A, Bernstein E (1988) Orienting of visual attention in progressive supranuclear palsy. Brain 111: 267–280

Ruberg M, Javoy-Agid F, Hirsch E, Scatton B, Lheureux R, Hauw JJ, Duyckaerts C, Gray F, Morel-Maroger A, Rascol A, Serdaru M, Agid Y (1985) Dopaminergic and cholinergic lesions in progressive supranuclear palsy. Ann Neurol 18: 523–529

Ruberg M, Hirsch E, Javoy-Agid F (1992) Neurochemistry. In: Litvan I, Agid Y (eds) Progressive supranuclear palsy: clinical and research approaches. Oxford University Press, New York, pp 89–109

Steele JC, Richardson JC, Olszewski J (1964) Progressive supranuclear palsy: a heterogeneous degeneration involving the brain stem, basal ganglia, and cerebellum, with vertical gaze and pseudobulbar palsy, nuclear dystonia, and dementia. Arch Neurol 10: 333–359

Thal LJ, Fuld PA, Masur DM, Sharpless NS (1983) Oral physostigmine without lecithin improves memory in Alzheimer's disease. J Am Geriatr Soc 37: 42–48

Tagliavini F, Pilleri G, Gemignani F, Lechi A (1983) Neuronal loss in the basal nucleus of Meynert in progressive supranuclear palsy. Acta Neuropathol 61: 157–160

Weiner DM, Levey AI, Brann MR (1990) Expression of muscarinic acetylcholine and dopamine receptor mRNAs in rat basal ganglia. Proc Natl Acad Sci USA 87: 7050–7054

Author's address: Dr. I. Litvan, Neuroepidemiology Branch, National Institutes of Neurological Disorders and Stroke, Federal Bldg., Rm. 714, Bethesda, MD 20892, U.S.A.

# Therapy for progressive supranuclear palsy: past and future

## D. G. Cole and J. H. Growdon

Department of Neurology, Massachusetts General Hospital, Charlestown, Massachusetts, U.S.A.

**Summary.** Dysfunction of multiple brain systems in progressive supranuclear palsy (PSP) has complicated attempts to treat the disease. Neurotransmitter replacement strategies targeting the dopaminergic, cholinergic, and serotonergic systems have been unsuccessful. In order to bypass the degenerated corticostriato-pallidal loop, we adminstered the adrenergic agonist idazoxan (IDA) to treat PSP in two randomized double-blind, placebo controlled, crossover studies. Approximately one half of patients enrolled in these studies showed statistically significant improvement in balance and manual dexterity while taking IDA compared to placebo. These results suggest that new therapies that target structures outside of the basal ganglia may be useful for symptomatic treatment of PSP. Applying this strategy and developing treatments that arrest or reverse clinical deterioration in PSP will require improved understanding of the process underlying the illness.

## Introduction

The clinical syndrome of progressive supranuclear palsy (PSP) is distinguished by postural imbalance, eye movement abnormalities, symmetrical axial rigidity, and bulbar dysfunction. Late stages of the illness are also characterized by the loss of manual dexterity. Because the etiology of PSP is unknown, therapeutic efforts have been directed at treating the symptoms. Currently, however, effective symptomatic therapy for PSP is unavailable. This chapter discusses the difficulties in developing treatment for PSP, reviews a recent approach that has provided encouraging results, and considers future possibilities for therapy.

Two major factors that hamper efforts to treat PSP are the complexity of the disorder and the gaps in knowledge regarding the disease. The complexity of PSP is demonstrated by neuropathological studies. Classical histological techniques have revealed dramatic degeneration of the globus pallidus (GP), subthalamic nucleus, substantia nigra, superior colliculus, dentate nucleus, locus ceruleus, periaqeuductal structures, and various brainstem nuclei (Steele, 1964); Recent investigations have noted that cortical structures may be involved as well (Jellinger, 1980; Hauw, 1990). The widespread degeneration of different areas of the brain limits the number of structures that can be targeted for therapy with a reasonable chance of success. For instance, because virtually all striatal output is funneled through the GP (Albin, 1989), treatments that restore normal striatal function are likely to be of little clinical benefit. The GP is simply unable to transmit these normal signals downstream to the thalamus.

Pharmacological data indicate another level of complexity in PSP: even single systems in the brain are abnormal at multiple levels. For example, biochemical studies have revealed a significant loss of dopamine (DA) from the mesostriatal pathway (Ruberg, 1985; Kish, 1985), and receptor assays have demonstrated that D2 receptors are depleted in the striatum (Pascual, 1992). Simply restoring DA does not address the loss of receptors and is therefore of limited value.

While present knowledge is sufficient to explain some of the problems in treating PSP, many issues remain unresolved. Despite evidence for pathologic involvement of the cholinergic nucleus basalis of Meynert (Tagliavini, 1984), and striatal cholinergic interneurons (Ruberg, 1985), and brainstem cholinergic nuclei (Hirsch, 1987; Juncos, 1991; Malessa, 1991; Zweig, 1987), biochemical studies have found inconsistent changes in the levels of acetylcholine (Kish, 1985 and Ruberg, 1985). Similarly, levels of norepinephrine (NE) and serotonin are reportedly normal (Kish, 1985) despite degeneration of the locus ceruleus and raphe neurons in the periaqueductal gray region (Steele, 1964). It is important to reconcile these apparent contradictions in order to ascertain precisely which brain systems are involved in PSP and whether these systems would be amenable to therapy.

Despite the complex pathologic and pharmacological abnormalities in PSP and the inconsistencies in the data regarding these abnormalities, straightforward neurotransmitter replacement therapy has been tried extensively to treat the disease. DA replacement with the precursor levodopa improves idiopathic Parkinson's disease but produces little or no benefit in PSP. In fact, a good therapeutic response to levodopa calls the diagnosis of PSP into question, although some patients may improve transiently early in the course of the illness (Agid, 1986). At least two factors account for this poor response: 1) loss of striatal D2 receptors (Pascual, 1992) and 2) degeneration of the GP, which is the ultimate limiting factor. Other approaches to treatment have included attempts to augment cholinergic neurotransmission with either the anticholinesterase drug physostigmine (Litvan, 1989) or the direct cholinergic agonist RS-86 (Foster, 1989). Neither agent has been found to be useful. Efforts to use the 5HT agonist lisuride (Neophytides, 1982) and the 5HT antagonist methysergide (Rafal, 1981 and Paulson, 1981) have been equally unsuccessful. Anecdotal report of benefits from tricylcic antidepressants (Newman, 1985) has not been supported by controlled therapeutic trials.

All of these palliative treatments have sought to enhance transmission in neuronal systems that pass through the basal ganglia despite the pathoanatomic evidence militating against the success of this approach. In view of the degeneration of the GP, we adopted the alternative approach of bypassing the striato-pallidal loop and increasing neurotransmission in motor systems outside of the basal ganglia. Given previous experience demonstrating poor response to DAergic, AChergic, and 5HTergic medications, we adminstered the noradrenergic agent idazoxan (IDA) to treat the disease (Ghika, 1991). IDA is an α-2 antagonist that increases noradrenergic neurotransmission by blocking presynaptic autoreceptors and therefore preventing feedback inhibition of NE release. This approach has the theoretical advantage over prior treatment attempts in that a large proportion of ascending projections from the locus ceruleus terminate in the cortex rather than the striatum. Because the cortex is relatively unaffected in PSP, and its subcortical noradrenergic input does not depend on normal function of the GP, we reasoned that this strategy could restore, at least partially, tonic activation of cortical structures.

## Methods

Nine patients with a clinical diagnosis of PSP (Golbe, 1988) were enrolled in a double blind, placebo-controlled, crossover study according to the provisions of a protocol approved by the Massachusetts General Hospital Human Studies Committee. There were 2 men and 7 women whose mean age was 70.3 ± 1.7 (SD) years old. The mean duration of disease was 4 ± 1.2 years, and the mean Hoehn and Yahr stage was 4.2 ± 0.2. Cognitive impairment, as revealed by history and performance on the Blessed dementia Scale mental status examination (Blessed, 1968), was commonly present, and 3 patients were demented by DSM-III-R criteria (1987). Patients with hypertension (≥160 mm Hg systolic or ≥90 mm Hg diastolic), patients on antihypertensive medication (except for diuretics or angiotensin converting enzyme inhibitors), and patients needing antidepressant medication were disqualified. The use of traditional antiparkinsonian medications was permitted.

Patients were randomized to receive either placebo or IDA during the first evaluation period with crossover to the other agent during the second period. They were evaluated at baseline and during drug wash-in with medication at weekly or fortnightly intervals over 4 weeks. Patients took 10 mg t.i.d. of IDA (or placebo) for the first week, 20 mg t.i.d. for the second week, and 40 mg t.i.d. for the last two weeks. After a 3 week washout, the protocol was repeated with crossover to the second agent.

At each visit, patients underwent examination, which included a modified United Parkinson's Disability Rating Scale (UPDRS; Lang, 1989), examination of eye movements, and assignment of a Hoehn and Yahr score by the examining physician (Hoehn, 1967), as well as evaluation of performance on the step second test (Lang, 1989) and the Purdue pegboard test (Bass, 1951). The patients and their families and the examining physician also gave a global assessment of the patient's status. Tests of psychiatric status included the Hamilton Depression scale (Hamilton, 1967) and the Geriatric Depression scale (Yesavage, 1983). Cognitive tests were administered at baseline and maximal dosage for each of the assessment periods. These tests included the Blessed Dementia Scale, digit span, New York University Stories (Randt, 1980), simple reaction time, Odd Man Out (Flowers, 1985), Stroop Test (Stroop, 1935), Verbal Fluency, Luria Mental Rotation (Golden, 1980), and Picture Arrangement (Weschler, 1980). The examining physician was blinded to the therapy being adminstered.

Scores for each variable at baseline were subtracted from scores after 2 weeks on maximal dosage for each of the two treatment periods. The value of the difference on placebo was compared with the value of the difference on IDA.

### Results

The results of the study revealed a significant ($p < 0.05$) decrease (improvement) in the global score and motor subscore of the UPDRS on IDA 40 t.i.d. Although overall mean scores for the group as a whole improved, benefit was pronounced in 5 patients but negligible or absent in 4 patients. Compared to a mean baseline global UPDRS of 82.3 (±6.1, SD), patients on IDA had a mean global score of 75.2 (±6.7). The UPDRS motor subscore decreased from 52.2 (±3.6) to 44.8 (±4.1). Features of the motor subscore that showed the most marked improvement included the ability to arise from a chair, gait, postural stability, digital dexterity, and movement speed. These objective changes were reflected in improvements in the Purdue pegboard, in functional improvement in mobility and tasks such as using eating utensils and writing, and in patients' self-evaluations of balance, gait, agility, stiffness, and sense of energy. There was no significant improvement in any of these measures when patients were taking placebo. There was excellent concordance among the objective data, the patients' impressions, and the subjective impressions of the examiner. No objective changes were observed in other areas of motor function (speech, facial expression, rigidity, finger tapping, posture, bradykinesia, eye movements), mood, or cognition during IDA or placebo administration.

### Discussion

This study demonstrated that IDA significantly improved balance, mobility, and dexterity for patients with PSP. These results were encouraging because they provided the first demonstration of effective drug therapy for PSP from a double-blind placebo controlled, crossover study.

In order to confirm this result, we conducted a second similar study with 11 patients according to a similar, but abbreviated protocol. These patients were demographically and clinically comparable to the group enrolled in the original study. Preliminary review of these data confirm a moderate benefit of IDA in limited spheres of motor function for approximately one half of the patients tested. Patients who completed either study were given the option of taking IDA chronically on a compassionate-use basis, and we continue to follow 8 patients on a long term basis. During the course of these studies, 7 patients who enrolled in the studies have died. Autopsies were performed on 4, and the clinical diagnosis of PSP was confirmed neuropathologically in each case.

Our results should not obscure two facts: 1) improvement with IDA is modest and 2) not all patients benefit. The magnitude of improvement during IDA administration generally represented <30% change in the UPDRS score. Only certain symptoms of the disease responded; for example, no significant improvement was seen in eye movements, bradykinesia, or bulbar function. Moreover, only about one half of the patients showed significant improvement. Because IDA is contraindicated in patients with hypertension or severe anxiety, and because it may interact with psychotropic medication, its potential use is limited, even within the sphere of PSP. In spite of these limitations, however, patients who did respond often felt that the medication had a noticeable impact on their lives. This impression was shared by family members and corroborated by objective improvements in motor skills and functional independence.

The discrepancy between the small magnitude of change as defined by the rating scales and the importance of the change to patients' function suggests that rating scales that have been devised for Parkinson's disease may be insufficient to describe impairment in PSP or to quantitate functionally important responses to drug therapy. Developing a scale specific for PSP or preferably objective quantitative measures of motor performance would improve the precision and quality of clinical drug trials for this illness.

The experience with IDA validates the concept of bypassing the striatopallidal loop to treat PSP. Because basal ganglia output is transmitted primarily through the ventroanterior/ventrolateral nuclei of the thalamus (Albin, 1989), identification of drugs that target these thalamic nuclei may be the basis for a more powerful application of this approach. In lieu of this possibility, the use of other cortically active drugs, or drugs that specifically influence the cerebellum or spinal cord, should be considered.

Evidence for the notion that IDA exerts its effects primarily by improving noradrenergic neurotransmission derives from the observations that the drug is highly selective for $\alpha$-2 adrenergic receptors and does not interact with DA, ACh, or 5HT receptors (Doxey, 1984; Freedman, 1984). This argument is substantiated by the fact that IDA consistently produces a transient hyperadrenergic state consisting of tremulousness, flushing, elevated pulse, and increased blood pressure. There are also parallels with a previous report that the noradrenergic precursor L-threo-dihydroxyphenylserine (L-threo-DOPS) improved mobility in patients with predominantly axial parkinsonism (Narabayashi, 1986).

These observations are consistent with the view that the noradrenergic nervous system plays a role in motor function, especially gait, balance, stability, and dexterity. A final conclusion in this regard, however, must be withheld for several reasons. All patients did not respond to these agents, and predictors of response have not been identified. Moreover, no patient obtained complete relief, suggesting that NE makes, at most, a partial contribution to these functions. Also, the anatomic basis for noradrenergic control of motor function is unclear. The nature of the improvements on IDA suggests, however, a diffuse rather than focal localization.

## Future directions

The positive response to IDA underscores the difficulty of interpreting available data regarding the nature of the brain abnormalities underlying PSP. This response, in spite of putatively normal levels of NE (Kish, 1985), suggests, for example, that assays of neurotransmitter levels in PSP inadequately characterize the possible involvement of neurotransmitter systems in the disease. Deficiencies limited to small subregions of the brain may contribute to clinical dysfunction. Such deficiencies would probably escape detection by available assay techniques. Alternatively, the timing of release of neurotransmitters may be abnormal in the face of grossly normal levels. Quantitative abnormalities of specific receptor subtypes may also be a factor, as may disruptions in postreceptor signal transduction. All of these possibilities must be kept in mind when evaluating studies of the brains of PSP patients and devising strategies to treat the disease.

The IDA data sustain hope that neurotransmitter replacement therapy, although palliative and not curative, may yet provide relief to patients with PSP. Two modifications of neurotransmitter therapy merit rigorous evaluation. First, drugs that interact with multiple neurotransmitter systems should be tested, based on the premise that significant symptomatic relief will require restoration of function in all of the neurotransmitter systems that may be involved in the disease. Second, the increasing availability of drugs that target specific receptor subtypes opens the possibility of designing more specific treatments. A theoretical advantage of receptor-specific drugs would be to minimize unwanted side effects. Rational application of receptor subtype-selective therapy, however, will depend on identifying which receptor subtypes are spared and which are damaged in PSP.

Our findings with IDA support the concept of administering drugs that bypass striatal circuits as a fruitful strategy for treating PSP. On the basis of this experience, the use of other agents that target structures such as the thalamus and the cortex, which maintain relative functional integrity during the course of the illness, must be considered. Of course, studies of brain pathology in PSP emphasize the changes that characterize the end stages of the disease. It is likely that, during earlier stages, the striato-pallidal outflow remains viable. Therefore, treatment strategies that focus on these structures remain a valid approach.

While the current emphasis in treatment of PSP is on neurotransmitter replacement, this strategy is unlikely to provide relief of the magnitude afforded by L-DOPA in Parkinson's disease. Ideally, strategies that provide consistent relief, retard the progression of the illness, reverse it, or even prevent it must be sought. Along these lines, preliminary data from our longitudinal study of IDA suggest that patients who originally respond to IDA continue to respond for at least several months, but inevitably display evidence of clinical and functional decline. This course is analogous to that of PD patients on L-DOPA, but much more rapid. Transplantation of DA producing cells into the brains of patients with PSP has been attempted with little success (Koller, 1989). The use of MAO-B inhibitors, such as selegeline, that reportedly slow the progression of PD, has not been tested in PSP. The development of other approaches to treating PSP will depend on new insights into the pathophysiology and pathogenesis of the illness. Fundamental questions remain unanswered. Much work remains to be done.

## References

Agid Y, Javoy-Agid F, Ruberg M, Pillon B, Dubois B, Duyckaerts C, Hauw JJ, Baron JC, Scatton B (1986) Progressive supranuclear palsy: anatomoclinical and biochemical considerations. In: Yahr MD, Bergmann KJ (eds) Advances in neurology, vol 45. Raven Press, New York, pp 191–206

Albin RL, Young AB, Penney JB (1989) The functional anatomy of basal ganglia disorders. TINS 12: 366–375

Bass BM, Stucki RE (1951) A note on a modified Purdue pegboard. J Appl Psychol 35: 312–313

Blessed G, Tomlinson BE, Roth M (1968) The association between quantitative measures of dementia and of senile changes in the grey matter of elderly subjects. Br J Psychiatry 114: 797–811

Diagnostic and statistical manual of mental disorders, 3rd edn (1987) American Psychiatric Association, Washington DC

Doxey JC, Lane AC, Roach ET, Virdee NK (1984) Comparison of the alpha adrenoceptor antagonist profiles of idazoxan (RX 781094), yohimbine, rauwolscine, and coryanthine. Naunyn Schmiedebergs Arch Pharmacol 325: 136–144

Flowers KA, Robertson C (1985) The effect of Parkinson's disease on the ability to maintain a mental set. JNNP 48: 517–529

Foster NL, Aldrich MS, Bluemlein L, White RF, Berent S (1989) Failure of cholinergic agonist RS-486 to improve cognition and movement in PSP despite effects on sleep. Neurology 39: 257–261

Freedman JE, Aghahanian GK (1984) Idazoxan (RX 781094) selectively antagonizes alpha-2 adrenoceptors on rat central neurons. Eur J Pharmacol 105: 205–265

Ghika J, Tennis M, Hoffman E, Schoenfeld D, Growdon J (1991) Idazoxan treatment in progressive supranuclear palsy. Neurology 41: 986–991

Golbe LI, Davis PH, Schoenberg BS, Duvoisin RC (1988) Prevalence and natural history of progressive supranuclear palsy. Neurology 38: 1031–1034

Golden CJ, Hammeke TA, Purisch AD (1980) Manual for the Luria-Nebraska neuropsychological battery. Western Psychological Services, Los Angeles

Hamilton M (1967) A rating scale for depression. JNNP 5: 135–140

Hauw JJ, Verny M, Delaere P, Cervera P, He Y, Duyckaerts C (1990) Constant neurofibrillary changes in the neocortex in progressive supranuclear palsy. Basic differences with Alzheimer's disease and aging. Neurosci Lett 19: 182–186

Hirsch EC, Graybiel AM, Duyckaerts C, Javoy-Agid F (1987) Neuronal loss in the pedunculopontine tegmental nucleus in Parkinson disease and in progressive supranuclear palsy. PNAS 84: 5976–5980

Hoehn MM, Yahr MD (1967) Parkinsonism: onset, progression, and mortality. Neurology 17: 427–442

Jellinger K, Riederer P, Tomonoga M (1980) Progressive supranuclear palsy: clinicopathological and biochemical studies. J Neural Transm [Suppl] 16: 111–128

Juncos JL, Hirsch EC, Malessa S, Duyckaerts C, Hersh LB, Agid Y (1991) Mesencephalic cholinergic nuclei in progressive supranuclear palsy. Neurology 41: 25–30

Kish SJ, Chang LJ, Mirchandani L, Shannak K, Hornykiewicz O (1985) Progressive supranuclear palsy: relationship between extrapyramidal disturbances, dementia, and brain neurotransmitter markers. Ann Neurol 18: 530–536

Koller WC, Morantz R, Vetere-Overfield B, Waxman M (1989) Autologous adrenal medullary transplant in progressive supranuclear palsy. Neurology 39: 1066–1068

Lang AE, Fahn S (1989) Assessment of Parkinson's disease. In: Munsat TL (ed) Quantification of neurological deficit. Butterworths, London, pp 271–284

Litvan I, Gomez C, Atack J, Gillespie M, Kask AM, Mouradian MM, Chase TN (1989) Physostigmine treatment of progressive supranuclear palsy. Ann Neurol 26: 404–407

Malessa, Hirsch EC, Cervera P, Javoy-Agid F, Duyckaerts C, Hauw JJ, Agid Y (1991) Progressive supranuclear palsy: loss of choline acetyl-transferase-like immunoreactive neurons in the pontine reticular formation. Neurology 41: 1593–1597

Narabayashi H, Timohoshi K, Yokochi F, Nagatsu T (1986) Clinical effects of L-threo-3,4-dyhydroxyphenylserine in cases of parkinsonism and pure akinesia. In: Yahr MD, Bergmann KJ (eds) Advances in neurology, vol 45. Raven Press, New York, pp 550–559

Newman GC (1985) Treatment of progressive supranuclear palsy with tricyclic antidepressants. Neurology 35: 1189–1193

Neophytides A, Lieberman AN, Goldstein M, Gopinathan G, Leibowitz M, Bock J, Walker R (1982) The use of lisuride, a potent dopamine and serotonin agonist, in the treatment of progressive supranuclear palsy. JNNP 45: 261–263

Paulson GW, Lowery HW, Taylor GC (1981) Progressive supranuclear palsy: pneumoencephalography, electronystagmography and treatment with methysergide. Eur Neurol 20: 13–16

Pascual J, Berciano J, Grijalba B, del Olmo E, Gonzalez AM, Figols J, Pazos A (1992) Dopamine D1 and D2 receptors in progressive supranuclear palsy: an autoradiographic study. Ann Neurol 32: 703–707

Rafals RD, Grimm RJ (1981) Progressive supranuclear palsy: functional analysis of response to methysergide and antiparkinsonian agents. Neurology 31: 1507–1518

Randt CT, Brown ER, Osborne DP (1980) A memory test for longitudinal measurement of mild to moderate deficits. Clin Neuropsychol 2: 184–194

Ruberg M, Javoy-Agid F, Hirsch E, Scatton B, LHeureux R, Hauw JJ, Duyckaerts C, Gray F, Morel-Maroger A, Rascol A, Serdaru M, Agid Y (1985) Dopaminergic and cholinergic lesions in progressive supranuclear palsy. Ann Neurol 18: 523–529

Steele JC, Richardson JC, Olszewski J (1964) Progressive supranuclear palsy. Arch Neurol 10: 333–359

Stroop JR (1935) Studies of interference in serial verbal reactions. J Exp Psychol 18: 643–662

Tagliavini F, Pilleri G, Bouras C, Constantinidis J (1984) The basal nucleus of Meynert in patients with progressive supranuclear palsy. Neurosci Lett 44: 37–42

Weschler D (1981) Weschler adult intelligence scale, revised. Psychological Corporation, New York

Yesavage JA, Brink TL (1983) Development and validation of a geriatric depression screening scale: a preliminary report. J Psychiatr Res 17: 37–49

Zweig RM, Whitehouse PJ, Casanova MF, Walker LC, Jankel WR, Price DL (1987) Loss of pedunculopontine neurons in progressive supranuclear palsy. Ann Neurol 22: 18–25

Authors' address: Dr. D. G. Cole, Laboratory of Molecular and Developmental Neurobiology, Massachusetts General Hospital East, Building 149 13th Street, Charlestown, MA 02129, U.S.A.

# Subject Index

K. F. Tipton, M. B. H. Youdim, C. J. Barwell,
B. A. Callingham, G. A. Lyles (eds.)

# Amine Oxidases: Function and Dysfunction

Proceedings of the 5th International Amine Oxidase Workshop,
Galway, Ireland, August 22-25, 1992

1994. 113 figures. XII, 457 pages.
Soft cover DM 220,–, öS 1540,–
Reduced price for subscribers to "Journal of Neural Transmission":
Soft cover DM 198,–, öS 1386,–
ISBN 3-211-82521-5
(Journal of Neural Transmission, Supplement 41)

Monoamine oxidase plays a major role in the pathogenesis of neuropsychiatric disorders including depressive illness, Parkinson's disease and Alzheimer's disease. The new generation of selective monoamine oxidase inhibitors, devoid of major side effects, has found a prominent place in the treatment of these diseases. Some of these drugs may have neuroprotective activity with prospects for treating progressive neurodegenerative diseases. The volume presents a collection of research papers on monoamine oxidase and its inhibitors. The topic is treated from the point of view of chemistry, biochemistry, pharmacology, physiology, neurology and psychiatry. The book serves as a quick and comprehensive reference source for obtaining the most up to date information.

*Prices are subject to change without notice*

Springer-Verlag Wien New York

Sachsenplatz 4–6, P.O.Box 89, A-1201 Wien · 175 Fifth Avenue, New York, NY 10010, USA
Heidelberger Platz 3, D-14197 Berlin · 37-3, Hongo 3-chome, Bunkyo-ku, Tokyo 113, Japan